Molecular Bases of Anesthesia

Pharmacology and Toxicology: Basic and Clinical Aspects

Mannfred A. Hollinger, Series Editor
University of California, Davis

Published Titles

Biomedical Applications of Computer Modeling, 2001, Arthur Christopoulos
Molecular Bases of Anesthesia, 2001, Eric Moody and Phil Skolnick
Manual of Immunological Methods, 1999, Pauline Brousseau, Yves Payette, Helen Tryphonas, Barry Blakley, Herman Boermans, Denis Flipo, Michel Fournier
CNS Injuries: Cellular Responses and Pharmacological Strategies, 1999, Martin Berry and Ann Logan
*Infectious Diseases in Immunocompromised Hosts,*1998, Vassil St. Georgiev
Pharmacology of Antimuscarinic Agents, 1998, Laszlo Gyermek
Basis of Toxicity Testing, Second Edition, 1997, Donald J. Ecobichon
Anabolic Treatments for Osteoporosis, 1997, James F. Whitfield and Paul Morley
Antibody Therapeutics, 1997, William J. Harris and John R. Adair
Muscarinic Receptor Subtypes in Smooth Muscle, 1997, Richard M. Eglen
Antisense Oligodeonucleotides as Novel Pharmacological Therapeutic Agents, 1997, Benjamin Weiss
Airway Wall Remodelling in Asthma, 1996, A.G. Stewart
Drug Delivery Systems, 1996, Vasant V. Ranade and Mannfred A. Hollinger
Brain Mechanisms and Psychotropic Drugs, 1996, Andrius Baskys and Gary Remington
Receptor Dynamics in Neural Development, 1996, Christopher A. Shaw
Ryanodine Receptors, 1996, Vincenzo Sorrentino
Therapeutic Modulation of Cytokines, 1996, M.W. Bodmer and Brian Henderson
Pharmacology in Exercise and Sport, 1996, Satu M. Somani
Placental Pharmacology, 1996, B. V. Rama Sastry
Pharmacological Effects of Ethanol on the Nervous System, 1996, Richard A. Deitrich
Immunopharmaceuticals, 1996, Edward S. Kimball
Chemoattractant Ligands and Their Receptors, 1996, Richard Horuk
Pharmacological Regulation of Gene Expression in the CNS, 1996, Kalpana Merchant
Experimental Models of Mucosal Inflammation, 1995, Timothy S. Gaginella
Human Growth Hormone Pharmacology: Basic and Clinical Aspects, 1995, Kathleen T. Shiverick and Arlan Rosenbloom
Placental Toxicology, 1995, B. V. Rama Sastry
Stealth Liposomes, 1995, Danilo Lasic and Frank Martin
TAXOL®: Science and Applications, 1995, Matthew Suffness
Endothelin Receptors: From the Gene to the Human, 1995, Robert R. Ruffolo, Jr.
Alternative Methodologies for the Safety Evaluation of Chemicals in the Cosmetic Industry,1995, Nicola Loprieno
Phospholipase A_2 in Clinical Inflammation: Molecular Approaches to Pathophysiology, 1995, Keith B. Glaser and Peter Vadas
Serotonin and Gastrointestinal Function, 1995, Timothy S. Gaginella and James J. Galligan
Chemical and Structural Approaches to Rational Drug Design, 1994, David B. Weiner and William V. Williams
Biological Approaches to Rational Drug Design, 1994, David B. Weiner and William V. Williams

Pharmacology and Toxicology: Basic and Clinical Aspects

Published Titles (*Continued*)

Direct and Allosteric Control of Glutamate Receptors, 1994, M. Palfreyman, I. Reynolds, and P. Skolnick
Genomic and Non-Genomic Effects of Aldosterone, 1994, Martin Wehling
Peroxisome Proliferators: Unique Inducers of Drug-Metabolizing Enzymes, 1994, David E. Moody
Angiotensin II Receptors, Volume I: Molecular Biology, Biochemistry, Pharmacology, and Clinical Perspectives, 1994, Robert R. Ruffolo, Jr.
Angiotensin II Receptors, Volume II: Medicinal Chemistry, 1994, Robert R. Ruffolo, Jr.
Beneficial and Toxic Effects of Aspirin, 1993, Susan E. Feinman
Preclinical and Clinical Modulation of Anticancer Drugs, 1993, Kenneth D. Tew, Peter Houghton, and Janet Houghton
In Vitro *Methods of Toxicology*, 1992, Ronald R. Watson
Human Drug Metabolism from Molecular Biology to Man, 1992, Elizabeth Jeffreys
Platelet Activating Factor Receptor: Signal Mechanisms and Molecular Biology, 1992, Shivendra D. Shukla
Biopharmaceutics of Ocular Drug Delivery, 1992, Peter Edman
Pharmacology of the Skin, 1991, Hasan Mukhtar
Inflammatory Cells and Mediators in Bronchial Asthma, 1990, Devendra K. Agrawal and Robert G. Townley

Molecular Bases of Anesthesia

Edited by
Eric Moody
and Phil Skolnick

CRC Press
Boca Raton London New York Washington, D.C.

Library of Congress Cataloging-in-Publication Data

Molecular bases of anesthesia / edited by Eric Moody and Phil Skolnick.
p. cm.-- (Pharmacology and toxicology)
Includes bibliographical references and index.
ISBN 0-8493-8555-5 (alk. paper)
1. Anesthetics. 2. Molecular pharmacology. I. Moody, Eric. II. Skolnick, Phil. III. Pharmacology & toxicology (Boca Raton, Fla.)

RD82 .M633 2000
617.9′6—dc21 00-064136

Direct all inquiries to CRC Press LLC, 2000 N.W. Corporate Blvd., Boca Raton, Florida 33431.

International Standard Book Number 0-8493-8555-5
Library of Congress Card Number 00-064136
Printed in the United States of America 1 2 3 4 5 6 7 8 9 0
Printed on acid-free paper

Preface

The concept of this monograph began with the search for a comprehensive, focused source of information on anesthetic mechanisms. Here, we have gathered a number of contributions from researchers with diverse points of view, and have sought to cover the most important areas of current interest concerning anesthetic mechanisms. The focus of this monograph is squarely on neuroreceptors, with deliberate emphasis on ion channels. This reflects the area of greatest current interest. Lipid interactions with general anesthetics have not been covered separately, but information relevant to this topic is covered in the introductory chapter. This bias is fully consistent with a rapidly accumulating body of evidence, including the recent identification of targets for anesthetics on protein ion channels.

The field of anesthetic mechanisms is far different from what it was a decade or two ago. The unique pharmacology and low potencies of general anesthetics complicated the search for appropriate targets, and our meager understanding of the basic foundations of consciousness led to myriad anesthetic theories. Likewise, basic conceptual issues, such as whether lipids or proteins were the primary targets of anesthetics, resulted in little common ground between these two groups of researchers. The seminal observation that the effects of anesthetics on lipids were very modest and mimicked by other conditions (such as small changes in temperature) caused many researchers to focus on proteins as targets of anesthetic action. At the same time, the basic physiology and structure of receptors and ion channels were elucidated. For example, insight into benzodiazepine actions at the $GABA_A$ receptors led to studies with barbiturates, alcohols, and volatile anesthetics at this family of ligand-gated ion channels. Simultaneously, studies were being conducted with anesthetics at other potential loci of action.

The use of molecular biological techniques has also been applied to this study of anesthetic mechanisms. As can be seen from many of the contributions to this monograph, recombinant receptors have been very useful in elucidating anesthetic actions. Several researchers have used mutagenesis techniques to identify crucial segments of receptors involved in anesthetic action. Putative loci for volatile anesthetics and alcohols, as well as intravenous anesthetics have been described. These are exciting findings providing a framework to evaluate theories of anesthesia and allowing a better understanding of anesthetic actions at the molecular level. Substitution of various amino acids at crucial loci on a receptor provide an opportunity to determine the size of potential binding sites. Moreover, the use of transgenic animals may provide the opportunity to determine if the *in vitro* effects of anesthetics cause *in vivo* changes in anesthetic effects. Such experiments, if successful, may provide a method for determining the pharmacological relevance of experimental observations and resolving discrepancies between opposing theories of anesthesia. Many of the opposing concepts relating to anesthetic mechanisms are not fully resolved. It is our hope that this volume will provide the information and framework for readers to review and evaluate these issues. In addition, we trust that this book will be a useful reference for those with an interest in anesthesiology, and will stimulate further study.

Dedication

This work is dedicated to the postdoctoral fellows whose hard work and commitment resulted in much of the data contained in this book, and to the Intramural Program at the National Institutes of Health for creating an exciting research environment which encouraged risk taking.

Editors

Phil Skolnick is currently a Lilly Research Fellow (Neuroscience) at Lilly Research Laboratories. He is also a research professor of anesthesiology at Johns Hopkins University, an adjunct professor of pharmacology and toxicology at Indiana University School of Medicine, and has served as a research professor of psychiatry at the Uniformed Services University of the Health Sciences (1989–1998). Dr. Skolnick was senior investigator and chief, Laboratory of Neuroscience, at the National Institutes of Health from 1986–1997. He earned a B.S. (summa cum laude) from Long Island University in 1968 and a Ph.D. from George Washington University in 1972. That year, Dr. Skolnick joined the NIH as a staff fellow working under Dr. John W. Daly. He was appointed a senior investigator in 1977.

Dr. Skolnick is a member of the American Society for Pharmacology and Experimental Therapeutics, International Society for Neurochemistry, Society for Neuroscience, Society for Biological Psychiatry, and a fellow of the American College of Neuropsychopharmacology. Dr. Skolnick was twice named a Wellcome visiting professor in the basic medical sciences.

He is an editor of *Current Protocols in Neuroscience* and serves on the editorial advisory boards of *The European Journal of Pharmacology; The Journal of Molecular Neuroscience*; and *Pharmacology, Biochemistry and Behavior.* His principal research interest is the physiology and pharmacology of ligand-gated ion channels. Dr. Skolnick has co-authored more than 480 articles and holds several patents.

Eric Moody attended Brown University for his undergraduate and medical school training. During this time, he was also a researcher with the National Science Foundation's U.S. Antarctic Research Program. After a residency in anesthesiology at the Harvard Medical School, Massachusetts General Hospital, he spent several years performing postdoctoral research in the Laboratory of Neuroscience at the National Institutes of Health. He has maintained an interest in anesthetic mechanisms since that time. Dr. Moody is currently an associate professor of anesthesiology at Johns Hopkins Hospital, where he was the recipient of the Merck Clinician Scientist award in 1995. His principal research interest is the mechanisms of general anesthesia.

Contributors

Stephen Daniels
Welsh School of Pharmacy
University of Wales
Wales, United Kingdom

Jo Ellen Dildy-Mayfield
Department of Pharmacology
University of Colorado Health Science Center and Denver Veterans Administration Medical Center
Boulder, Colorado

James P. Dilger
Department of Anesthesiology/Pharmacology
State University of New York at Stony Brook
Stony Brook, New York

Daniel S. Duch
Department of Anesthesiology and Physiology
Cornell University Medical Center
New York, New York

Roderic G. Eckenhoff
Department of Anesthesiology, Biochemistry and Biophysics
University of Pennsylvania
Philadelphia, Pennsylvania

Pamela Flood
Department of Anesthesiology
Columbia University
New York, New York

R. Adron Harris
Department of Molecular Biology
University of Texas at Austin
Austin, Texas

Hugh C. Hemmings, Jr.
Department of Anesthesiology/Pharmacology
New York Hospital
Cornell University Medical Center
New York, New York

Piotr K. Janicki
Department of Anesthesia/Medicine
Vanderbilt University School of Medicine
Nashville, Tennessee

Jonas S. Johansson
Department of Anesthesiology, Biochemistry and Biophysics
University of Pennsylvania
Philadelphia, Pennsylvania

Donald D. Koblin
Department of Anesthesia
University of California–San Francisco
Department of Anesthesiology
V.A. Administration Hospital
San Francisco, California

Eric J. Moody
Department of Anesthesiology and Critical Care Medicine
Johns Hopkins University
Baltimore, Maryland

Philip G. Morgan
Department of Anesthesiology
University Hospitals of Cleveland
Cleveland, Ohio

Robert A. Pearce
Department of Anesthesiology
University of Wisconsin School of Medicine
Madison, Wisconsin

Margaret M. Sedensky
Department of Anesthesiology
University Hospitals of Cleveland
Cleveland, Ohio

Phil Skolnick
Department of Anesthesiology and Critical Care Medicine
Johns Hopkins University
and Neuroscience Discovery
Eli Lilly & Company
Indianapolis, Indiana

Tatyana N. Vysotskaya
Department of Anesthesiology and Physiology
Cornell University Medical Center
New York, New York

Table of Contents

Acknowledgment

The members of the Department of Anesthesiology at Johns Hopkins University are recognized for their long-term career support of Dr. Moody, which has enabled him to investigate anesthetic mechanisms and contribute to this publication.

1 Basic Pharmacology of Volatile Anesthetics

James P. Dilger

CONTENTS

1.1 INTRODUCTION

What is (are) the site(s) of action of volatile anesthetics that produce general anesthesia? The answer to this fundamental question would constitute a great advance toward determining the mechanisms by which anesthetics operate. Alas, the answer has eluded anesthetic mechanicians since they started asking it in 1899. There are two serious experimental obstacles that impede progress. One is that general anesthetics have a relatively low potency. The most potent volatile anesthetics are effective at a concentration of several hundred micromolar. Nonspecific binding of anesthetics in the central nervous system is widespread, making it difficult to clearly separate specific and nonspecific binding. The second problem is that there is no known compound that specifically antagonizes general anesthesia. If we had such

0-8493-8555-5/01/$0.00+$.50

a marvelous substance, we could use it to bait our hooks and go on a fishing expedition for anesthetic binding sites in the brain.

In the absence of a potent, chemical antagonist for general anesthesia, we must rely on alternative ways of assessing the suitability of putative anesthetic targets. A number of different approaches have been and are being taken. We classify them into three areas: chemistry, thermodynamics, and genetics. In this chapter, we discuss and criticize the usefulness of these approaches. The goal is to determine which, if any, of them might provide reliable information about anesthetic binding sites. In addition, we suggest what additional experimental evidence is needed to strengthen the approaches. The last section deals with hypotheses of anesthetic mechanisms. We review the history and current status of these hypotheses.

These are controversial issues, not only in terms of basic science but also in terms of politics. Funds for basic research are limited. It would be useful to have a rational set of criteria to identify those research projects that propose to study only the most promising putative anesthetic targets. Therefore, a second reason for analyzing the experimental approaches is to answer the question, "Is it appropriate for the field of anesthetic mechanisms to adopt such criteria at this time?" My conclusion is that none of the available criteria are supported by enough experimental evidence to justify their adoption at this time.

1.2 ANESTHETIC POTENCY

Minimum alveolar concentration (MAC) is the most widely used index for determining whether a patient or experimental animal is anesthetized.[1,2] One MAC is defined as the concentration of inhalational anesthetic required to blunt the muscular response to surgical skin incision of 50% of a population of unparalyzed patients. Figure 1.1a shows data for the determination of the MAC for halothane in humans, plots the fraction of patients who did not move in response to surgical incision at each halothane concentration. Thus, the figure is a quantal concentration–response curve for halothane anesthesia.[2,3] One MAC equals 0.20 m*M* halothane. Sixty-eight percent of patients are anesthetized at halothane concentrations between 0.185 m*M* and 0.215 m*M* (vertical lines in Figure 1.1a), so the standard deviation of MAC is 0.015 m*M* (7.5% of 1.0 MAC).

The numerical values of MAC for different anesthetics are indispensable to anesthesiologists, but the concept of MAC is not so useful because surgical patients are often paralyzed by muscle relaxants. The concept of MAC also has limited usefulness to anesthetic mechanicians. MAC provides a single number, the ED_{50} (1.0 MAC), to characterize the effects of anesthetics. Depth of anesthesia measured in this way is a discrete or "all-or-none" measurement; the subject either moves or remains still. The concept of MAC contains no information about the depth of anesthesia for a subject exposed to 0.5 or 2.0 MAC.

Figures 1.1b and c illustrate how MAC is related to a hypothetical curve for halothane binding to a site in the central nervous system (CNS). Figure 1.1b considers the case where binding of a single molecule of halothane is sufficient to affect some process; Figure 1.1c considers the case where the cooperative binding of five molecules of halothane is needed. Anesthesia occurs when a binding threshold is

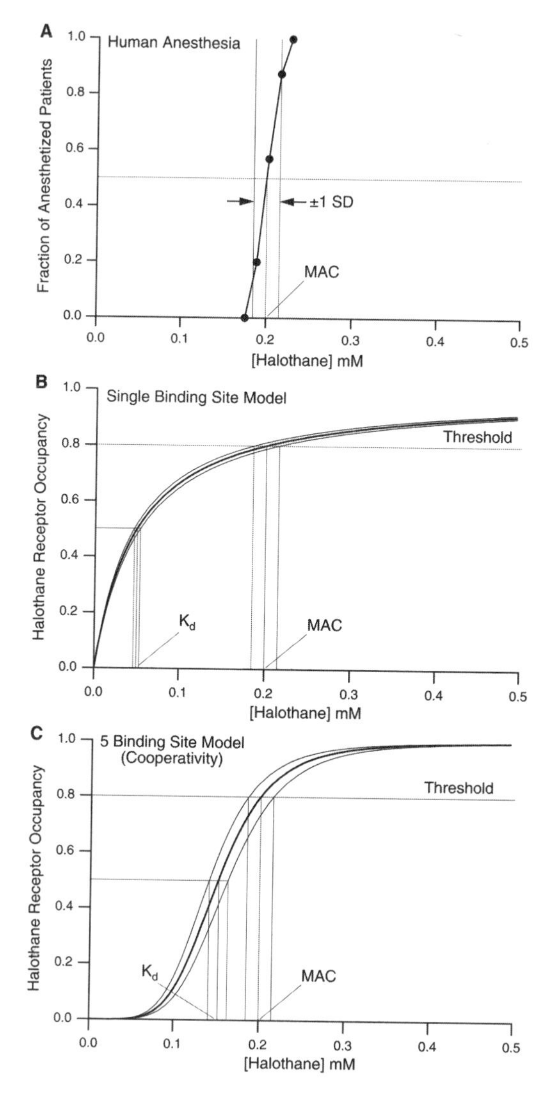

FIGURE 1.1 (A) The quantal concentration–response curve for halothane anesthesia. Human MAC of halothane is determined from the cumulative distribution of patients who did not show a muscle response to surgical incision. One MAC is the concentration at which 50% of the patients did not respond. The vertical lines indicate ±1 standard deviation of MAC; MAC = 0.20 ± 0.015 m*M* halothane. (B, C) Hypothetical molecular binding curves for anesthesia. These curves illustrate the origin of the quantal concentration–response curve for halothane anesthesia. The binding curves are calculated according to the Hill equation: fraction bound = $(c/K_d)/1 + (c/K_d)^n$, where c = anesthetic concentration, K_d = anesthetic dissociation constant, and n = Hill coefficient. For the three curves in B, K_d = 0.0465, 0.05, and 0.0535 m*M*, n = 1. For the three curves in C, K_d = 0.141, 0.152, and 0.163 m*M*, n = 5. The threshold, the level of binding required to produce general anesthesia, is assumed to be 80%. A population variability in K_d of ±7.5% leads to a population variability in MAC of ±7.5%. This determines the slope of the quantal concentration–response curve for anesthesia, as shown in A.

reached (the safety margin is exceeded). Here, we use 80% as the binding threshold but the argument is valid for any threshold level. The concentration of halothane at which the binding curve crosses 80% would be 1.0 MAC. For the single binding site model (Figure 1.1b), the affinity for halothane binding to the site (K_d) would have to be 0.05 m*M* in order to achieve 80% binding at 0.2 m*M* halothane. There will be some variation in the binding affinity from subject to subject, say ±7.5%. This gives rise to a ±7.5% variability in the concentration of halothane at which the binding curve crosses 80% and, hence, a ±7.5% variability in MAC. This variability describes the entire quantal concentration–response curve for halothane anesthesia (Figure 1.1a). The steepness of Figure 1.1a may simply reflect the population variability in halothane sensitivity.

The steepness of Figure 1.1a is sometimes interpreted to imply that the site of action of general anesthetics has a steep dependence on concentration or that anesthesia is a highly cooperative phenomenon. This is false. Figure 1.1c shows that if the hypothetical binding curve for halothane anesthesia requires the binding of five molecules of halothane, a variability in the binding affinity of ±7.5% still gives rise to a ±7.5% variability in the concentration of halothane at which the binding curve crosses 80%. There is no additional variability in MAC due to halothane binding cooperativity. The quantal concentration–response curve for anesthesia provides no information about the underlying mechanism of anesthetic action.[4] (See Ref. 5 for another opinion regarding the significance of the steepness of the concentration–response curve.)

Some conceptual difficulties of MAC will be discussed in Section 1.3. There is, so far, no viable alternative to MAC, so it remains as the benchmark for anesthetic potency. We must ask, then, how measurements of MAC for other animals compare to that for humans. For experimental animals, another noxious stimulus (e.g., tail clamp, hot plate) may be used instead of surgical incision to assess anesthesia. The type of stimulus may not be important* provided the intensity is supramaximal.[2,6] The muscular response should be a "gross purposeful muscular movement";[2] a movement that clearly indicates that the subject is experiencing pain. The definition of anesthesia has been extended to permit studies with animals that cannot easily be administered a noxious stimulus, but undergo spontaneous movement. The reversible cessation of that movement is taken to be an indication of anesthesia.

The ED_{50} for anesthesia varies among animals. Figure 1.2 shows that the ED_{50} for halothane is within 0.18 to 0.26 m*M* (0.72 to 1.1% atm) for mammals. (A previously published version of this data[10] did not take into account the different temperatures at which ED_{50} determinations were done.) Most nonmammalian species are less sensitive to halothane than are mammals; the exceptions being fruit flies and tadpoles, which are as sensitive as mammals. Nematodes are the least sensitive of the animals that have been tested; nematodes require about 10 times more

* The importance of the stimulus intensity also can be seen by comparing halothane MAC awake (0.41% atm),[7] where the stimulus is a verbal command, and MAC (0.75% atm), where the stimulus is surgery. Animal studies show a similar effect. Fruit flies will stop flying when they are exposed to 0.15% atm halothane, but they will still move when stimulated by heat until the concentration reaches 0.41% atm halothane.[8,9]

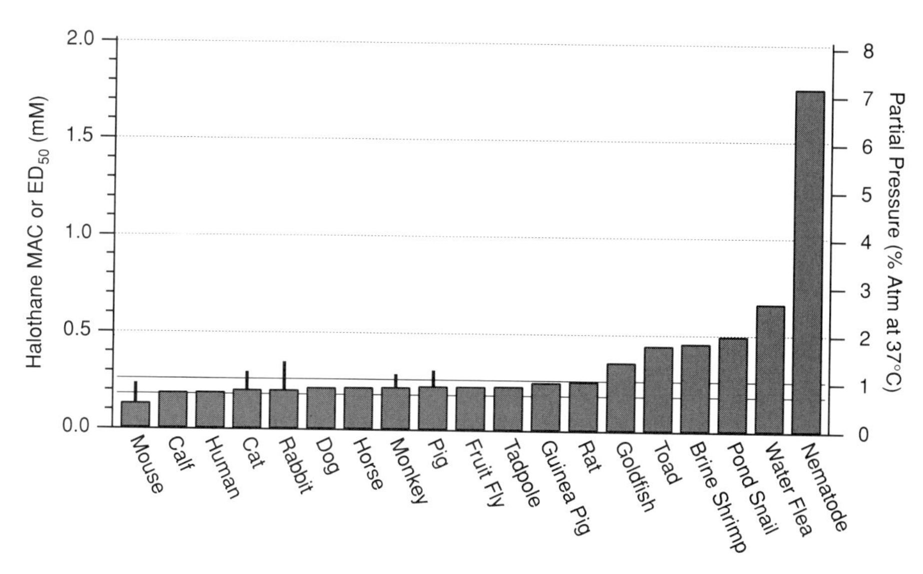

FIGURE 1.2 MAC or ED_{50} of halothane for 19 animal species. Values from the literature that were expressed as partial pressures (all except tadpoles), were converted to aqueous concentrations (left-hand axis) using the halothane Bunsen gas:water partition coefficient for the temperature of the experiment (0.62 at 37°C, 1.2 at 25°C, 1.4 at 20°C). The right-hand axis shows the partial pressure of halothane at 37°C for the corresponding aqueous concentration. The lines extending from the tops of the histogram bars for mouse, cat, rabbit, monkey, and pig indicate the upper range of published MAC values for these animals. The narrow horizontal lines at 0.18 and 0.26 m*M* halothane indicate the extent of MAC values of mammals. (References: mouse,[101,106] calf,[107] human,[12] cat,[98,108] rabbit,[102,109] dog,[12] horse,[99] monkey,[110,111] pig,[112,113] fruit fly,[9] tadpole,[114] guinea pig,[98] rat,[12] goldfish,[60] toad,[115] brine shrimp,[54] pond snail,[116] water flea,[103] nematode.[104]) (Adapted from Reference 10.)

halothane than humans. Nematodes are lower on the evolutionary scale than any other animal in Figure 1.2. Their resistance to anesthetics may be due to the anesthetic end point (cessation of the spontaneous movement), to the simplicity of their nervous system (302 neurons), or to evolutionary adaptation to a harsh environment.[11] The trend in Figure 1.2 is that the more primitive and simpler organisms are less sensitive to halothane. It could be argued, however, that environment plays the biggest role in anesthetic sensitivity; consider the ecological niches occupied by brine shrimp, snails, and nematodes!

Although the differences in halothane sensitivity among mammals are small, the same relative differences are consistently seen with other anesthetics. The most thorough comparative studies have been done with humans, dogs, and rats. For seven volatile anesthetics, the MAC for both rats and dogs is consistently 1.2 to 2.0 times greater than MAC for humans.[12]

MAC values for 13 volatile anesthetics administered to humans are given in Table 1.1. The anesthetics are ranked from highest to lowest potency as determined by the MAC values. Among the most commonly used anesthetics, there is an eightfold range of MAC values from halothane to sevoflurane.

TABLE 1.1
MAC in Humans and Partition Coefficients for Some Volatile General Anesthetics[12,95-97]

Anesthetic	MAC (% atm)	MAC (m*M*)	Partition Coefficients Water/Gas	Partition Coefficients Oil/Gas	Partition Coefficients Oil/Water
Methoxyflurane	0.16	0.26	4.2	850	200
Halothane	0.77	0.19	0.63	200	320
Isoflurane	1.2	0.24	0.54	91	170
Enflurane	1.7	0.52	0.78	97	120
Diethylether	1.9	8.22	11	57	5.2
Sevoflurane	2.0	0.29	0.37	47	130
Fluroxene	3.4	0.95	0.71	67	94
Desflurane	6.0	0.52	0.22	19	86
Cyclopropane	9.2	0.72	0.20	9.7	49
Butane	20	0.15	0.019	15	790
Ethylene	67	2.2	0.085	1.1	13
Xenon	71	2.1	0.075	1.8	24
Nitrous oxide	101	15	0.39	1.3	3.3

Source: Adapted from Reference 10.

Knowing the water/gas partition coefficient ($PC_{water/gas}$) for an anesthetic, we can convert each of the MAC values (partial pressure, p, in % atm) to an aqueous concentration (c_{aq} in m*M*) using the following equation:*

$$c_{aq}(mM) = \frac{1000}{224} * \frac{273}{273 + \text{temp}(°C)} * \frac{p(\%\ \text{atm})}{100} * PC_{water/gas} \quad (1.1)$$

$$\text{at } 37°C,\ c_{aq}(mM) = 0.39 * p(\%\ \text{atm}) * PC_{water/gas}$$

After the conversion, the rank order of anesthetic potency changes to butane, halothane, isoflurane, methoxyflurane, sevoflurane, enflurane, desflurane, cyclopropane, fluroxene, xenon, ethylene, diethylether, nitrous oxide. There is no contradiction here. It is simply a question of where the anesthetic concentration is measured. Consider diethylether and sevoflurane, which both have MAC values of about 2%. Diethylether vapor partitions 30 times more strongly into water than does sevoflurane vapor. As a result, the aqueous concentration of 1 MAC diethylether will be 30 times greater than that of 1 MAC sevoflurane. When expressed as aqueous concentrations, the range of anesthetizing concentrations among the commonly used anesthetics is only 2.7-fold from halothane to enflurane and desflurane. Potency is a confusing concept if we neglect to specify how the anesthetic is measured. (Temperature introduces a further complication. This will be discussed in Section 1.5.2.)

* The origins of the terms in Equation 1 are as follows: 1000 to convert *M* to m*M*; 22.4 mol/l for an ideal gas; 273/(273 + T) corrects the Bunsen partition coefficient, which is based on the volume of 1 mol of gas at standard temperature and pressure (STP); 100 to convert % atm to atm.

1.3 ASSESSING ANESTHETIC HYPOTHESES AND CRITERIA

As we evaluate theories of anesthesia and the usefulness of experimental tools in anesthetic research, it will be convenient to have a conceptual model to describe consciousness and anesthesia. The model should be as simple and as general as possible. We will consider two such models; the first model is shown in Figure 1.3. Model I consists of a neuron, the "consciousness center" that receives both excitatory and inhibitory inputs. The brain exerts regulatory control over the level of activity of this neuron by adjusting the degree of excitatory and inhibitory inputs (indicated by EX and IN, respectively). We assume that consciousness is represented by a high level of activity of the neuron (the level of activity is represented on a gray scale with darker levels indicating higher activity) due to a high degree of excitatory input and a low degree of inhibitory input (Figure 1.3a). We also assume that the action

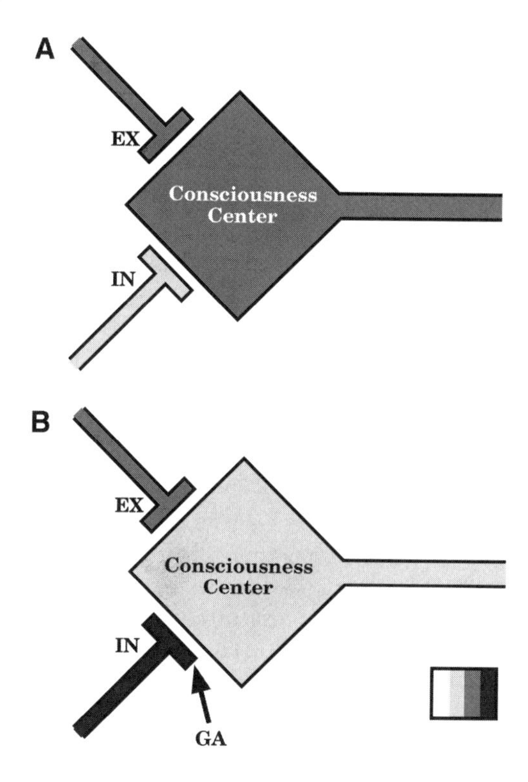

FIGURE 1.3 Model I. A simple, general unitary model of consciousness and anesthesia. The consciousness center neuron has both excitatory (EX) and inhibitory (IN) inputs. (A) In the conscious subject, the neuron has a high level of activity. (B) Anesthetics are assumed to increase the inhibitory input to the neuron, rendering it less active. Neuronal activity is indicated by the gray scale, with the highest activity indicated by the darkest shade of gray. The model can be used to test hypotheses and criteria of anesthetic action (see text). GA, general anesthetic.

of a general anesthetic (indicated by GA) is to increase the degree of inhibitory input while leaving the excitatory input unchanged (Figure 1.3b). The net effect is to decrease the level of activity of the neuron and, thereby, decrease consciousness. Perhaps sleep and/or coma in the absence of anesthetics results from endogenous factors acting to increase the degree of inhibitory input to the neuron.

Model I is quite general. It does not require specification of the location of the anesthetic effect (e.g., lipid vs. protein, axonal vs. pre- or postsynaptic). The action of an anesthetic could be to decrease the excitatory input rather than to increase inhibitory input. Moreover, the model is not restricted to neurons and synapses. For example, the consciousness center could be considered to be an intracellular enzyme that can be phosphorylated (excitatory input) and dephosphorylated (inhibitory input). The important features are that the consciousness center has positive and negative regulatory mechanisms and that anesthetics interfere with one of these mechanisms.

We will use Model I as a first step in evaluating hypotheses of anesthetic action and criteria for anesthetic relevance. We will ask the question, "Is there a unique way to interpret an hypothesis or a criterion in terms of Model I?" If the answer is "Yes," then the hypothesis or criterion will be considered "robust" at this level of scrutiny. If the answer is "No," then the hypothesis or criterion will be considered "ambiguous."

An implicit assumption of Model I is that MAC faithfully reports the status of consciousness; Model I does not distinguish between loss of consciousness and lack of purposeful movement due to painful stimuli. Several studies indicate that the spinal cord processes painful sensory input and initiates a motor response independent from the brain. For example, the MAC of isoflurane in rats is unchanged after either acute decerebration[13] or spinal cord transection.[14] In addition, the MAC of isoflurane is increased when the anesthetic (GA) is preferentially applied to the brain of goats.[15] To incorporate this into our conceptual model of anesthesia, we consider Model II (Figure 1.4). Model II consists of two neurons. The first neuron represents the consciousness center and the second neuron, which may be in the spinal cord, represents the movement center. If the two neurons are connected in series (Figure 1.4a), anesthetics acting on the consciousness center might transmit this information to the movement center, decreasing its activity. A direct action of anesthetics on the movement center would not be necessary. In this case, Model II is equivalent to Model I. Alternatively, there may be no functional link between the consciousness and movement centers (Figure 1.4b); the two neurons are in parallel. In order to suppress movement, anesthetics would have to increase inhibitory input to the movement center.

A criterion that is deemed robust in terms of Model I may be ambiguous in terms of Model II. To remove this ambiguity, it is necessary to determine the direct relationships of the criterion on the movement center. Thus, Model II might suggest ways in which criteria can be strengthened experimentally. Alternatively, if MAC is independent of the consciousness center (the parallel neuron version of Model II, Figure 1.4b), then all of the criteria discussed below are actually probing the interactions of anesthetics with the movement center rather than the consciousness center.

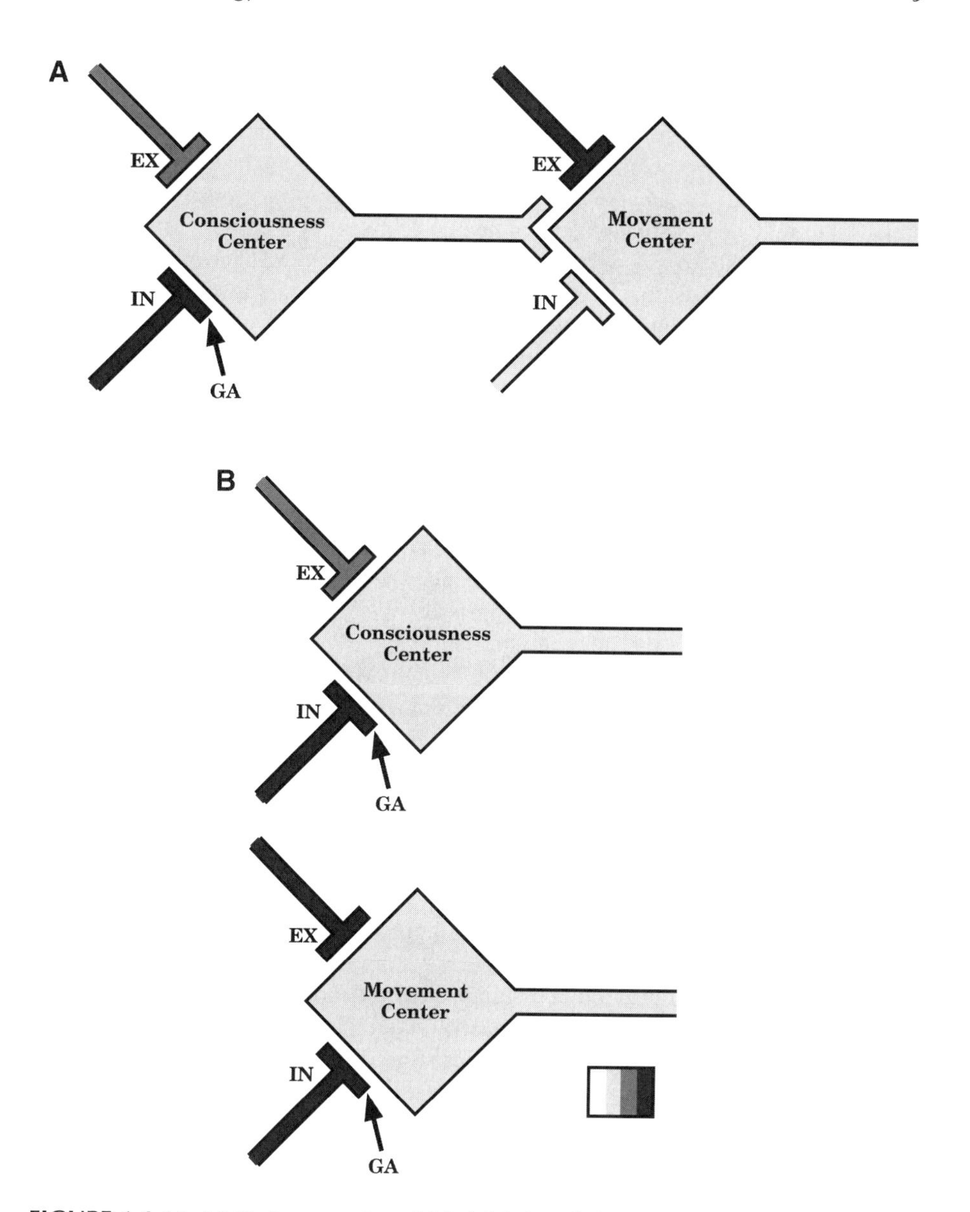

FIGURE 1.4 Model II. An extension of Model I that distinguishes the consciousness and movement components of anesthesia. The second neuron, the movement center, is also subject to excitatory (EX) and inhibitory (IN) regulation. (A) Neurons in series. The movement center neuron receives synaptic input from the consciousness center neuron. The effect of anesthetics is transmitted via this synapse only. (B) Neurons in parallel. The movement center neuron does not receive synaptic input from the consciousness center neuron. Anesthetics inhibit movement by increasing the inhibitory input to the movement center. GA, general anesthetic.

Both Models I and II are "unitary" in that all general anesthetics (at least the volatile anesthetics under discussion here) are assumed to act at the same site (or the same pair of consciousness and movement sites) and produce anesthesia in the same way. There have been some recent challenges to the unitary viewpoint[16] (see also Section 1.6 in this chapter). These do not necessarily denounce all aspects of unitarity. It may be possible to identify several "classes" of anesthetics (even among the volatile anesthetics) such that anesthetics within each class have common sites of action. Our bias is that our knowledge has too many gaps to permit us to completely abandon unitary hypotheses. We should acknowledge the primitive nature of our understanding of neuronal signaling. We still must identify all of the ion channels and intracellular enzymes that participate, elucidate their kinetic mechanisms, measure the time dependence of neurotransmitter concentrations at synapses, and learn the wiring diagram for the nervous system. Some day the experimental evidence may force us to abandon all unitary models, but until that time, we can learn much by constructing, testing, and refining unitary models.

Another important factor to consider when evaluating anesthetic criteria is generality. A criterion that has been tested in only one or two animals, is not as robust as one that shows consistent results in tests on many experimental animals. *Homo sapiens* are, of course, the ultimate species of interest, but ethical considerations limit experiments on humans. High priority should probably be given to mammals whose MAC values most closely resemble humans.

The final factor we will consider is the discriminatory power of an anesthetic criterion. Although a criterion may have a unique interpretation in terms of Models I and II, and be generally valid for many animals, it will not be useful if it gives many false positives. Thus, the criterion should be tested in systems that are unlikely to be relevant to anesthesia as well as in systems that might be relevant. An example of an irrelevant system is firefly luciferase.[17] Although anesthetics interact specifically with this enzyme, it is found only in lightning bugs. So far, it has not been detected in the CNS of any animal. Other examples of irrelevant systems are a nonspecific protein such as bovine serum albumin and a pure, single-component lipid bilayer. Most anesthetic mechanicians would agree that anesthetic effects on the latter system are not significant enough to be relevant to anesthesia. However, the inclusion of lipids here is not meant to dismiss all lipid systems from the list of possible relevant sites of anesthetic action.

The four factors that we will use to evaluate anesthetic criteria are listed in Table 1.2.

TABLE 1.2
The Four Factors Used in This Chapter to Evaluate Anesthetic Criteria and Mechanisms

1. Is there a unique interpretation in terms of Model I?
2. Is there a unique interpretation in terms of Model II?
3. Is it generally valid among animals?
4. Does it have discriminatory power?

1.4 CHEMISTRY

1.4.1 Structure of Anesthetics

The most striking thing about the structure of general anesthetics is the variety of chemical compounds that produce anesthesia. The inert gases xenon, argon, and krypton are general anesthetics, as are diatomic gases such as hydrogen and nitrogen. Simple organic compounds such as chloroform and cyclopropane have been used to anesthetize both human patients and animal subjects. Then, there are the more familiar clinical anesthetics: ether, six synthetic halogenated ethers (isoflurane, enflurane, methoxyflurane, desflurane, sevoflurane, and fluroxene) and the polyhalogenated alkane, halothane. Finally, there are a host of compounds that have been used on animals including alkanes, alcohols, diols, acetone, sulfur hexafluoride, hydrogen, nitrogen, argon, krypton, and thiomethoxyflurane. All of these compounds are relatively small; their molecular weight is less than 200 Da and their molar volumes range between 28 and 240 ml. However, they do not have any chemical groups in common.

1.4.2 Meyer–Overton Correlation

The first chemical clue relating the structure of anesthetics to their potency was discovered in 1899 by a pharmacologist, Hans Horst Meyer,[18] and an anesthetist, Charles Ernst Overton.[19] Working independently, Meyer and Overton found a strong correlation between the polarity of a compound and its potency as an anesthetic. The polarity is expressed as the oil/gas partition coefficient, and anesthetic potency is expressed as the partial pressure in atmospheres. Figure 1.5 is a Meyer–Overton correlation for 18 anesthetics used on mice. The correlation is plotted on a log–log scale to accommodate the large range of partition coefficients and anesthetic potencies (more than 4 orders of magnitude). The slope of the regression line fit to the data is –1.02, implying that MAC is inversely proportional to partition coefficient or potency is directly proportional to partition coefficient. The Meyer–Overton correlation also can be plotted with potency expressed as aqueous concentrations. In this case, the abscissa of the graph must be the oil:water partition coefficient (see Table 1.1).

Another way of expressing the Meyer–Overton correlation is to say that the product of MAC (in atm) and the oil/gas partition coefficient (dimensionless) is a constant. The constant for human anesthesia is 1.3 ± 0.4 (mean ± standard deviation) atm.[12] Alternatively, the product of the aqueous concentration equivalent of MAC and the oil/water partition coefficient gives a constant value of 50 ± 15 m*M*. This can be interpreted as the concentration of anesthetic in olive oil corresponding to anesthesia. If we can represent the site of anesthetic action as a bulk oil phase, the concentration of any anesthetic at 1.0 MAC is 50 m*M*.

Meyer and Overton used olive oil in their partition coefficient measurements and this has become the most commonly used reference solvent (the data in Figure 1.5 and Table 1.1 are based on olive oil). The partition coefficients of anesthetics into hexadecane,[20] octanol,[12,20] lipid bilayers,[21] benzene,[12] and lipids[12] also have been measured. Amphipathic solvents such as octanol and lecithin (phosphatidylcholine)

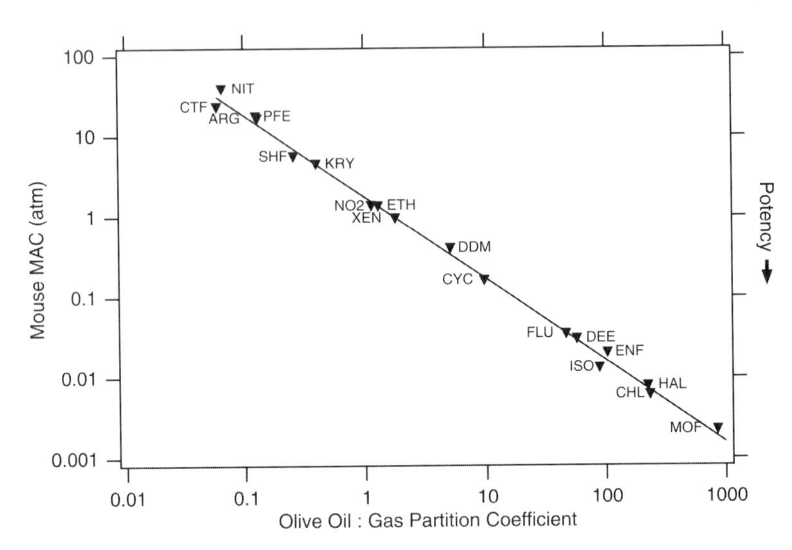

FIGURE 1.5 Meyer–Overton correlation for volatile general anesthetics in mice.[95] The slope of the regression line is –1.02 and the correlation coefficient, $r^2 = 0.997$. CTF, carbon tetrafluoride; NIT, nitrogen; ARG, argon; PFE, perfluoroethane; SHF, sulfur hexafluoride; KRY, krypton; NO2, nitrous oxide; ETH, ethylene; XEN, xenon; DDM, dichlorodifluoromethane; CYC, cyclopropane; FLU, fluroxene; DEE, diethylether; ENF, enflurane; ISO, isoflurane; HAL, halothane; CHL, chloroform; MOF, methoxyflurane.

produce stronger Meyer–Overton correlations than olive oil. In fact, octanol and lecithin are the only solvents that correctly predict that isoflurane is more potent than enflurane.[12] Purely hydrophobic solvents such as hexadecane and benzene produce weaker correlations than olive oil.

The Meyer–Overton correlation suggests that the site at which anesthetics bind is primarily a hydrophobic environment. The fact that the correlation is stronger with amphipathic solvents than with nonpolar solvents indicates that the environment is amphipathic. This provides a limited amount of information about the site of anesthetic action. The lipid bilayer component of cell membranes is amphipathic. Most proteins, both membrane proteins and water-soluble proteins are amphipathic. It is clear that we need more clues to isolate the sites in the CNS that are relevant to general anesthesia. The Meyer–Overton correlation provides the first criterion for assessing a putative site of anesthetic action.

Criterion 1. The site of action of general anesthetics must obey the Meyer–Overton correlation

We cannot evaluate Criterion 1 in terms of either Model I or II because the consciousness and movement centers are defined in terms of MAC, which is our only measure of anesthetic potency. The discriminatory power of Criterion 1 is weak. The inhibition of firefly luciferase[17] and the depression of the phase transition temperature of phosphatidylcholine[22] by anesthetics obey the Meyer–Overton

TABLE 1.3
Comparison of MAC or ED_{50} Values for Enflurane and Isoflurane in Various Animal Species

Animal	MAC Enflurane (% atm)	MAC Isoflurane (% atm)	Enf/Iso	Ref.
Guinea pig	2.2	1.2	1.9	98
Horse	2.1	1.3	1.6	99
Dog	2.2	1.4	1.6	12
Rat	2.2	1.5	1.5	12
Cat	2.4	1.6	1.5	100
Mouse	2.0	1.3	1.5	101
Human	1.7	1.2	1.4	12
Rabbit	2.9	2.1	1.4	102
Fruit fly	0.51	0.38	1.3	9
Water flea	1.4	1.2	1.2	103
Nematode	5.9	7.2	0.8	104

Note: Anesthetic potencies are given at the temperature of the experiment (37°C for mammals and 20 to 22°C for nonmammals). If we assume that the temperature dependence of the water/gas partition coefficient is the same for both anesthetics, then the ratio of MAC values expressed as partial pressures will be the same as those expressed in aqueous concentrations.

correlation. The binding of anesthetics to bovine serum albumin, however, follows the Meyer–Overton correlation on a coarse scale only; a 100-fold range of partition coefficients leads to a 100-fold range of albumin binding affinities, but the rank order of binding affinities is not predicted by the partition coefficients.[23] What about the generality of Criterion 1? All animals appear to have the same rank order of anesthetic potency, at least on a gross scale. Mammals can distinguish the subtle difference between the isomers isoflurane and enflurane; mammals are less sensitive to enflurane by a factor of 1.4 to 1.9 (Table 1.3). For insects, the ratio of MAC values for enflurane and isoflurane is closer to unity (1.2 to 1.3). Nematodes, however, exhibit the reverse pattern of sensitivity; they are more sensitive to enflurane than to isoflurane. This suggests the following criterion.

Criterion 2. The site of action of general anesthetics should be about 1.5 times more sensitive to isoflurane than to enflurane

A greater sensitivity for isoflurane than enflurane seems to be generally valid for many mammals but not for insects or nematodes. The discriminatory power of Criterion 2 has not been tested.

A second type of deviation from the Meyer–Overton correlation is also possible. The rank order of potency might be correct, but the concentration of anesthetic required to affect the test site differs from the concentration of anesthetic needed to produce anesthesia.

Criterion 3. The site of action of general anesthetics should have a sensitivity to anesthetics close to MAC

This may appear to be an obvious requirement. However, the relationship between the anesthetic dissociation constant (K_d) and MAC depends strongly on the model used to describe the anesthetic binding site. Figures 1.1b and c show that MAC represents the K_d for anesthetic binding to this site only if the margin of safety is 50%. If the margin of safety is 80%, MAC overestimates K_d for anesthetic binding to this site: $K_d = 0.25 \times \text{MAC}$. If the margin of safety is 20%, MAC underestimates binding: $K_d = 4 \times \text{MAC}$. The existence of multiple binding sites results in a closer equivalence between K_d and MAC (Figure 1.1c).

The actions of depolarizing and nondepolarizing muscle relaxants demonstrate how the margin of safety helps determine the clinical potency of a drug. Nondepolarizers are competitive inhibitors of the acetylcholine receptor. Neuromuscular transmission is said to have a high safety margin because there is an excess of postsynaptic receptors. As a result, nondepolarizing muscle relaxants must bind to (inhibit) 80% of the receptors before there is any diminution in the twitch response. Thus, the concentration of muscle relaxant needed to block the twitch response by 50% may be 10 or more times the K_d for drug binding to the receptors.[24] In contrast, depolarizing muscle relaxants exert their effect by opening acetylcholine receptor channels and producing a prolonged depolarization of the muscle cell. They can achieve this by binding to only 20% of the postsynaptic receptors.[25] These drugs are clinically effective at concentrations much lower than their binding K_d. Clearly, knowledge of the K_d without some mechanistic information is insufficient to predict the "MAC" of a drug.

So far, this analysis has assumed that anesthesia follows directly from the binding of anesthetic to some site. If anesthetic binding is followed by a conformational change in a protein and this is what produces anesthesia, then the equilibrium between the inactive and active conformational states (L) also enters into the determination of MAC. Experimentally, binding and/or effect may be the measured variable. In a three-state model (unbound/inactive, bound/inactive, bound/active) the fractional occupancy of the active conformational state, [AR*], is

$$R \xleftrightarrow{K_d} AR \xleftrightarrow{L} AR^* \quad [AR^*] = \frac{\dfrac{c}{K_d L}}{1 + \dfrac{c}{K_d}\left(1 + \dfrac{1}{L}\right)} \tag{1.2}$$

where c is the concentration of anesthetic. If $L = 0.1$ (the active state is favored by a factor of 10), then the K_d necessary for an 80% occupancy of the active conformation at c = MAC is $K_d = 1.5 \times \text{MAC}$. If the active state is favored by a factor of 100, then $K_d = 32 \times \text{MAC}$. In the absence of a model, K_d has no definite relationship to MAC.

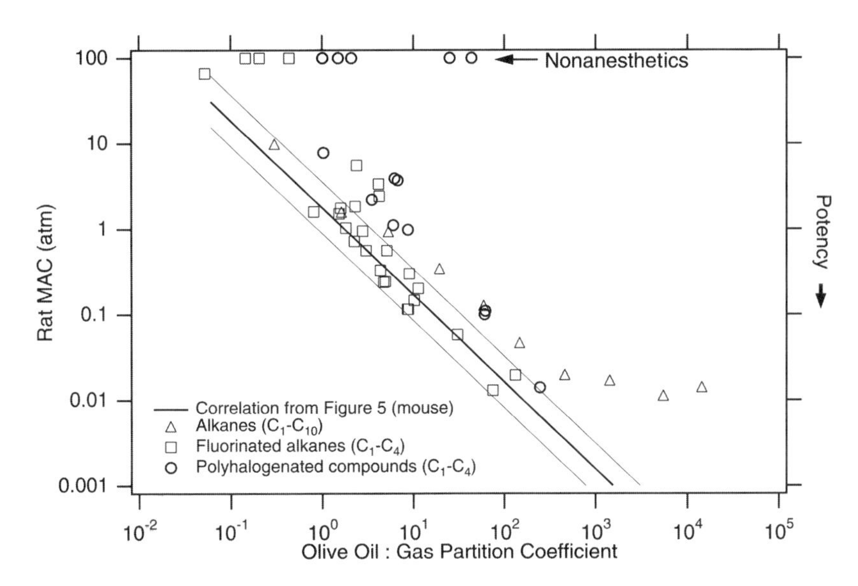

FIGURE 1.6 Deviations from the Meyer–Overton correlation as determined in rats. The dark line is the correlation found in Figure 1.5 for mice; the light lines correspond to the correlation shifted twofold to higher and lower potency values. (The mouse data were used because there are no published MAC values for rats with the less common anesthetics used in Figure 1.5. For the commonly used anesthetics, the MAC values for rats and mice are within a factor of 1.5.) △, alkanes methane (least potent) through decane;[28] ❑, fluorinated alkanes (1 to 4 carbons;[30] ❍, polyhalogenated alkanes (1 to 4 carbons).[31] See text for more information about the compounds.

1.4.3 Exceptions to the Meyer–Overton Correlation

Although a wide variety of compounds lie on the Meyer–Overton correlation line, there are many compounds that do not. An early observation was that there was a cut-off in the potency of n-alcohols. Alcohols longer than 12 or 13 carbons in length are not anesthetics despite their high lipid solubility.[26,27] Figure 1.6 shows the relationship between anesthetic potency and the oil/gas partition coefficient for 51 volatile compounds: alkanes (1 to 10 carbons),[28] fluorinated alkanes (1 to 4 carbons, containing one or more F substitutions for H),[29,30] and polyhalogenated compounds (1 to 4 carbons, in which all hydrogens are replaced by different combinations of F, Cl, and Br).[31] Of these compounds, 22 lie within a factor of 2 of the Meyer–Overton correlation, 21 compounds are more than a factor of 2 less potent than predicted, and 8 compounds are nonanesthetics (they have no anesthetic activity at all).

Experimentally, if a compound shows no anesthetic activity by itself, it is administered to the rat along with a subanesthetizing dose of another, more potent, anesthetic (usually desflurane). The ability of the compound to decrease the MAC of desflurane is tested. It is assumed that the MAC level of two co-applied anesthetics can be calculated by adding the MAC level of each anesthetic separately.[32] In this way, the MAC of an unknown compound can be determined from how much it lowers the MAC of desflurane. The eight nonanesthetics in Figure 1.6 did not lower the MAC of desflurane.[29,31] The MAC of six polyhalogenated compounds was determined by their ability to decrease the MAC of desflurane, because they did not have anesthetic activity by themselves.[31] The MAC of perfluoromethane was determined by its ability to decrease the MAC of desflurane, because the MAC value is very close to a lethal concentration of perfluoromethane.[29]

Figure 1.6 shows that the Meyer–Overton correlation does not completely describe all anesthetics. This means that the chemical properties of the anesthetic site differ from those of olive oil. Looking at the exceptions, we can try to determine the important differences. Most of the alkanes are less potent than would be expected from their partition coefficient. Moderate fluorination increases potency but complete fluorination is disastrous. Of the four perfluorinated alkanes (CF_4, C_2F_6, C_3F_8, C_4F_{10}), only perfluoromethane is an anesthetic. Polyhalogenated compounds are generally not very potent anesthetics.

The low potency of alkanes is not surprising considering the previous indications that the site of anesthetic action is amphipathic rather than purely hydrophobic. Short chain-length alkanes may bind to the anesthetic site in spite of their being nonpolar because their small size may allow them to avoid the polar region of the binding site. As chain length increases, the potency of alkanes deviates increasingly more than predicted. A moderate amount of fluorination increases the polarity of the compound and makes it a more potent anesthetic, but complete fluorination decreases the polarity of the compound again. The fact that perfluorinated alkanes are always less potent than the alkanes themselves supports the idea that hydrogen bonding at the anesthetic binding site is important.[33]

In the discussion above, we have assumed that the low (or absent) potency of the Meyer–Overton exceptions is due to their having a lower binding affinity to the anesthetic binding site than what is predicted from their oil:gas partition coefficient. An alternative explanation also exists. Assume that the binding site is on a protein. These compounds may have a normal binding affinity, but a low efficacy. They may bind to the protein just as well as a potent anesthetic, but may not allow some critical conformational change to occur. The potent anesthetic does promote this conformational change. Thus, nonanesthetics may actually be competitive antagonists (or partial agonists) at the anesthetic binding site. (For an analogy, compare the actions of acetylcholine and d-tubocurarine on the acetylcholine receptor. When acetylcholine is bound, the closed-to-open channel conformational change can take place. When d-tubocurarine is bound, the conformational change does not occur and the channel remains closed.) In this scenario, we could not deduce any information about the structure of the anesthetic binding site from the low potency of some compounds. The structural information they provide would be more subtle because it would involve the transition rates of protein conformational changes rather than equilibrium binding affinities.

The low (or absent) potency of the Meyer–Overton exceptions is complicated by the fact that many of these compounds, including the nonanesthetics, cause convulsions or other hyperactivity in rats.[31] Our evaluation of the usefulness of these compounds must account for this.

Criterion 4. The site of action of general anesthetics should discriminate between anesthetics and nonanesthetic compounds

Figure 1.7a illustrates one interpretation of the interaction of nonanesthetics with Model I; they don't interact act all. This would be true for the "ideal" nonanesthetic. It does not adequately describe the available nonanesthetics because it does not account for the convulsive and hyperactivity effects of these compounds. Figures 1.7b and c show two ways of accounting for hyperactivity by nonanesthetics within Model I. Nonanesthetics may suppress the normal inhibitory input to the consciousness center (Figure 1.7b). Alternatively, nonanesthetics may stimulate excitatory input to the consciousness center (Figure 1.7c). In both of these scenarios, it is possible that nonanesthetics have two effects, anesthesia and hyperactivity, but the degree of hyperactivity overwhelms the anesthetic effect. If this were so, nonanesthetics should also decrease the ability of potent anesthetics to produce anesthesia.

Some of the nonanesthetics have been found to increase significantly the MAC of desflurane,[31,34] isoflurane, and halothane[29] by as much as 50%. However, the increases are not consistently seen with each potent anesthetic tested and are not correlated with the concentration of the nonanesthetic. These results suggest that nonanesthetics are neither agonists nor antagonists at the anesthetic binding site of Model I; they do not interact with this site at all (Figure 1.7c). Thus, Criterion 4 is robust in terms of Model I. The status of nonanesthetics in terms of Model II is ambiguous. Nonanesthetics may decrease the activity of the consciousness neuron, but this information may not be communicated to the movement center because either (a) the coupling between the consciousness center and the movement center is impaired by nonanesthetics (Figure 1.4a) or (b) the consciousness center and the movement center are not physically coupled (Figure 1.4b).

The generality of Criterion 4 has not been tested. The rat is the only animal to have been administered the nonanesthetic fluorinated alkanes and polyhalogenated compounds. Testing these compounds for convulsant and anesthetic effects on other animals is crucial for determining the suitability of Criterion 4 in assessing putative sites of action of general anesthetics. The discriminatory power of nonanesthetics has not yet been tested on luciferase, bovine serum albumin, or simple lipid bilayer systems. However, the fact that these compounds act as convulsants indicates that they can discriminate between the convulsant site of action and the anesthetic site of action. Nonanesthetics also suppress learning and/or memory in rats[35] and must, therefore, bind to sites in the brain that involve these processes. Tests of the discriminatory power of long-chain alcohols have produced mixed results. Luciferase does show a cut-off in potency for long-chain-length alcohols.[36] Alcohols up to 15 carbons in length do not exhibit a cut-off in partitioning into lipid vesicles containing egg lecithin, cholesterol, and phosphatidic acid.[37]

1.4.4 Stereoisomers of General Anesthetics

Four clinically used volatile general anesthetics — isoflurane, enflurane, halothane, and desflurane — possess a chiral carbon atom. These anesthetics are synthesized as racemic mixtures of (+) and (–) stereoisomers. In the case of isoflurane and

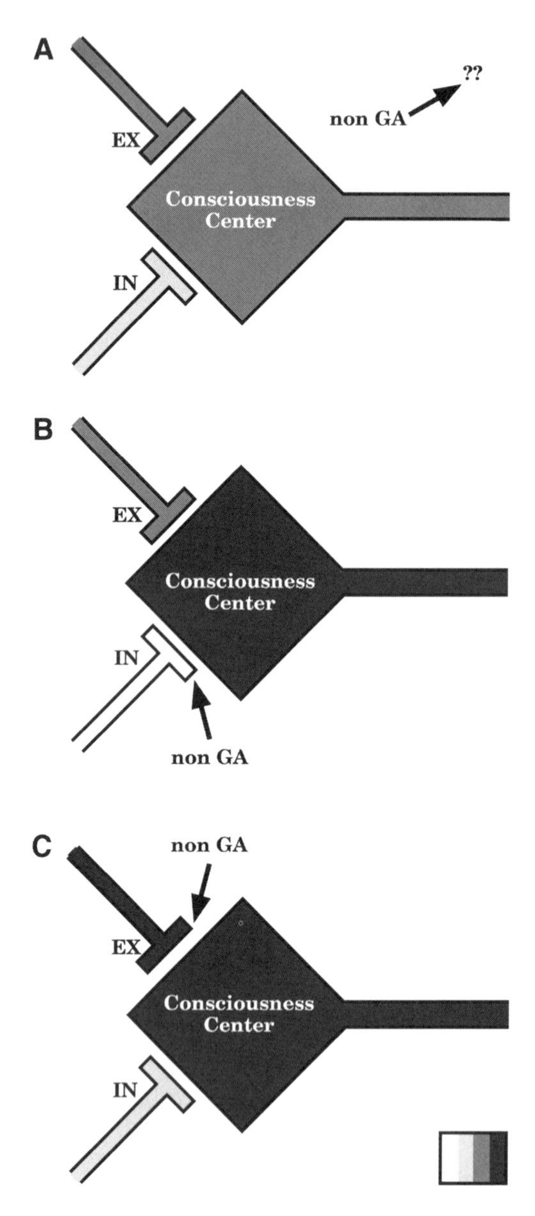

FIGURE 1.7 Possible ways in which nonanesthetics may affect the consciousness center in Model I. (A) Nonanesthetics do not interact with any component of Model I. (B) Nonanesthetics act to decrease inhibitory input (IN). (C) Nonanesthetics act to increase excitatory input (EX). GA, general anesthetic.

halothane, the isomers have been separated and studied to a limited extent on whole animals, isolated organs, and ion channel preparations. Rats are more than 50% more sensitive to the (+) isomer than to the (–) isomer of isoflurane[38] (Figure 1.8). This suggests a new criterion.

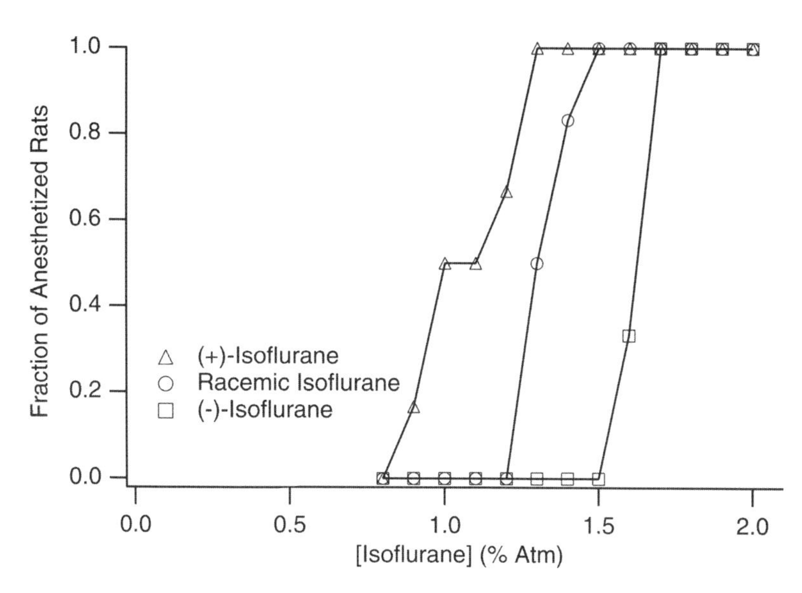

FIGURE 1.8 The determination of MAC for the two stereoisomers of isoflurane, plotted using data from Ref. 38.

Criterion 5. The site of action of general anesthetics should discriminate between the optical isomers of chiral general anesthetics

Criterion 5 is robust in terms of Model I because the stereoisomers of an anesthetic bind to the same anesthetic binding site in a unitary model. However, Criterion 5 is ambiguous in terms of Model II. The movement center could receive information about anesthetic binding either via input from the consciousness center or from anesthetic effects on the inhibitory input of the movement center. It is not clear whether the consciousness center or the movement center distinguishes the optical isomers.

The generality of Criterion 5 has not been tested on very many animals. A second study on rats suggested a much smaller and nonsignificant difference between the isoflurane enantiomers.[39] Some strains of nematodes can distinguish optical isomers of isoflurane; other strains cannot.[11] Moreover, the ability of a strain to distinguish optical isomers of isoflurane does not predict its ability to distinguish optical isomers of halothane.[11,40] Tadpoles cannot distinguish optical isomers of isoflurane.[41] Tadpoles also cannot distinguish stereoisomers of the 2-alcohols: 2-butanol through 2-octanol.[42] These compounds have not been tested in other animals. Studies of the potency of anesthetic stereoisomers in other animals are essential before Criterion 5 can be considered a suitable test for anesthetic relevance.

The discriminatory power of stereoisomers is mostly positive. Luciferase does not distinguish the two isomers of isoflurane,[43] and the effects of (+) and (–) isoflurane on the phase transition temperature of phosphatidylcholine are identical.[44] The (–) isomer of isoflurane is 2.5-fold more potent than the (+) isomer in inhibiting the decrease in adenosine triphosphate (ATP) concentration in anoxic rat hepatocytes,[45] an effect that is presumably unrelated to anesthesia. The equilibrium binding of isoflurane to bovine serum albumin is not stereoselective, but the association and dissociation rates for (+) isoflurane binding are slower than those for (–) isoflurane.[46]

1.5 THERMODYNAMICS

1.5.1 Pressure

In 1950, Johnson and Flagler made the observation that tadpoles anesthetized with ethanol resume swimming when the pressure is increased to 200 to 300 atm.[47] Pressure reversal is often considered a critical test for relevance to anesthesia. A complicating feature is that high pressure, by itself, usually has an excitatory effect on animals. There is an excellent review of the effects of high pressure and anesthetics.[48]

Criterion 6. The site of action of general anesthetics should exhibit pressure reversal of anesthetic effects

Figure 1.9 illustrates the two general ways in which high pressure might antagonize the effects of anesthetics in terms of Model I.[48] Pressure could displace anesthetic molecules from their site of action (Figure 1.9a). In this scheme, the excitatory effect of pressure by itself might occur by displacement of endogenous factors from their sites on the inhibitory input to the consciousness center. Alternatively, pressure could cause an excitation that counteracts the effects of anesthetics (Figure 1.9b). In order to assess the robustness of Criterion 6, we must first examine its generality. One difficulty is that few studies have been performed with volatile anesthetics; we must consider ethanol, barbiturates, and other agents as well. Among the animals that exhibit pressure reversal are tadpoles,[47,49] newts,[50] mice,[51] and rats.[52] Pressure reversal is not observed in a marine amphipod,[53] freshwater shrimp,[54] and nematodes.[55] Although it is tempting to ignore the negative results seen with some nonmammalian species in favor of the positive results seen with mammals, an additional observation makes this difficult. High pressure applied by itself has an immobilizing effect in the three nonmammalian animals. This could explain why high pressure does not reverse anesthesia in these animals.

Whether or not we place importance on the experiments showing the lack of pressure reversal in some species, Criterion 6 is weak. If we disregard these experiments, the criterion is ambiguous in terms of Model I. The two possible explanations for pressure reversal are equally likely. If, on the other hand, we emphasize these experiments, Criterion 6 is robust (Figure 1.9b), but it is not a useful indicator of anesthetic relevance. Pressure would be acting at a site that does not participate in general anesthesia. The site of action of general anesthetics would not be influenced by high pressure. Although the question of the discriminatory power of Criterion 6 is probably moot, for completeness we mention that anesthetic inhibition of firefly luciferase is not reversed by high pressure.[56]

1.5.2 Temperature

MAC, measured as a partial pressure, decreases as the temperature of the animal is lower. This has been observed in dogs,[57] rats,[58] goats,[59] and fish.[60] For example, halothane MAC in dogs is 0.86% at 37°C but only 0.47% at 28°C.[57]

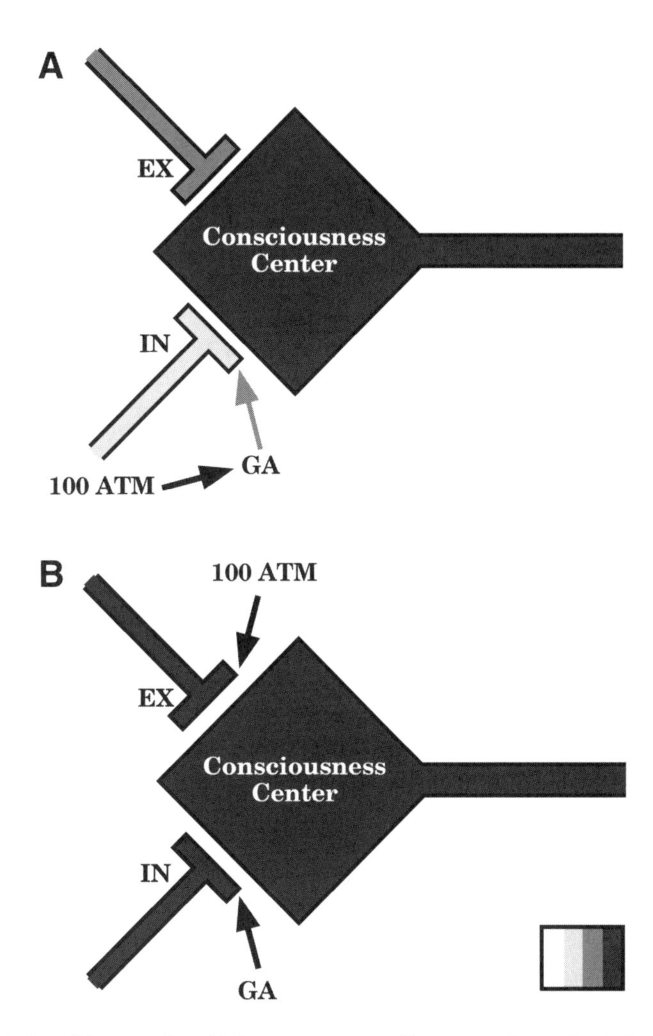

FIGURE 1.9 Possible ways in which pressure may affect general anesthesia in Model I. (A) Pressure displaces anesthetics (GA) so that they no longer act to increase inhibitory input (IN) to the consciousness neuron. The activity of the neuron returns to unanesthetized levels. (B) Pressure stimulates excitatory input (EX) to the consciousness neuron. This counteracts the effects of anesthetics on the activity level of the neuron.

Criterion 7. The site of action of general anesthetics should be less sensitive to anesthetics at lower temperatures

Several factors may contribute to the temperature dependence of MAC: (1) an increase in the oil/gas and water/gas partition coefficients with decreasing temperature, (2) changes in organ physiology and cell biochemistry with decreasing temperature, and (3) a genuine change in efficacy of the anesthetic with decreasing temperature. The first factor is, by far, the most important.[61] For halothane, the oil:gas and water:gas partition coefficients increase about 1.5-fold between 37 and 28°C

TABLE 1.4
Temperature Dependences of Halothane MAC in Dogs and of Halothane Partition Coefficients

Temp. (°C)	MAC (% atm)	Oil/Gas PC	Water/Gas PC	$[Hal]_{oil}$ (m*M*)	$[Hal]_{aq}$ (m*M*)
37	0.86	200	0.63	70	0.22
28	0.47	290	1.0	55	0.19

Note: The MAC values were obtained from Ref. 57.

(Table 1.4). These values are used to calculate the concentration of halothane in an oil and an aqueous phase. The aqueous concentration of halothane needed for anesthesia decreases only 15% between 37 and 28°C. This can be characterized by an enthalpy of transfer from the aqueous phase to the animal of –10 to –20 kJ/mol.[62]

To what can we ascribe this remaining temperature dependence of the anesthetic requirement? The effects of temperature on animal physiology and biochemistry are too numerous to reduce to a single factor. An extreme, but useful view is that physiology and biochemistry do not contribute to the temperature dependence of MAC. Rather, temperature dependence is a consequence of a temperature-dependent binding affinity. The partitioning of anesthetics into lipid bilayers composed of phosphatidylcholine, phosphatidic acid, and cholesterol is less at lower than at higher temperatures[21] and is characterized by a positive enthalpy of transfer.[62] Most studies on proteins have not included the important temperature range of 28 to 37°C. Anesthetic binding to luciferase is stronger at 4 than at 25°C; the enthalpy of transfer is about –20 kJ/mol.[63] Over the same temperature range, anesthetic inhibition of a neuronal acetylcholine receptor is characterized by an enthalpy of transfer of –20 to –30 kJ/mol.[62] The temperature dependence of the potentiation of $GABA_A$ receptors by anesthetics has been examined between 10 and 37°C.[64] Anesthetics are more potent at lower temperatures, but the enthalpy cannot be calculated because it is not clear how to relate potentiation to a binding constant.

In summary, it is difficult to assess the validity of Criterion 7. Changes in anesthetic potency with temperature are relatively small but anesthetic specific.[64] More studies will be needed to judge whether the criterion will be useful as a means of evaluating putative sites of anesthetic action.

1.6 GENETICS

The application of genetics to the study of anesthetic mechanisms uses selective breeding to produce mice,[65] fruit flies,[66] and nematodes[67] with altered sensitivities to volatile anesthetics. The goal of such experiments is to use mutant animals to help map the genetic locus of sites that govern an animal's sensitivity to anesthetics. This should point to the physical location of anesthetic targets in the nervous system.

An interesting finding in genetic studies is that animals selected for altered sensitivity to one anesthetic do not exhibit altered sensitivity to all anesthetics. Table

TABLE 1.5
Differential Sensitivity of Animal Strains to Volatile Anesthetics

Animal	Strain	Anesthetic	MAC Ratio	Anesthetic	MAC Ratio	Ref.
Mouse	Long–Evans vs. Sprague–Dawley	Isoflurane	0.77	Nitrous oxide	1.03	105
Mouse	129/SvJ vs. Spret/Ei	Desflurane	0.77	Halothane	0.95	68
		Isoflurane	0.79			
Fruit fly	har63 vs. wild-type	Trichloroethylene	0.6	Methoxyflurane	1.3	9
Nematode	unc-79 vs. N2 (wild type)	Halothane	0.31	Enflurane	1.06	69

1.5 shows some examples of this. Several other examples can be found in a recent study[68] of five anesthetics and 11 mouse strains (including inbred, hybrid, outbred, and knock-out mice). The largest deviation from a MAC ratio of 1.0 in Table 1.5 is for the unc-79 mutant and the wild-type nematode.[69] This mutant was selected for its high sensitivity (for a nematode) to halothane; the ED_{50} values are 3.2% (wild type) and 0.98% (unc-79). However, the unc-79 mutant is *less* sensitive to enflurane anesthesia than is the wild-type nematode: 5.9% (wild type), 6.2% (unc-79). These data may be interpreted as evidence that halothane and enflurane have different molecular sites of action. This, of course, challenges the unitary hypothesis of anesthetic action.

Examples for which the one animal strain lacks a protein that is expressed by the other strain (e.g., unc-79 vs. wild-type nematodes and the PKCγ knock-out mouse[68]) constitute strong evidence that different anesthetics are acting at different sites. Further studies of knock-out animals are needed to tell us whether these results pass the generality criterion.

In other cases, it is possible for differential sensitivity of strains or mutants to anesthetics to be interpreted in terms of unitary mechanisms of anesthetic action. The only additional assumption needed is that the different anesthetics may bind at different locations within a single molecular binding site. Figure 1.10 illustrates a hypothetical anesthetic binding site modeled as an amphipathic pocket within a protein. Enflurane and halothane both bind within the pocket but not at identical locations. The oxygen atom of enflurane (gray circle in Figure 1.10) keeps it close to the aqueous end of the pocket. In the wild-type binding site, halothane is prevented from penetrating further into the pocket by a kink in the pocket. In the mutant binding site, the kink is straightened, allowing halothane to bind more tightly by penetrating further into the pocket. The binding of enflurane is unaffected by the mutation because it is constrained by the requirement to remain close to the aqueous interface.

These two explanations for differential sensitivity to anesthetics — (1) two anesthetic binding sites controlled by separate genes and (2) two anesthetic binding locations on a single gene product (a protein, in this example) — both require one additional assumption over and above our simple model of anesthetic action (Model I). We conclude, then, that observations of differential sensitivity to anesthetics do not force us to abandon unitary models of anesthetic action.

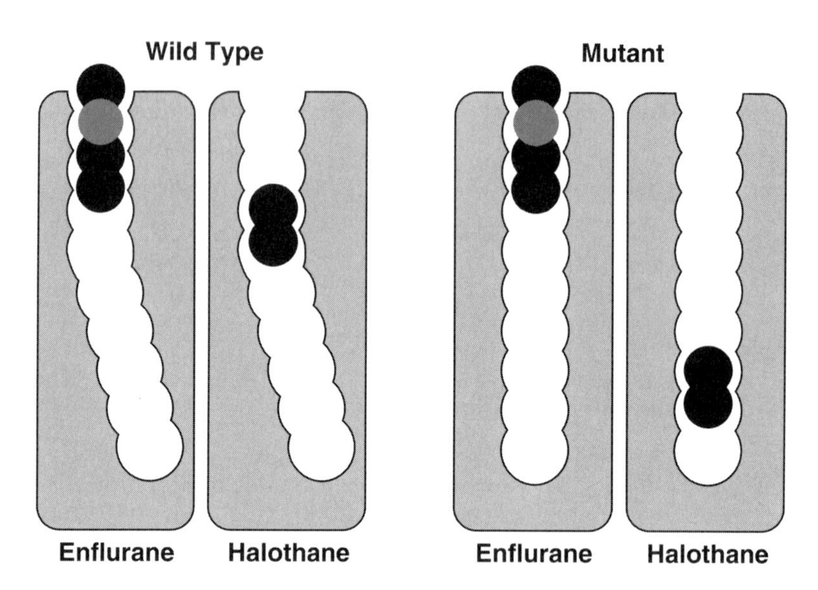

FIGURE 1.10 A genetic change within an anesthetic binding pocket can explain the differential sensitivity of strains and mutants to anesthetics. The binding pocket is assumed to be within a protein. The wild-type binding pocket has a kink but the mutant does not. Enflurane, represented by three (black) carbon atoms and one (gray) oxygen atom, binds far enough away from the kink so as to be unaffected by the mutation. Halothane, two (black) carbon atoms, binds near the kink in the wild type, but can penetrate deeper into the pocket of the mutant protein. Therefore, the mutation affects the binding of halothane more than the binding of enflurane.

A variation of the model in Figure 1.10 was used to interpret the differential sensitivity of two forms of firefly luciferase for anesthetics.[70] The affinity of the ATP–luciferase complex for anesthetics is greater than the affinity of free luciferase for anesthetics. In the homologous series of n-alcohols, the difference in affinity between the two forms of luciferase increases as the chain length increases. The interpretation is that the binding of ATP to luciferase causes a conformational change that results in an increase in the length of an amphipathic pocket on luciferase. This increases the pocket's binding affinity for long alcohols more than for short alcohols. ATP can be considered as a modulator of the anesthetic sensitivity of luciferase.[70] Similarly, we can consider the idea that some mutations are not affecting the site of action of anesthetics per se, but rather, are affecting some modulator of the anesthetic site.

Are there other results from genetic studies of anesthetics that might support the concept of multiple molecular sites of anesthetic action? There are some nematode mutants (e.g., fc23) that differ from wild type in their sensitivity to enflurane but not halothane.[71] This mutation codes for a gene product that is distinct from that of unc-79. Thus, two different protein (or lipid) molecules must be involved. One possibility is that the fc23 mutation affects a modulator of the anesthetic binding site. This explanation is just as valid as explanations involving multiple sites of anesthetic action.

1.7 HYPOTHESES

1.7.1 LIPID VERSUS PROTEIN

The lipid/protein controversy among anesthetic mechanicians has existed for many years. But it is far from being an old and tired debate topic. The controversy has, and continues to stimulate critical thinking about the interactions of general anesthetics with biological molecules. In addition, the controversy continues to evolve as our understanding of the structure and function of lipids and proteins improves. The center of the debate is whether lipids or proteins are the primary target for general anesthetics. In the fluid mosaic model of the cell membrane, proteins have active functions and lipids have passive, supporting functions. A reasonable hypothesis is that during anesthesia the normal function of some critical membrane protein (or proteins) is disrupted. This could be due to either a direct effect of anesthetics on the *protein* or an effect of anesthetics on the *lipids* surrounding the protein which, in turn, modifies the function of the protein.

Recent developments in cell physiology question a simple dichotomy of the roles of lipids and proteins. Lipids such as the phosphoinositides actively participate in membrane signaling. Proteins such as spectrin and agrin have primarily structural roles. And intracellular enzymes such as kinases and phosphatases modulate information processing. As a result, the divisions in the lipid/protein controversy have become less clear.

Consider the impact of anesthetic molecules on lipids. At MAC, the concentration of anesthetic molecules in a hydrophobic phase is about 50 m*M*. If the anesthetic molecules were distributed uniformly throughout the lipid bilayer of a 50 Å thick cell membrane, there would be only one anesthetic molecule for every 60 lipid molecules (i.e., 1.5% of the molecules in the membrane[72] and only 0.5% of the membrane volume[73]). In this scenario, the anesthetic molecules would be distributed too diffusely to have a significant effect on membrane properties. If, on the other hand, anesthetic molecules partition preferentially to a certain region of lipid bilayers, within the polar region or adjacent to a protein, then the anesthetic could potentially have a significant effect on local properties of the membrane.

A similar argument can be made for anesthetic–protein interactions. Halothane has a molecular weight of 197. The $GABA_A$ receptor channel has a molecular weight of about 250,000 Da. Could the binding of a single molecule of halothane possibly have a significant effect on the behavior of the protein? There are probably many sites on the protein where halothane could bind and have no effect. (The function of many proteins is unaffected by anesthetics, but it is not known if anesthetics bind to these proteins.) But there may be some sites where the binding of a single molecule of halothane could critically disrupt the function of the protein. Proteins are designed to undergo global structural changes upon the binding of small substrate molecules. γ-Aminobutyric acid, the neurotransmitter that allows the $GABA_A$ receptor channel to undergo the closed-to-open channel transition, has a molecular weight of only 103 Da.

Whether anesthetics have their primary effect on lipids or on proteins, the key features of the interaction must be specificity and selectivity. As discussed previously, only if anesthetics bind to a specific, localized region will they have a significant effect. Furthermore, the fact that general anesthesia is relatively safe and readily reversible supports the idea that anesthetics selectively interfere with a few important groups of nerve cells and allow the rest of the animal to function normally.

1.7.2 Lipid Hypotheses

Lipid hypotheses of anesthetic action postulate that the physical chemistry of lipid bilayers is modified by anesthetics. Some sensitive membrane proteins would react to this change in their environment. Lipid hypotheses are usually stated as unitary hypotheses; all general anesthetics have the same effect on the lipid bilayer membrane when a critical concentration of about 50 m*M* is reached. The membrane properties that have been considered to be important for anesthesia include thickness,[74] area per lipid molecule or volume,[50] curvature,[75] fluidity,[76,77] dielectric properties,[78,79] phase transition temperature,[80,81] and ionic permeability.[82]

Lipid hypotheses of anesthetic action flourished up until the mid-1970s mainly because they provided conceptually simple explanations for the Meyer–Overton correlation and the diversity of anesthetic structures. There was very little quantitative experimental information that either supported or refuted the idea that the lipid bilayer of cell membranes was the primary site of action of anesthetics. When data of the effects of anesthetics on lipids began to accumulate and anesthetic mechanicians began to examine it critically,[20,72,73,83,84] lipid hypotheses became less tenable. In the case of membrane thickness, volume, fluidity, permeability, and dielectric constant the problem is that there are simply too few anesthetic molecules in a bulk lipid phase to have an appreciable effect on these properties. (The effects of membrane curvature on protein function have not been adequately investigated.) Although one MAC of an anesthetic lowers the phase transition temperature of some lipids by several degrees, there are no known examples of biological membranes that are close to a thermal phase transition.[20]

It is possible that we have not yet identified the appropriate lipid property that is relevant to anesthesia. It is also possible that there exists a particularly anesthetic-sensitive combination of lipids somewhere in the nervous system. One relatively new idea is that anesthetics disrupt lipid domains in biological membranes.[85] Lipid domains can arise from the intrinsic properties of lipids, from electrostatic interactions between lipids and calcium, and from long-range organization by proteins.[86] Lipid domains are involved in vesicle formation[87] and in the association of membrane-bound substrates with intracellular enzymes such as myristoylated alanine-rich C kinase substrate with protein kinase C.[88]

What kind of experimental evidence would be needed to support disruption of lipid domains (using this as an example of a lipid hypothesis) as a possible anesthetic mechanism? It is easy to see how lipid domains might be sensitive to disruption by low concentrations of anesthetics because the adsorption of a small number of anesthetic molecules might act as a "defect" that destroys the integrity of a lipid domain. We would also expect that the effect would be a steep function of anesthetic concentration due to cooperativity. It would be necessary to demonstrate that the effect of the anesthetic was to disrupt the domain via interactions with the lipids rather than via interactions with proteins responsible for domain formation and stability. This would be less of a problem if domain formation occurred spontaneously in the absence of proteins. Finally, we must consider the assay for domain integrity. The best assay would be one that does not rely on the function of a protein, such as the microscopy of fluorescently labeled lipids.[86]

1.7.3 Amphipathic Pocket Protein Binding Site Hypothesis

The finding that the water-soluble enzyme, luciferase, is inhibited by general anesthetics[17,89] sparked an interest in considering direct interactions between anesthetics and proteins. General anesthetics act as competitive antagonists at the luciferin binding site on luciferase. The number of anesthetic molecules that can bind to this site depends on the size of the molecule. Thus, the binding site can accommodate a single molecule of decanol or two molecules of halothane. These results suggest a picture of the luciferin binding site of luciferase as an amphipathic pocket (see Figure 1.10). In this model, the dimensions of the pocket on luciferase are approximately the size of a molecule of decanol; the volume of the pocket is 250 ml/mol.[17]

The amphipathic pocket protein binding site hypothesis suggests a unitary mechanism of action. An amphipathic pocket could exist anywhere on a protein. For an amphipathic pocket to be a good candidate as a site of action for general anesthetics, it would either have some functional role (e.g., as a binding site for a natural substrate) or have importance for the structural stability of the protein in one or more of the protein's conformational states. The pore of an ion channel protein is an example of an amphipathic pocket. A pore contains hydrophobic residues that help anchor it to the cell membrane. It also has polar residues that stabilize partially hydrated ions as they pass through the pore. Moreover, there is an obvious functional consequence if an anesthetic molecule binds within the pore — the normal flow of ions through the pore is impeded.[90]

What sort of experimental evidence might be used to support the idea that anesthetics bind to an amphipathic pocket in a protein? First of all, the binding curve should saturate and its steepness should correspond to binding by one or two anesthetic molecules. The steepness of the binding curve for a small anesthetic molecule might be greater than that for a large anesthetic molecule. The kinetics of binding should indicate that the association rate is slower for larger anesthetic molecules because diffusion of large molecules into a narrow amphipathic pocket will be restricted. The dissociation rate should be slower for more hydrophobic molecules because of the strong correlation of anesthetic potency and hydrophobicity. If it is not possible to determine binding directly, then some information could be obtained by performing a functional assay at different anesthetic concentrations (the light reaction in the case of luciferase). A next step would be to use site-directed mutagenesis to produce proteins that have mutations in a region that might constitute part of the binding pocket. Changing the size and polarity of such regions should change anesthetic binding in a predictable way. Labeling experiments may also help identify anesthetic binding sites. Direct photoaffinity labeling of C^{14} halothane, followed by protein digestion and sequencing, can be used to locate amino acids labeled by halothane.[91] One difficulty is that there may be multiple halothane binding sites on the protein and there is no way to determine the sites for which halothane binding has a functional consequence. Combining photoaffinity labeling with mutagenesis could potentially overcome this drawback.[92]

The strongest evidence would be to obtain atomic-resolution structural information of the protein in the absence and presence of anesthetics. This is beginning to be possible for membrane proteins such as ion channels. A voltage-gated potassium channel has been imaged at 3.2 Å resolution.[93] The structure of the nicotinic acetylcholine receptor channel[94] has been determined at 4.6 Å. It may not be long before such studies can be used to reveal the location of small anesthetic molecules.

1.8 SUMMARY

Consider the chemical tools for identifying the sites of action of general anesthetics: Meyer–Overton correlation, nonanesthetics, stereoisomer anesthetics. Any realistic candidate for an anesthetic site must show a strong correlation between the effects of anesthetics on this site and the partition coefficient of the anesthetic. The poor discriminatory power of this criteria means that its usefulness is limited. Both nonanesthetics and stereoisomer anesthetics are potentially powerful tools, but they have not been extensively tested for generality. These two tools are also limited by their expense (although the enantiomers of the 2-alcohols are commercially available) or their special experimental requirements (pressure chamber for some of the nonanesthetics). It will be interesting to see if the discriminatory power of comparing sensitivity to isoflurane and enflurane is great enough to allow them to be considered as an inexpensive alternative to the stereoisomers to isoflurane.

The main problem with using pressure reversal as a test for relevance to anesthesia is that high pressure, in the absence of anesthetics, causes excitation in most animals. It is not easily determined, then, whether pressure reversal is due to the actions of high pressure on the anesthetic site or to an increased level of excitation elsewhere in the nervous system.

Many *in vitro* studies of the effects of anesthetics (especially those on ion channels) are conducted at room temperature rather than body temperature. Although temperature control can be implemented in many cases, higher temperatures can introduce experimental problems. For example, temperature-dependent kinetic processes may speed up to the point of being faster than the time resolution of the recording device. Nevertheless, the validity of experiments performed at nonphysiologic temperatures can be questioned. The question of choosing appropriate anesthetic concentrations is the easiest to answer: calculate aqueous concentrations rather than partial pressures.[61] Questions of physiologic relevance can be answered only by performing experiments at body temperature. It is important to remember, though, that temperature is only one of several parameters that must be considered to achieve appropriate physiological conditions. The ideal preparation would consist of intact neurons with functional synapses, exposed to normal ion concentrations and subject to the proper time-dependent neurotransmitter and neuromodulator concentrations — and, of course, at 37°C. Progress in understanding the actions of anesthetics can and will be made using less-than-ideal conditions. We cannot sit back and wait for the ultimate recording device to become commercially available.

The determination of genetic control of anesthetic sensitivity is a potentially powerful tool. However, even at this early stage of the application of these tools, the complexity of genetic control is apparent. The results obtained so far do not discount unitary theories of anesthetic action.

The lipid/protein controversy will probably continue for a long time. If consciousness is dependent on a membrane protein, then it will require a combination of structural and functional techniques to determine if the primary site for anesthetics is the protein, the lipid, or the interface between lipid and protein.

REFERENCES

1. Merkel, G. and Eger, E. I., II, A comparative study of halothane and halopropane anesthesia. Including method for determining equipotency, *Anesthesiology* 24, 346, 1963.
2. Eger, E. I., Saidman, L. J. and Brandstater, B., Minimum alveolar anesthetic concentration: A standard of anesthetic potency, *Anesthesiology* 26, 756, 1965.
3. Quasha, A. L., Eger, E. I. and Tinker, J. H., Determination and applications of MAC, *Anesthesiology* 53, 315, 1980.
4. Waud, B. E. and Waud, D. R., On dose-response curves and anesthetics, *Anesthesiology* 33, 1, 1970.
5. Eckenhoff, R. G. and Johansson, J. S., On the relevance of "clinically relevant concentrations" of inhaled anesthetics in *in vitro* experiments, *Anesthesiology* 91, 856, 1999.
6. Eger, E. I., Johnson, B. H., Weiskopf, R. B., Holmes, M. A., Yasuda, N., Targ, A. and Rampil, I. J., Minimum alveolar concentration of I-653 and isoflurane in pigs: Definition of a supramaximal stimulus, *Anesth. Analg.* 67, 1174, 1988.
7. Stoelting, R. K., Longnecker, D. R. and Eger, E. I., II, Minimum alveolar concentrations in man on awakening from methoxyflurane, halothane, ether and fluroxene anesthesia: MAC awake, *Anesthesiology* 33, 5, 1970.
8. Allada, R. and Nash, H. A., Drosophila-melanogaster as a model for study of general anesthesia — the quantitative response to clinical anesthetics and alkanes, *Anesth. Analg.* 77, 19, 1993.
9. Campbell, D. B. and Nash, H. A., Use of *Drosophila* mutants to distinguish among volatile general anesthetics, *Proc. Natl. Acad. Sci. USA* 91, 2135, 1994.
10. Dilger, J. P., Basic pharmacology of inhalational anesthetic agents, in *The Pharmacological Basis of Anesthesiology*, Bowdle, T. A., Horita, A. and Kharasch, E. D., Eds., Churchill Livingstone, New York, 1994, 497.
11. Morgan, P. G., Usiak, M. F. and Sedensky, M. M., Genetic differences affecting the potency of stereoisomers of isoflurane, *Anesthesiology* 85, 385, 1996.
12. Taheri, S., Halsey, M. J., Liu, J., Eger, E. I., Koblin, D. D. and Laster, M. J., What solvent best represents the site of action of inhaled anesthetics in humans, rats, and dogs?, *Anesth. Analg.* 72, 627, 1991.
13. Rampil, I. J., Mason, P. and Singh, H., Anesthetic potency (MAC) is independent of forebrain structures in the rat, *Anesthesiology* 78, 707, 1993.
14. King, B. S. and Rampil, I. J., Anesthetic depression of spinal motor neurons may contribute to lack of movement in response to noxious stimuli, *Anesthesiology* 81, 1484, 1994.

15. Antognini, J. F. and Schwartz, K., Exaggerated anesthetic requirements in the preferentially anesthetized brain, *Anesthesiology* 79, 1244, 1993.
16. Antkowiak, B., Different actions of general anesthetics on the firing patterns of neocortical neurons mediated by the $GABA_A$ receptor, *Anesthesiology* 91, 500, 1999.
17. Franks, N. P. and Lieb, W. R., Do general anaesthetics act by competitive binding to specific receptors?, *Nature* 310, 599, 1984.
18. Meyer, H. H., Theorie der Alkoholnarkose, *Arch. Exp. Pathol. Pharmakol.* 42, 109, 1899.
19. Overton, E., *Studies on Narcosis,* Lipnick, R. L., Ed., Routledge, Chapman and Hall, New York, 1990.
20. Franks, N. P. and Lieb, W. R., Where do general anaesthetics act?, *Nature* 274, 339, 1978.
21. Smith, R. A., Porter, E. G. and Miller, K. W., The solubility of anesthetic gases in lipid bilayers, *Biochim. Biophys. Acta* 645, 327, 1981.
22. Hill, M. W., The effect of anaesthetic-like molecules on the phase transition temperature in smectic mesophases of dipalmitoyllecithin. I. The normal alcohol up to C-9 and three inhalation anaesthetics, *Biochim. Biophys. Acta* 356, 117, 1974.
23. Eckenhoff, R. G. and Shuman, H., Halothane binding to soluble proteins determined by photoaffinity labeling, *Anesthesiology* 79, 96, 1993.
24. Paton, W. D. M. and Waud, D. R., The margin of safety of neuromuscular transmission, *J. Physiol. (Lond.)* 191, 59, 1967.
25. Marshall, C. G., Ogden, D. C. and Colquhoun, D., The actions of suxamethonium (succinyldicholine) as an agonist and channel blocker at the nicotinic receptor of frog muscle, *J. Physiol.* 428, 155, 1990.
26. Seeman, P., The membrane actions of anesthetics and tranquilizers, *Pharmacol. Rev.* 24, 583, 1972.
27. Pringle, M. J., Brown, K. B. and Miller, K. W., Can the lipid theories of anesthesia account for the cutoff in anesthetic potency in homologous series of alcohols?, *Mol. Pharmacol.* 19, 49, 1981.
28. Taheri, S., Laster, M. J., Liu, J., Eger, E. I., Halsey, M. J. and Koblin, D. D., Anesthesia by n-alkanes not consistent with the Meyer-Overton hypothesis — determinations of the solubilities of alkanes in saline and various lipids, *Anesth. Analg.* 77, 7, 1993.
29. Liu, J., Laster, M. J., Taheri, S., Eger, E. I., Koblin, D. D. and Halsey, M. J., Is there a cutoff in anesthetic potency for the normal alkanes?, *Anesth. Analg.* 77, 12, 1993.
30. Eger, E. I., Liu, J., Koblin, D. D., Laster, M. J., Taheri, S., Halsey, M. J., Ionescu, P., Chortkoff, B. S. and Hudlicky, T., Molecular properties of the "Ideal" inhaled anesthetic: Studies of fluorinated methanes, ethanes, propanes, and butanes. *Anesth. Analg.* 79, 245, 1994.
31. Koblin, D. D., Chortkoff, B. S., Laster, M. J., Eger, E. I., Halsey, M. J. and Ionescu, P., Polyhalogenated and perfluorinated compounds that disobey the Meyer-Overton hypothesis, *Anesth. Analg.* 79, 1043, 1994.
32. Murray, D. J., Mehta, M. P. and Forbes, R. B., The additive contribution of nitrous oxide to isoflurane MAC in infants and children, *Anesthesiology* 75, 186, 1991.
33. Abraham, M. H., Lieb, W. R. and Franks, N. P., Role of hydrogen bonding in general anesthesia, *J. Pharm. Sci.* 80, 719, 1991.
34. Fang, Z., Laster, M. J., Ionescu, P., Koblin, D. D., Sonner, J., Eger, E. I., II, and Halsey, M. J., Effects of inhaled nonimmobilizer, proconvulsant compounds on desflurane minimum alveolar anesthetic concentration in rats, *Anesth. Analg.* 85, 149, 1997.

35. Kandel, L., Chortkoff, B. S., Sonner, J., Laster, M. J. and Eger, E. I., II, Nonanesthetics can suppress learning, *Anesth. Analg.* 82, 321, 1996.
36. Franks, N. P. and Lieb, W. R., Mapping of general anaesthetic target sites provides a molecular basis for cutoff effects, *Nature* 316, 349, 1985.
37. Franks, N. P. and Lieb, W. R., Partitioning of long-chain alcohols into lipid bilayers: Implications for mechanisms of general anesthesia, *Proc. Natl. Acad. Sci. USA* 83, 5116, 1986.
38. Lysko, G. S., Robinson, J. L., Casto, R. and Ferrone, R. A., The stereospecific effects of isoflurane isomers in vivo, *Eur. J. Pharmacol.* 263, 25, 1994.
39. Eger, E. I., Koblin, D. D., Laster, M. J., Schurig, V., Juza, M., Ionescu, P. and Gong, D., Minimum alveolar anesthetic concentration values for the enantiomers of isoflurane differ minimally, *Anesth. Analg.* 85, 188, 1997.
40. Sedensky, M. M., Cascorbi, H. F., Meinwald, J., Radford, P. and Morgan, P. G., Genetic differences affecting the potency of stereoisomers of halothane, *Proc. Natl. Acad. Sci. USA* 91, 10054, 1994.
41. Firestone, S., Ferguson, C. and Firestone, L., Isoflurane's optical isomers are equipotent in Rana pipiens tadpoles, *Anesthesiology* 77, A758, 1992.
42. Alifimoff, J. K., Firestone, L. L. and Miller, K. W., Anesthetic potencies of secondary alcohol enantiomers, *Anesthesiology* 66, 55, 1987.
43. Franks, N. P., Dickinson, R. and Lieb, W. R., Effects of isoflurane enantiomers on general anaesthetic target sites, in *Fluorine in Medicine in the 21st Century*, Banks, R. E. and Lowe, K. C., Eds., University of Manchester Institute of Science & Technology, UK, Paper 16, 1994, 1.
44. Franks, N. P. and Lieb, W. R., Stereospecific effects of inhalational general anesthetic optical isomers on nerve ion channels, *Science* 254, 427, 1991.
45. Pohorecki, R. A., Howard, B. J. A., Matsushita, M. A., Stemmer, P. M. A., Becker, G. L. A. and Landers, D. F. T., Isoflurane isomers differ in preservation of ATP in anoxic rat hepatocytes, *J. Pharmacol. Exp. Ther.* 268, 625, 1994.
46. Xu, Y., Tang, P., Firestone, L. and Zhang, T. T., ^{19}F nuclear magnetic resonance investigation of stereoselective binding of isoflurane to bovine serum albumin, *Biophys. J.* 70, 532, 1996.
47. Johnson, F. H. and Flagler, E. A., Hydrostatic pressure reversal of narcosis in tadpoles, *Science* 112, 91, 1950.
48. Wann, K. T. and MacDonald, A. G., Actions and interactions of high pressure and general anaesthetics. *Prog. Neurobiol.* 30, 271, 1988.
49. Halsey, M. J. and Wardley-Smith, B., Pressure reversal of narcosis produced by anaesthetics, narcotic, and tranquillisers, *Nature* 257, 811, 1975.
50. Miller, K. W., Paton, W. D. M., Smith, R. A. and Smith, E. B., Pressure reversal of general anesthesia and the critical volume hypothesis, *Mol. Pharmacol.* 9, 131, 1973.
51. Miller, K. W. and Wilson, M. W., The pressure reversal of a variety of anesthetic agents in mice, *Anesthesiology* 48, 104, 1978.
52. Halsey, M. J., Wardley-Smith, B. and Green, C. J., Pressure reversal of general anaesthesia — a multi-site expansion hypothesis, *Br. J. Anaesth.* 50, 1091, 1978.
53. Youngson, A. F. and MacDonald, A. G., Interaction between halothane and hydrostatic pressure, *Br. J. Anaesth.* 42, 801, 1970.
54. Simon, S. A., Parmentier, J. L. and Bennett, P. B., Anesthetic antagonism of the effects of high hydrostatic pressure on locomotory activity of the brine shrimp Artemia, *Comp. Biochem. Physiol.* 75A, 193, 1983.
55. Eckenhoff, R. G. and Yang, B. J., Absence of pressure antagonism of ethanol narcosis in *C. elegans, NeuroReport* 6, 77, 1994.

56. Moss, G. W. J., Lieb, W. R. and Franks, N. P., Anesthetic inhibition of firefly luciferase, a protein model for general anesthesia, does not exhibit pressure reversal, *Biophys. J.* 60, 1309, 1991.
57. Eger, E. I., Saidman, L. J. and Brandstater, B., Temperature dependence of halothane and cyclopropane anesthesia in dogs: Correlation with some theories of anesthetic action, *Anesthesiology* 26, 764, 1965.
58. Vitez, T. S., White, P. F. and Eger, E. I., Effects of hypothermia on halothane MAC and isoflurane MAC in the rat, *Anesthesiology* 41, 80, 1974.
59. Antognini, J. F., Hypothermia eliminates isoflurane requirements at 20°C, *Anesthesiology* 78, 1152, 1993.
60. Cherkin, A. and Catchpool, J. F., Temperature dependence of anesthesia in goldfish, *Science* 144, 1460, 1964.
61. Franks, N. P. and Lieb, W. R., Temperature dependence of the potency of volatile general anesthetics: Implications for *in vitro* experiments, *Anesthesiology* 84, 716, 1996.
62. Dickinson, R., Lieb, W. R. and Franks, N. P., The effects of temperature on the interactions between volatile general anaesthetics and a neuronal nicotinic acetylcholine receptor, *Br. J. Pharmacol.* 116, 2949, 1995.
63. Dickinson, R., Franks, N. P. and Lieb, W. R., Thermodynamics of anesthetic/protein interactions. Temperature studies on firefly luciferase, *Biophys. J.* 64, 1264, 1993.
64. Jenkins, A., Franks, N. P. and Lieb, W. R., Effects of temperature and volatile anesthetics on $GABA_A$ receptors, *Anesthesiology* 90, 484, 1999.
65. Koblin, D. D., Dong, D. E., Deady, J. E. and Eger, E. I., Selective breeding alters murine resistance to nitrous oxide without alteration in synaptic membrane composition, *Anesthesiology* 62, 401, 1980.
66. Gamo, S., Nakashima-Tanaka, E. and Ogali, M., Strain differences in minimum anesthetic concentrations in *Drosophila menlanogaster*, *Anesthesiology* 54, 289, 1981.
67. Morgan, P. and Cascorbi, H. F., Effect of anesthetics and a convulsant on normal and mutant *Caenorhabditis elegans*, *Anesthesiology* 62, 738, 1985.
68. Sonner, J. M., Gong, D., Li, J., Eger, E. I., II and Laster, M. J., Mouse strain modestly influences minimum alveolar anesthetic concentration and convulsivity of inhaled compounds, *Anesth. Analg.* 89, 1030, 1999.
69. Morgan, P. G., Sedensky, M. and Meneely, P. M., Multiple sites of action of volatile anesthetics in *Caenorhabditis elegans*, *Proc. Nat. Acad. Sci. USA* 87, 2965, 1990.
70. Moss, G. W. J., Franks, N. P. and Lieb, W. R., Modulation of the general anesthetic sensitivity of a protein: A transition between two forms of firefly luciferase, *Proc. Natl. Acad. Sci. USA* 88, 134, 1991.
71. Morgan, P. G. and Sedensky, M. M., Mutations conferring new patterns of sensitivity to volatile anesthetics in *Caenorhabditis elegans*, *Anesthesiology* 81, 888, 1994.
72. Franks, N. P. and Lieb, W. R., The structure of lipid bilayers and the effects of general anesthetics. An X-ray and neutron diffraction study, *J. Mol. Biol.* 133, 469, 1979.
73. Franks, N. P. and Lieb, W. R., Is membrane expansion relevant to anaesthesia?, *Nature* 292, 248, 1981.
74. Haydon, D. A., Hendry, B. M., Levinson, S. R. and Requena, J., The molecular mechanism of anesthesia, *Nature* 268, 356, 1977.
75. Gruner, S. and Shyamsunder, E., Is the mechanism of general anesthesia related to lipid membrane spontaneous curvature?, *Ann. NY Acad. Sci.* 625, 685, 1991.
76. Metcalfe, J. C., Seeman, P. and Burgen, A. S. V., The proton relaxation of benzyl alcohol in erythrocyte membranes, *Mol. Pharmacol.* 4, 87, 1968.

77. Ueda, I., Hirakawa, M., Arakawa, K. and Kamaya, H., Do anesthetics fluidize membranes?, *Anesthesiology* 64, 67, 1986.
78. Gage, P. W., McBurney, R. N. and Schneider, G. T., Effects of some aliphatic alcohols on the conductance change caused by a quantum of acetylcholine at the toad endplate, *J. Physiol.* 244, 409, 1975.
79. Enders, A., The influence of general, volatile anesthetics on the dynamic properties of model membranes, *Biochim. Biophys. Acta* 1029, 43, 1990.
80. Trudell, J., A unitary theory of anesthesia based on lateral phase separations in nerve membranes, *Anesthesiology* 46, 5, 1977.
81. Suezaki, Y., Tamura, K., Takasaki, M., Kamaya, H. and Ueda, I., A statistical mechanical analysis of the effect of long-chain alcohols and high pressure upon the phase transition temperature of lipid bilayer membranes, *Biochim. Biophys. Acta* 1066, 225, 1991.
82. Bangham, A. D. and Mason, W. T., Anaesthetics may act by collapsing pH gradients, *Anesthesiology* 53, 135, 1980.
83. Katz, Y. and Simon, S. A., Physical parameters of the anesthetic site, *Biochim. Biophys. Acta* 471, 1, 1977.
84. Akeson, M. A. and Deamer, D. W., Steady-state catecholamine distribution in chromaffin granule preparations: A test of the pump-leak hypothesis of general anesthesia, *Biochemistry* 28, 5120, 1989.
85. Janes, N., Hsu, J. W., Rubin, E. and Taraschi, T. F., Nature of alcohol and anesthetic action on cooperative membrane equilibria, *Biochemistry* 31, 9467, 1992.
86. Glaser, M., Lipid domains in biological membranes, *Curr. Opin. Struct. Biol.* 3, 475, 1993.
87. Brown, D. A. and Rose, J. K., Sorting of GPI-anchored proteins to glycolipid-enriched membrane subdomains during transport to the apical cell surface, *Cell* 68, 533, 1992.
88. Yang, L. and Glaser, M., Membrane domains containing phosphatidylserine and substrate can be important for the activation of protein kinase C, *Biochemistry* 34, 1500, 1995.
89. DeLuca, M., Hydrophobic nature of the active site of firefly luciferase, *Biochemistry* 8, 160, 1969.
90. Dilger, J. P., Vidal, A. M., Mody, H. I. and Liu, Y., Evidence for direct actions of general anesthetics on an ion channel protein — a new look at a unified mechanism of action, *Anesthesiology* 81, 431, 1994.
91. Eckenhoff, R. G., An inhalational anesthetic binding domain in the nicotinic acetylcholine receptor, *Proc. Natl. Acad. Sci. USA* 93, 2807, 1996.
92. Husain, S. S., Forman, S. A., Kloczewiak, M. A., Addona, G. H., Olsen, R. W., Pratt, M. B., Cohen, J. B. and Miller, K. W., Synthesis and properties of 3-(2-hydroxyethyl)-3-n-pentyldiazirine, a photoactivable general anesthetic, *J. Med. Chem.* 42, 3300, 1999.
93. Doyle, D. A., Morais Cabral, J., Pfuetzner, R. A., Kuo, A., Gulbis, J. M., Cohen, S. L., Chait, B. T. and MacKinnon, R., The structure of the potassium channel: Molecular basis of K+ conduction and selectivity, *Science* 280, 69, 1998.
94. Miyazawa, A., Fujiyoshi, Y., Stowell, M. and Unwin, N., Nicotinic acetylcholine receptor at 4.6 Å resolution: Transverse tunnels in the channel wall, *J. Mol. Biol.* 288, 765, 1999.
95. Firestone, L. L., Miller, J. C. and Miller, K. M., Tables of physical and pharmacological properties of anesthetics, in *Molecular and Cellular Mechanisms of Anesthetics,* Roth, S. H. and Miller, K. W., Eds., Plenum Press, New York, 1986, 267.

96. Eger, E. I., II, Partition coefficient of I-653 in human blood, saline and olive oil, *Anesth. Analg.* 66, 971, 1987.
97. Strum, D. P. and Eger, E. I., II, Partition coefficients for sevoflurane in human blood, saline and olive oil, *Anesth. Analg.* 66, 654, 1987.
98. Seifen, A. B., Kennedy, R. H., Bray, J. P. and Seifen, E., Estimation of minimum alveolar concentration (MAC) for halothane, enflurane and isoflurane in spontaneously breathing guinea pigs, *Lab. Anim. Sci.* 39, 579, 1989.
99. Steffey, E. P., Howland, D., Jr., Giri, S. and Eger, E. I., Enflurane, halothane, and isoflurane potency in horses, *Am. J. Vet. Res.* 38, 1037, 1977.
100. Drummond, J. C., Todd, M. M. and Shapiro, H. M., Minimal alveolar concentrations for halothane, enflurane, and isoflurane in the cat, *J. Am. Vet. Med. Assoc.* 182, 1099, 1983.
101. Mazze, R. I., Rice, S. A. and Baden, J. M., Halothane, isoflurane, and enflurane MAC in pregnant and nonpregnant female and male mice and rats, *Anesthesiology* 62, 339, 1985.
102. Drummond, J. C., MAC for halothane, enflurane and isoflurane in the New Zealand white rabbit, and a test for the validity of MAC determinations, *Anesthesiology* 62, 336, 1985.
103. McKenzie, J. D., Calow, P. and Nimmo, W. S., Effects of inhalational general anaesthetics on intact *Daphnia-magna* (Cladocera, Crustacea), *Comp. Biochem. Physiol.* 101C, 9, 1992.
104. Morgan, P. G., Sedensky, M. M., Meneely, P. M. and Cascorbi, H. F., The effect of two genes on anesthetic response in the nematode *Caenorhabditis elegans*, *Anesthesiology* 69, 246, 1988.
105. Russell, G. B. and Graybeal, J. M., Differences in anesthetic potency between Sprague-Dawley and Long-Evans rats for isoflurane but not nitrous oxide, *Pharmacology* 50, 162, 1995.
106. Miller, K. W., Paton, W. D. M., Smith, E. B. and Smith, R. A., Physiochemical approaches to the mode of action of general anesthetics, *Anesthesiology* 36, 339, 1972.
107. Steffey, E. P. and Howland, D., Jr., Halothane anesthesia in calves, *Am. J. Vet. Res.* 40, 372, 1979.
108. Brown, B. R. and Crout, J. R., A comparative study of the effects of five general anesthetics on myocardial contractility. I. Isomeric conditions, *Anesthesiology* 34, 236, 1971.
109. Davis, N. L., Nunnally, R. L. and Malinin, T. I., Determination of the minimum alveolar concentration (MAC) of halothane in the white New Zealand rabbit, *Br. J. Anaesth.* 47, 341, 1975.
110. Steffey, E. P., Gillespie, J. R., Berry, J. D., Eger, E. I., II and Munson, E. S., Anesthetic potency (MAC) of nitrous oxide in the dog, cat and stump-tail monkey, *J. Appl. Physiol.* 36, 530, 1974.
111. Tinker, J. H., Sharbrough, F. W. and Michenfelder, J. D., Anterior shift of the dominant EEG rhythm during anesthesia in the Java monkey, Correlation with anesthetic potency, *Anesthesiology* 46, 252, 1977.
112. Weiskopf, R. B. and Bogetz, M. S., Minimum alveolar concentrations (MAC) of halothane and nitrous oxide in swine, *Anesth. Analg.* 63, 529, 1984.
113. Lerman, J., Oyston, J. P., Gallagher, T. M., Miyasaka, K., Volgyesi, G. A. and Burrows, F. A., The minimum alveolar concentration (MAC) and hemodynamic effects of halothane, isoflurane, and sevoflurane in newborn swine, *Anesthesiology* 73, 717, 1990.

114. Kita, Y., Bennett, L. J. and Miller, K. W., The partial molar volumes of anesthetics in lipid bilayers, *Biochim. Biophys. Acta* 647, 130, 1981.
115. Shim, C. Y. and Andersen, N. B., The effect of oxygen on minimal anesthetic requirements in the toad, *Anesthesiology* 34, 333, 1971.
116. Cruickshank, S. G. H., Girdlestone, D. and Winlow, W., The effects of halothane on the withdrawal response of *Lymnaea*, *J. Physiol.* 367, 8P, 1985.

2 Experimental Approaches to the Study of Volatile Anesthetic–Protein Interactions

Jonas S. Johansson and Roderic G. Eckenhoff

CONTENTS

2.1 INTRODUCTION

Inhalational anesthetics provide unique experimental challenges because of their volatility and the relatively weak energetics of their interactions with *in vivo* target site(s), based upon their measured EC_{50} values of 0.2 to 1 m*M*. These weak energetics translate into rapid target binding and dissociation rates, making traditional

0-8493-8555-5/01/$0.00+$.50

pharmacologic approaches to the study of ligand–receptor interactions, such as radioligand binding studies, impractical. Furthermore, the clinically used inhalational anesthetics are chemically inert and contain few functional groups that lend themselves readily to biochemical and biophysical analyses. These factors have made it difficult to study anesthetic–macromolecule interactions, contributing to the slow progress in our understanding of both the site(s) and the mechanism(s) of volatile general anesthetic action.

Despite these experimental difficulties, it will be essential to define how these drugs interact with biological macromolecules at the structural level, in order to understand mechanisms of action. We focus on interactions with protein, because current consensus favors proteins as a general target,[1,2] and also because the molecular interactions of anesthetics with proteins have been less studied than those with lipid. This chapter reviews the experimental tools currently being used and will point out their respective advantages and limitations. Although the focus of this chapter is on methodology, the implications to anesthetic mechanisms will be discussed where the data allow. Further, this is not intended to be an exhaustive review of each method as it relates to anesthetics; rather, we intend to briefly describe each approach, then present some examples of how it has been used to study the interactions of anesthetics with proteins. Approaches useful for characterizing binding,[3-9] such as X-ray crystallography, ^{19}F-NMR (nuclear magnetic resonance) spectroscopy, fluorescence quenching, and photoaffinity labeling, will be preceded by a brief review of the basic principles of anesthetic binding interactions. The methods for characterizing the structural changes that occur in the protein target after anesthetic binding will also be discussed, with the ultimate goal of laying the experimental framework for an understanding of how anesthetics might alter protein function and thereby cause the clinical state of anesthesia.

2.2 BASIC PRINCIPLES OF ANESTHETIC BINDING TO PROTEIN

Binding of an anesthetic (A) to a protein target (P) can be described as

$$A + P \underset{k_{-1}}{\overset{k_1}{\leftrightarrow}} A - P \underset{k_{-2}}{\overset{k_2}{\leftrightarrow}} A - P^*$$

where k_1 and k_{-1} are the on and off rate constants for anesthetic binding and decomplexation, respectively, and k_2 and k_{-2} are the rate constants for any ensuing structural change in the protein (P*), and its reversal, respectively, that leads to a functional change in protein activity. This is, of course, the most simplified scheme, because the path from A – P to A – P* may involve multiple discrete steps, each with its own kinetic signature. Although limited, the kinetic data for volatile anesthetic binding and decomplexation are reviewed in Section 2.5, which deals with ^{19}F-NMR spectroscopy.

From a free energy standpoint,[10] the transfer of a hydrophobic anesthetic molecule from water to a protein binding site can be broken down, for ease of analysis, into the following four discrete steps: (1) transfer of anesthetic from water to vacuum; (2) collapse of the water cavity; (3) creation of a cavity in the protein; and (4) transfer of the anesthetic from vacuum to the protein cavity. In the case where a cavity already exists in the protein, the third step (which is energetically unfavorable) is not required, and the overall free energy of transfer from water to protein is more favorable. Furthermore, if the protein cavity is filled with one or more water molecules in the native state, the free energy of transfer of anesthetic from bulk water to protein will be even more favorable, because of the entropic gain associated with transfering bound water to the bulk phase.[11,12] Thus, it is predicted, from an energetic point of view, that cavities (or packing defects) in proteins, which occur in the vast majority of proteins studied to date,[13-15] are likely to be occupied by volatile anesthetic molecules, if they exhibit suitable volumes and shapes. Experimental support for this is outlined in this chapter. However, it remains to be determined whether occupancy of such widely available protein sites by anesthetics has any effect on protein structure, dynamics, or function.

Finally, what is known about the nature of the *in vivo* inhalational anesthetic binding site(s)? Because the target(s) for these agents is unknown, investigators have historically approached this question by correlating *in vivo* potency data with solvation into essentially homogeneous organic solvents. Because of improved correlations with solvents such as n-octanol, these studies indicate that there is an important polar component to the anesthetic site of action.[16-18] One interpretation of this finding has been that anesthetics are able to form hydrogen bonds with macromolecular targets,[19-24] explaining not only binding energetics, but also the change in protein function through the concept of competitive hydrogen bonding. The action of the volatile anesthetics might then be related to their ability to compete with native protein hydrogen bonds, causing conformational and functional changes. The current structural understanding of anesthetic binding sites on proteins, determined using a variety of experimental approaches, will be discussed in several of the sections that follow.

2.3 X-RAY CRYSTALLOGRAPHY

X-ray diffraction analysis is a powerful method for determining the three-dimensional structure of molecules. In the case of proteins, a suitable crystal is first obtained, then irradiated with X-rays which are scattered in a characteristic pattern by the electrons present. Because the protein molecules orient in the same manner in a crystal, the diffraction pattern of each atom (or group) will add. Based upon the diffraction pattern obtained, it is possible to reconstruct a model of the protein structure.

The ability to obtain high-resolution crystal structures of proteins in the presence of various ligands allows direct determination of the role of specific side chains for recognition and stability.[25] Changes in the topology of the protein in the presence of the complexed ligand provide structural data for how binding might lead to an alteration in protein function. For example, X-ray diffraction analysis of hemoglobin crystals has helped our understanding of how the binding of one oxygen molecule causes the structural changes that facilitate binding of additional oxygen molecules.[26] More recently, X-ray crystallography has shown how the binding of the inducer (allolactose) to the *lac* repressor protein results in structural changes that prevent it from binding to DNA.[27] Thus, X-ray crystallography potentially can provide a precise structural description of the anesthetic binding sites, and the underlying interactions, and also how binding might alter protein conformation and therefore function.

X-ray crystallography was one of the first techniques used to probe the molecular features of inhalational anesthetic–protein complexes. Thus, xenon (which is a general anesthetic with a minimum alveolar concentration of 0.71 atm[28]) was shown to bind to a discrete site (at 2.5 atm), equidistant from the proximal histidine and one of the pyrrole rings of the heme moiety, in sperm whale myoglobin.[29] Further studies showed that both cyclopropane and dichloromethane bound to the same site in metmyoglobin.[30,31] On raising the xenon pressure to 7 atm, as many as four distinct binding sites for xenon, of varying affinity, were demonstrated in myoglobin.[32] As a model anesthetic target, myoglobin is unable to reproduce the clinical pharmacology of a number of different anesthetic molecules,[33] but the importance of this work lies in the demonstration that ligands may bind to hydrophobic regions of proteins, through weak interactions, to aliphatic residues and aromatic rings (pyrrole, phenylalanine, and tyrosine). Furthermore, as will become apparent from the remainder of the work described in this Chapter, the side-chain composition of the binding sites in myoglobin is a recurrent feature of such domains in other proteins. X-ray diffraction analysis has also revealed that xenon binds to hemoglobin.[34] Each α and β chain of hemoglobin binds a single xenon molecule, but at sites that are not analogous to the highest affinity site in myoglobin. However, the character of the hemoglobin xenon binding domains, lined by valine, leucine, and phenylalanine residues, is similar to that of the site in myoglobin.

The only X-ray diffraction study to incorporate a clinically relevant haloalkane anesthetic is that describing halothane binding to adenylate kinase.[35] Halothane inhibits this enzyme with a K_i (inhibition constant) of 2.5 m*M*, at low substrate (adenosine 5′-monophosphate, AMP) concentration (100 μM), suggesting a competitive interaction between halothane and AMP. Crystals of adenylate kinase soaked in saturated solutions of halothane (≈ 18 m*M* anesthetic[36]) showed localization of the anesthetic molecule in a discrete interhelical niche, lined by the aliphatic residues valine, leucine, and isoleucine, and also by the more polar residues tyrosine, arginine, and glutamine.[37] Such a site satisfies the suggestions made above that an anesthetic binding site, although necessarily hydrophobic (aliphatic), should also have some polar (a heading under which we include aromatic residues[12,38-41]) character. Interestingly, there was no indication of a structural consequence to adenylate kinase on binding halothane, but this may be due to insufficient resolution (6 Å).

Dichloroethane was one of the first haloalkane compounds to be shown to have useful anesthetic properties.[42] Binding of this halogenated alkane to crystals of bacterial haloalkane dehalogenase,[43] and insulin,[44] has been analyzed with X-ray diffraction. In the former protein, the dichloroethane is the native substrate that *Xanthobacter autotrophicus* GJ10 uses to derive energy. The dihaloalkane binding site is a predominantly hydrophobic pocket containing four phenylalanines, two

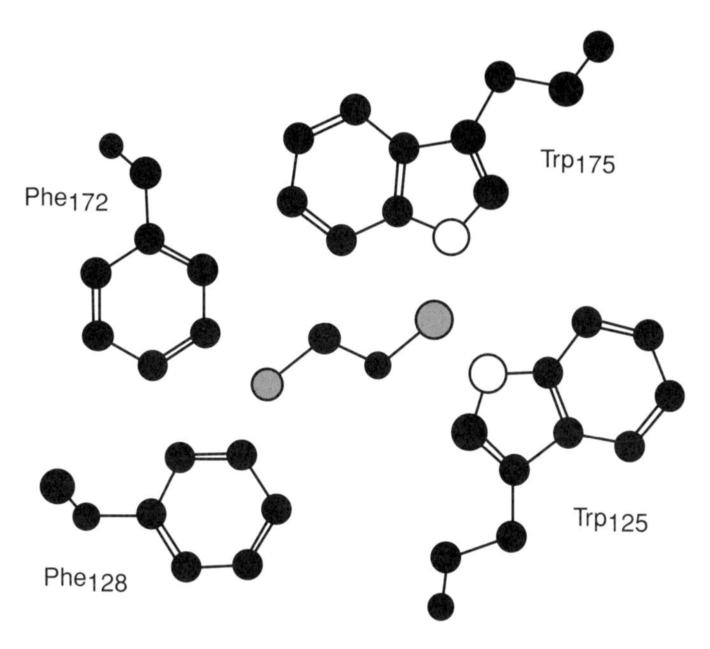

FIGURE 2.1 Binding of the substrate 1,2-dichloroethane to the active site of haloalkane dehalogenase from *Xanthobacter autotrophicus* GJ10. One of the chlorine atoms electrostatically interacts with the indole ring nitrogen protons of Trp125 and Trp175, whereas the other chlorine interacts with the aromatic ring protons of Phe128 and Phe172. Carbons are black, nitrogens are white, and chlorines are gray. (Adapted from Ref. 43.)

tryptophans, two leucines, a valine, a proline, and an aspartate.[43,45] Two of the phenylalanines and two of the tryptophans make electrostatic contacts with the dihaloalkane chlorines (Figure 2.1). The Michaelis constant (K_m) for dichloroethane dehalogenation by this enzyme is 0.7 to 1.1 m*M*. This site also binds and dehalogenates smaller haloalkanes such as methyl- and ethyl chloride,[46] but with higher apparent K_m values, suggesting the importance of the two additional electrostatic contacts on the two-carbon halogen, or loss of favorable van der Waals contacts. Site-directed mutagenesis studies involving the replacement of the two active site tryptophan residues by glutamines or phenylalanines indicate the importance of aromatic residues for substrate binding in this protein.[47] In contrast to this bacterial protein, the cavity in insulin is lined by serine, valine, glutamate, and tyrosine residues and binds only the *cis* conformation of 1,2-dichloroethane. This cavity is apparently too small and sterically hindered to bind the *trans* conformation. All of the current clinically used haloalkanes are too large to bind to this insulin cavity due to steric constraints. Smaller haloalkanes, such as dichloromethane, also bind poorly, presumably due to loss of stabilizing van der Waals contacts. Interestingly, the insulin cavity contains structured water molecules, some of which are displaced on haloalkane binding, suggesting an additional favorable entropic contribution to complex formation. Because of the small size of water molecules, and the inability to observe hydrogen atoms (because the single electron on hydrogen scatters X-rays poorly), high-resolution crystal structures are required in order to allow the position of water molecules to be inferred. This level of resolution has not been achieved in other protein–anesthetic systems to date, so it is unclear whether water displacement is a general feature of anesthetic binding.

Of related interest are studies on benzene binding to a hydrophobic cavity (or packing defect) engineered into T4 lysozyme.[48] Although toxic, benzene is also a volatile anesthetic with an EC_{50} of 1 to 2 m*M* in tadpoles.[49] Benzene does not bind appreciably to the native T4 lysozyme. Replacement of a leucine by an alanine in the hydrophobic core of the enzyme creates a cavity, lined by both aliphatic valine and leucine residues, and an aromatic phenylalanine residue, allowing benzene to bind with a K_d of 0.4 to 1.1 m*M*. Again, the structural consequences to the protein target are subtle in the presence of benzene, involving reorientation of some of the side chains in direct contact with the bound ligand and a 0.4 Å shift in the position of a short α-helical portion of the protein. Binding of benzene to this T4 lysozyme cavity mutant stabilizes the overall structure of the folded protein, as assessed by circular dichroism thermal denaturation spectroscopy (see Section 2.9). Such changes in side-chain orientation and global protein stability in the presence of anesthetic represent potential mechanisms whereby protein function may be altered.

This section has so far dealt exclusively with the results of X-ray diffraction studies on water-soluble proteins. Although X-ray diffraction studies of membrane proteins have been technically difficult, because these proteins do not readily form three-dimensional crystals, some progress has been made with novel crystallographic approaches using short chain detergents or antibody fragments,[50-53] or with highly ordered natural membrane proteins, such as bacteriorhodopsin.[24] In the latter case, low-resolution difference analysis of electron density maps with and without the haloalkane anesthetic diiodomethane showed preferential localization in the phospholipid-filled center of the naturally occurring protein trimers, suggesting lipid–protein interfacial binding. However, the natural organization of this membrane protein biases the interpretation, since the paucity of phospholipid molecules renders them all essentially interfacial.

The advantage of using X-ray crystallography to study protein–ligand complexes is that it provides a view of molecular interactions that is illuminating from both structural and mechanistic points of view. The types of residues, and the noncovalent forces involved in complexing the ligand, can be directly determined. The strength of various interactions resulting in binding can be predicted from the measured distances between individual atoms on ligand and protein. Finally, and perhaps most importantly, changes in the structure of the protein suggest mechanistically how protein function is altered following ligand binding. The principal drawback is the technical difficulty associated with growing suitable crystals, particularly for the membrane proteins. In addition, the unusual conditions necessary to grow crystals may make the ligand concentration difficult to control. For example, studies on the binding of inhaled anesthetics to protein crystals are typically performed with saturated solutions of anesthetic[31,35,44] so that it will remain unclear whether the site occupied by the anesthetic would also be filled at the concentrations of anesthetic used clinically. This is clearly important as demonstrated in the case of xenon binding to myoglobin, where additional sites are recruited as the anesthetic concentration is increased.[32] Furthermore, the assumptions and algorithms used to reduce the electron density map to a three-dimensional protein model have an inherent degree of subjectivity, suggesting that several structures are equally plausible.[54] Finally, given the harsh conditions under which protein crystals are grown,[55,56] it is perhaps not surprising that the structure assumed by the protein in the solid state may not always correspond to its native conformation in aqueous solution.[57]

2.4 GAS CHROMATOGRAPHIC PARTITION ANALYSIS

In usual practice, gas chromatographic partition analysis measures the gas phase concentration of an anesthetic in equilibrium with either a buffer or a protein solution. Partitioning of anesthetic from the gas phase into a solution of protein or lipid, in excess of that determined for buffer alone, is taken as evidence that a binding interaction exists. By plotting this binding as a function of anesthetic concentration, an estimate of both the dissociation constant(s) and the number of binding sites can be made. This method has proven useful for the study of anesthetic binding to macromolecular targets.

The method has been used to determine the apparent affinity (1.4 ± 0.2 m*M*) and the number of binding sites (4.2 ± 0.3) on bovine serum albumin (BSA) for isoflurane.[3] The measured dissociation constant using gas chromatographic partition analysis is in good agreement with that obtained using ^{19}F-NMR spectroscopy. Wishnia and Pinder[58,59] used partition analysis to determine the affinity of BSA and β-lactoglobulin for the alkanes butane and pentane. Specific alkane binding sites were noted on these proteins, whose affinity and number were conformation dependent. For example, four butane molecules were shown to bind to BSA with an average K_d of 1 m*M*. More recently, this approach has been used to show that the enantiomers of isoflurane solvate equally into phospholipid bilayers.[60]

Separation of specific and nonspecific binding can be a problem with this approach. Nonspecific binding may be approximated in some cases by measuring partitioning into denatured protein (by changing the pH or the temperature, or by adding a chaotropic agent such as guanidinium chloride). To study nonspecific binding of isoflurane to albumin, Dubois and Evers[3] measured anesthetic binding at a low pH (which partially unfolds the protein). This assumes that the specific binding domain is removed by the denaturant conditions, leaving only the nonspecific binding sites unchanged. Although such conditions appear to remove specific anesthetic binding domains, it is interesting that the secondary structure of albumin at low pH is altered only to a minor degree compared to the native state.[61] This suggests that secondary structure (α-helix in the case of albumin) *per se* is not sufficient to allow anesthetic binding.

Partition analysis has the disadvantage of requiring a high protein concentration,[3] on the order of 0.4 to 1.5 m*M*, because of the low-affinity binding that is characteristic of the volatile general anesthetics. In addition to the practical limitation of obtaining this much protein, the use of high protein concentrations means that there is the risk of aggregation, which might create intermolecular domains not present in more dilute solutions. Finally, the method provides little information concerning the structural nature of the anesthetic binding domains, although some insight can be gleaned through the addition of competitors or by altering the environment of the protein.

2.5 ^{19}F-NUCLEAR MAGNETIC RESONANCE SPECTROSCOPY

NMR spectroscopy can be used to gain information about the structure, interactions, and dynamics of biological systems.[62] In NMR spectroscopy, an external magnetic field is applied to a sample containing nuclei with nonzero spin (or angular momentum), such as ^{1}H, ^{13}C, ^{19}F, or ^{31}P. The external magnetic field acts to split (and align) the magnetic moment of the nuclei being studied. Transitions between these two energy levels can then be induced by the absorption of electromagnetic radiation (in the radiofrequency range), in the same way transitions between electronic levels are induced during the more familiar ultraviolet/visible absorption spectroscopy. Each chemically different nonzero spin nucleus in a molecule will have a unique NMR absorption frequency. The chemical shift is the variation in the NMR absorption frequency due to the variation in the chemical (i.e., electronic) environment of the nucleus. Chemical shifts are affected by intermolecular interactions, allowing the investigator to distinguish between molecules in different environments.

Modern clinical volatile anesthetics are heavily fluorinated to ensure minimal (1) flammability, (2) metabolism, and (3) side effects, in particular cardiac arrhythmias.[63] This characteristic is an advantage in the study of anesthetic interactions with macromolecules, because fluorine occurs infrequently in biological materials. The use of a method to monitor fluorine is therefore associated with an excellent signal-to-noise ratio. Based on the measured fluorine chemical shift (the position of resonance), NMR spectroscopy can be used to differentiate between distinct chemical environments experienced by the ^{19}F nucleus. This technique has been used to determine the energetics and the kinetics of volatile general anesthetic binding to BSA.[3,4,9] These studies reveal that the fluorinated volatile anesthetics isoflurane, halothane, methoxyflurane, and sevoflurane have discrete binding sites on albumin with K_d values of 1.4 ± 0.2 m*M*, 1.3 ± 0.2 m*M*, 2.6 ± 0.3 m*M*, and 4.5 ± 0.6 m*M*, respectively.[3,4] Inhibition of isoflurane binding by the other three volatile anesthetics occurred with inhibition constants (K_i) that approximated the measured K_d values for each individual anesthetic, suggesting that the same sites were being occupied competitively by the different anesthetic molecules. In addition, the average bound lifetime of isoflurane[3] was estimated to be 250 μs from the measured off rate (k_{-1}) of 4000 s^{-1}, a result confirmed by a more recent study.[9] This implies that the average residence time for a bound anesthetic molecule is less than 1 ms, directly pointing out the difficulty of studying anesthetic binding by more conventional approaches. For example, if radioactively labeled halothane were equilibrated with a target protein, and the investigator wished to separate bound from free ligand (by filtration, dialysis, or centrifugation), the allowable separation time[64] for an interaction with a dissociation constant of 1 m*M* would be only 0.1 ms.

In addition to the experimental challenge, these kinetic data also suggest that there are discrete binding sites for the general anesthetics on selected proteins. If anesthetics bound to macromolecules only by nonspecific interfacial (surface) binding, as suggested by some investigators,[65] then the on rate, k_1, should approach that for a strictly diffusion-controlled event, because almost all collisions between anesthetic and protein would result in complex formation. A discrete site, on the other hand, would be expected to comprise only a small fraction of the protein surface, significantly reducing the number of collisions resulting in binding.

Thus, using the measured K_d and k_{-1}, for isoflurane binding to BSA,[3,9] it is possible to calculate the on rate constant value, k_1, for anesthetic binding to albumin as $k_1 = 3 \cdot 10^6\ M^{-1}\ s^{-1}$. This value is two to three orders of magnitude less than that of a diffusion-controlled process, calculated to be $\approx 10^9\ M^{-1}\ s^{-1}$ for a protein and a small ligand.[66,67] In fact, the k_1 value for isoflurane binding to BSA is comparable to the binding rate constants of 10^4 to $10^7\ M^{-1}\ s^{-1}$, observed experimentally with other ligand–receptor interactions[68-70] known to occur at discrete sites. Similar on rate constants (10^6 to $10^7\ M^{-1}\ s^{-1}$) have been measured for benzene binding to a variety of engineered hydrophobic cavities in T4 lysozyme.[70] Interestingly, no clear pathway for benzene access to the cavity exists, indicating that transient protein conformational fluctuations are required to create a path for ligand entry. The same situation occurs in the case of myoglobin, where no obvious pathway for oxygen entry to the heme pocket exists,[71] and may also exist in BSA, although a higher resolution structure of the albumin–anesthetic complex will be required before a conclusion can be made.

Although a useful technique for monitoring binding energetics and kinetics for fluorinated volatile general anesthetics, the ^{19}F-NMR technique has the drawback of requiring relatively large amounts of protein in a pure form (25 to 90 μM).[3,4] It will probably have limited utility in the study of anesthetic interactions with natural membrane proteins, which are present in low relative concentrations in biological specimens. The technique provides little information on the location and characteristics of the anesthetic binding domains in the protein, although this may in some cases be approached using additional biophysical tools such as nitroxide spin labels in concert with ^{19}F-NMR spectroscopy.[72] As for partition analysis, more information may be accessible with the use of competitors or altered buffer conditions.

2.6 FLUORESCENCE SPECTROSCOPY

Fluorescence spectroscopy can be used to monitor ligand–protein interactions, providing information about equilibrium binding energetics, kinetics, and dynamic changes in protein structure after complex formation. Fluorophores may be intrinsic to the protein itself and may be incorporated via covalent attachment or simple partitioning. Intrinsic protein fluorescence is principally due to tryptophan residues, when present, but tyrosine[73] and phenylalanine[74] fluorescence can also be monitored, in selected cases. Because tryptophan is the least common amino acid in proteins, its study by fluorescence spectroscopy can provide structural information about the protein, because it allows the investigator to probe specific protein domains containing the tryptophan residues. Several features of the fluorescence spectrum can provide valuable information on ligand binding, including the fluorescence yield, the wavelength of emission, the fluorescence anisotropy, and the fluorescence lifetime. Thus, in a protein such as BSA, which contains only two tryptophan residues, changes in the fluorescence of the protein may reflect local perturbations in the protein structure in the vicinity of these residues. One of these tryptophan residues in BSA, the conserved Trp212, is located in the IIA binding

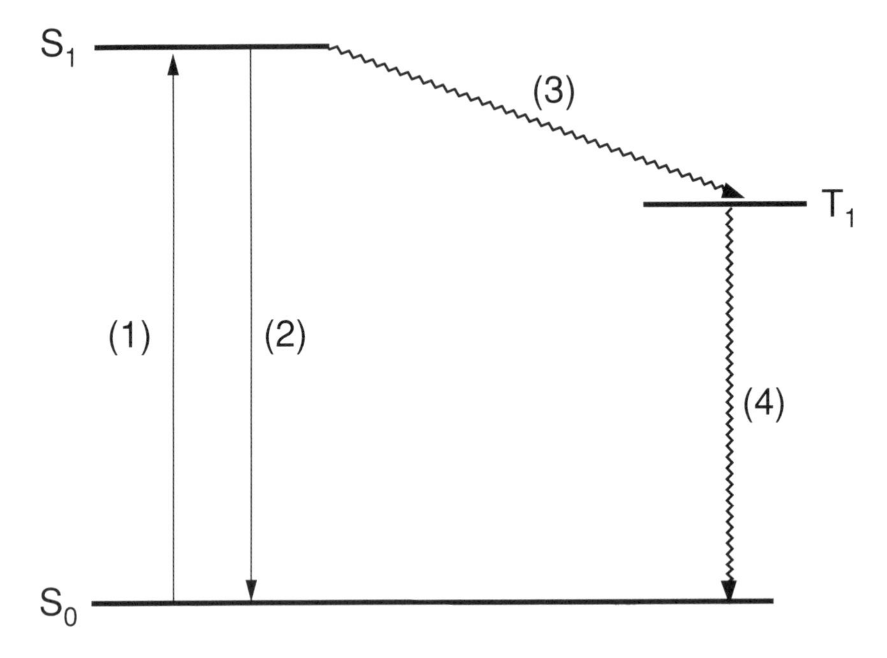

FIGURE 2.2 Schematic of Jablonski energy level diagram, showing (1) absorption of photon that promotes an electron from the ground state (S_0) to the excited singlet state (S_1). The excited singlet state may then be deactivated via (2) fluorescence or (3) intersystem crossing to the triplet excited state (T_1), which in turn may relax (4) via intersystem crossing to the ground state. Halothane quenching of tryptophan fluorescence is thought to occur by favoring step 3 over step 2. This mode of fluorescence quenching arises from increased spin orbit coupling resulting from the interaction of the excited electronic spin with the highly positively charged nucleus of the heavy atom,[76] which are the bromine and chlorine atoms in the case of halothane.

domain of albumin, which is known to bind small aromatic molecules such as warfarin and triiodobenzoic acid.[75]

An approach to determining the binding of halothane to water-soluble proteins has been developed.[6,8] This method relies on the ability of the heavier halogens, bromine and chlorine (i.e., not fluorine), to directly quench the fluorescence of selected tryptophan residues, or added fluorescent probes. In addition to providing information on the affinity of the anesthetic–protein interaction, and therefore the free energy of binding, the technique has the advantage that it reports the location of the anesthetic in the protein matrix. This is because the probable quenching mechanism is heavy atom (atoms of high atomic number) perturbation (Figure 2.2). This mode of fluorescence quenching requires close contact (less than 3 to 5 Å) between anesthetic and fluorophore, as shown for other heavy-atom-containing ligands and groups.[77,78] It is currently the only method available that provides direct solution information about the location of an anesthetic in the protein matrix, assuming that other quenching mechanisms (secondary to protein conformational changes) are not operative.

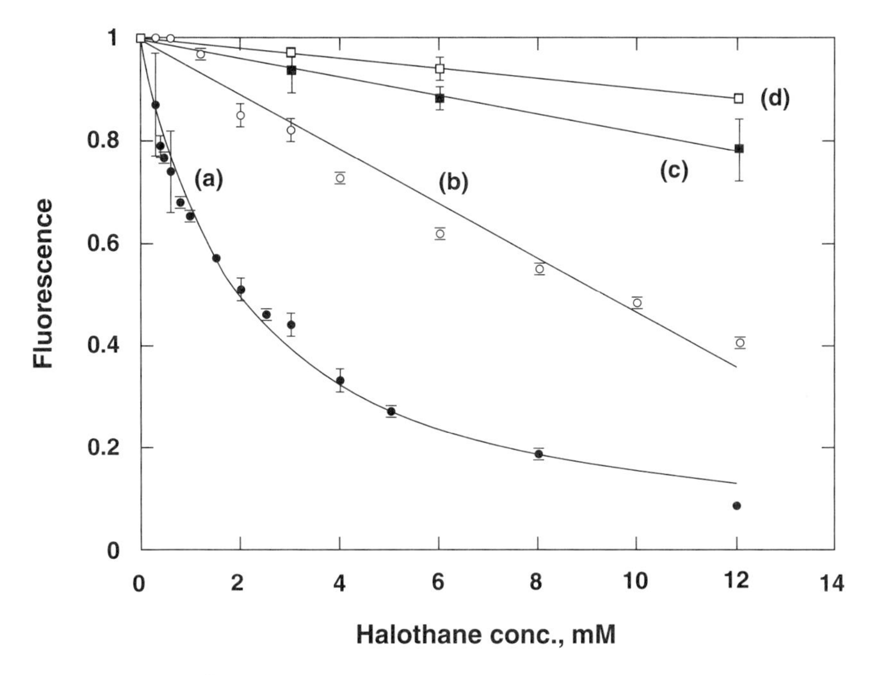

FIGURE 2.3 (a) Quenching of bovine serum albumin (BSA, 5 μ*M*) fluorescence by halothane at pH 7.0, in phosphate buffer. (b) Halothane quenching of BSA (5 μ*M*) fluorescence at pH 3.0. (c) Effect of halothane on free L-tryptophan (10 μ*M*) fluorescence. (d) Effect of halothane on equine apomyoglobin (5 μ*M*) fluorescence. (From Johansson, J. S., Eckenhoff, R. G., and Dutton, P. L., *Anesthesiology*, 83, 316–324, 1995. With permission.)

Our initial studies were performed with BSA because this protein has been shown to bind halothane using ^{19}F-NMR spectroscopy[4] and direct photoaffinity labeling.[5] Figure 2.3a shows that coequilibration of halothane with BSA at pH 7.0 in phosphate buffer causes a concentration-dependent decrease in the intrinsic protein tryptophan fluorescence. The line through the data points in Figure 2.3a yields a dissociation constant of 1.8 ± 0.2 m*M* and a maximum fluorescence quenching of 0.99 ± 0.04, indicating that the fluorescence of both of the tryptophan residues (134 and 212) is quenched in BSA. The results of the fluorescence experiments[6] agree well with the earlier studies and serve as a verification of the fluorescence quenching technique.

Halothane (at concentrations up to 12 m*M*) has only a small quenching effect on *free* L-tryptophan fluorescence, as shown in Figure 2.3c. However, at higher concentrations (up to 170 m*M*), halothane quenches tryptophan fluorescence to similar degrees as is observed in albumin, indicating that the presence of halothane in the vicinity of the indole rings in the protein is sufficient to explain the protein tryptophan fluorescence quenching, and that the albumin sites bind, and therefore "concentrate," the anesthetic.

To evaluate the importance of the native BSA conformation to halothane binding, experiments were performed at pH 3.0. At this pH, BSA changes from an ellipsoid (normal) to a fully uncoiled (expanded) shape.[61] Figure 2.3b shows that there is a large decrease in the amount of quenching of BSA tryptophan fluorescence compared to that at pH 7.0 (Figure 2.3a), and that there is apparent loss of saturable binding. Because the expanded pH 3.0 form of BSA retains the vast majority of the secondary structure present in the native pH 7.0 form,[61] tertiary rather than secondary structure must be the primary determinant producing suitable anesthetic binding domains in proteins. This conclusion is supported by studies using synthetic peptides which show that secondary structure per se is not sufficient to allow the formation of an anesthetic binding site.[8,79,80] Rather, supersecondary structure,[81] involving the complex geometrical relationships characteristic of adjacent secondary structural elements (helices, sheets, and reverse turns), appears to be necessary.

Further support for the importance of protein tertiary structure to anesthetic binding is emphasized in Figure 2.3d, which shows the effect of halothane on apomyoglobin tryptophan fluorescence at pH 7.0. Apomyoglobin has two tryptophan residues at positions 7 and 14.[82] Addition of halothane causes only a small linear decrease in the apomyoglobin tryptophan fluorescence, indicating that the presence of tryptophan residues alone is not sufficient to produce a halothane binding site, consistent with crystallographic data showing steric constraints.[13,32,83]

Thus far, we have considered quenching only through direct contact. However, fluorescence quenching might result from structural modifications of BSA upon halothane binding at allosteric sites. To examine this possibility, we used far-ultraviolet circular dichroism spectroscopy (see Section 2.9) to characterize the secondary structure of BSA, and we failed to detect protein conformational changes in the presence of halothane.[6] This suggests that anesthetic binding does not alter the protein secondary structure, and that the observed fluorescence quenching is most likely due to direct halothane interactions with the indole rings. In support of this, recent direct photoaffinity labeling experiments with halothane and BSA have demonstrated covalent labeling of these two tryptophan residues and adjacent amino acids.[84] It should be noted, however, that although these data suggest proximity as the cause of fluorescence quenching in BSA, the approaches used may not be sensitive to tertiary structural changes in the presence of halothane. For instance, Lopez and Kosk-Kosicka[85] have interpreted changes in plasma membrane Ca^{2+}-ATPase intrinsic tryptophan fluorescence in terms of anesthetic-induced conformational changes in the enzyme.

Measurement of the dynamics of fluorophores can provide further insight into the nature of anesthetic–protein interactions. For example, measuring fluorescence lifetimes allows classification of quenching interactions as static versus collisional.[86,87] Collisional quenching results from the random diffusional encounters between quencher and fluorophore as might occur between halothane and indole (for example) in a suitable solvent. The measured lifetime of the fluorophore, under these conditions, will be inversely proportional to the halothane concentration. On the other hand, a static interaction occurs when a complex is formed between the fluorophore and the quencher. Under these conditions, no change in the fluorescence lifetime is observed as the quencher concentration is increased. This is because the quencher molecule is already present in the vicinity of the fluorophore at the moment of excitation, and therefore causes instantaneous fluorescence quenching. This latter behavior was observed for the quenching of the tryptophan fluorescence of a four-α-helix bundle protein by halothane, demonstrating that binding to the protein in the vicinity of the indole rings has occurred.[79]

Fluorescence anisotropy measurements describe the rotational mobility of the entire protein and/or the local dynamics of individual fluorophores.[88] A decrease in anisotropy reflects an increase in probe mobility. The considerable mobility of tryptophan residues in proteins about the C_α-C_β bond can be quantified using anisotropy measurements and may have utility for studying anesthetic–protein interactions. Using this technique, Vanderkooi and colleagues[89] showed that general anesthetics decrease the fluorescence anisotropy of the hydrophobic probe 1-phenyl-6-phenylhexatriene, suggesting an increase in the fluidity of the membrane. Similarly, halothane has been shown to increase the mobility of skeletal muscle sarcoplasmic reticulum Ca^{2+}-ATPase labeled with a phosphorescent probe (erythrosin 5-isothiocyanate), also using anisotropy measurements.[90] Of importance was the fact that changes in enzyme mobility correlated with changes in Ca^{2+}-ATPase activity. Interestingly, halothane had the opposite effect on the phosphorescence anisotropy and the activity of the Ca^{2+}-ATPase from cardiac muscle sarcoplasmic reticulum,[91] indicating that these structurally different enzymes can be affected by volatile anesthetics in different ways. Therefore, changes in fluorescence or phosphorescence anisotropy may provide information on local changes in protein structure, and perhaps provide clues for how anesthetic binding alters protein function.

The principal advantages of fluorescence spectroscopy in characterizing anesthetic–protein interactions are that the instrumentation is readily available and that low concentrations of protein (1 to 5 μM) are typically sufficient. Methods based on fluorescence can be used to carry out thermodynamic and kinetic studies on anesthetic binding to proteins in solution, providing similar information to that obtained by ^{19}F-NMR spectroscopy, but also allowing the detection of changes in protein structure and dynamics. The main limitations with fluorescence quenching to study anesthetic–protein complex formation are that it fails to detect anesthetic binding to proteins at sites not containing aromatic residues and that anesthetics containing heavy atoms are required.

2.7 DIRECT PHOTOAFFINITY LABELING

The technique called photoaffinity labeling has been used extensively to study the structural and functional characteristics that underlie the interactions of receptors and enzymes with ligands and substrates.[92,93] Photolabile groups on the ligand allow reversible equilibrium binding to be converted to stable covalent linkages at target sites on the protein. Structural biochemical approaches can subsequently be used to narrow down the portion of the protein that comprises the binding site. The typical photoaffinity ligand is an already characterized ligand of interest that has been altered to include a highly photolabile group, such as an azido- or diazo group. Photolysis of these groups at relatively long ultraviolet wavelengths ($\approx$ 350 nm) leads to the production of a highly reactive carbene radical and a stable nitrogen molecule. Direct photoaffinity labeling, on the other hand, denotes the use of a chemically unmodified ligand that already includes a photolabile bond, and generally requires shorter

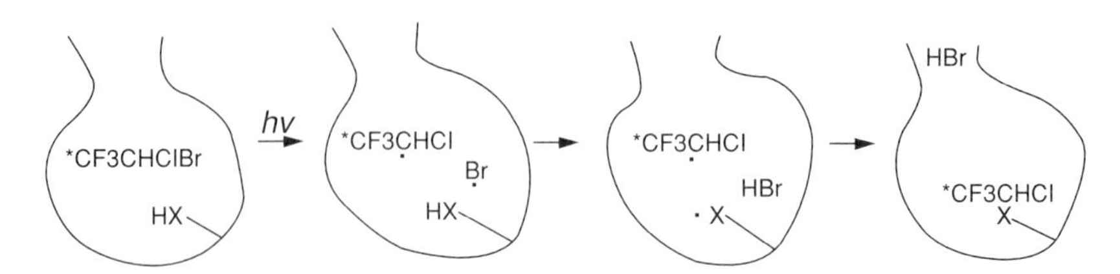

FIGURE 2.4 Probable mechanism for halothane photolabeling. Depicted is a protein cavity with a target group designated –HX. The X could be a carbon, nitrogen, oxygen, or sulfur atom on an amino acid side chain. Although unlikely, it is also possible that halothane may label main-chain atoms. The asterix on the 1-carbon (trifluoromethyl group) of halothane indicates a label, such as ^{14}C. Short UV light (<260 nm) causes homolytic cleavage of the C–Br bond to produce the two free radicals as shown in the second frame. The free Br· radical abstracts a hydrogen from the target group, leaving a target free radical, which then reacts (terminates) quickly with the chlorotrifluoroethyl radical containing the reporter (radioactive) label, while the HBr diffuses out of the presumably hydrophobic pocket. Alternatively, both halothane-derived radicals may insert directly into target bonds within the cavity, as bromine can be easily detected in irradiated proteins using X-ray fluorescence spectroscopy.[99]

wavelength ultraviolet light (≈ 250 nm) for photolysis. The use of photoaffinity labeling to determine the binding character and the targets for the inhaled anesthetics is particularly desirable because the rapid kinetics of these ligands preclude the use of the more conventional radioligand equilibrium binding assays.[94] Recently, direct photoaffinity labeling has been applied to the widely used inhalational anesthetic, halothane.[5,8,84,95,96]

Halothane includes a carbon–bromine bond that may be photolyzed at wavelengths shorter than about 260 nm (peak absorption ≈ 210 nm). Photolysis of this bond produces both bromine and chlorotrifluoroethyl (CTFE) radicals, which generally recombine to form the parent molecule in water.[97] This is likely the result of (1) confinement of halothane within a clathrate shell of approximately 23 water molecules,[98] (2) the influence of the high dielectric environment, and (3) the inability of the free radicals generated to insert into water. However, when a bound (immobilized) halothane molecule is photolyzed in an appropriate environment, other results may ensue (Figure 2.4). One possibility is that the reactive CTFE radical inserts directly into various target bonds in a single-step process. Alternatively, the bromine radical may abstract a hydrogen from the target site, producing a target radical with which the CTFE radical may, in turn, react. In either case, the two-carbon halothane backbone becomes rapidly incorporated into the target site, presumably in a distribution reflective of reversible, equilibrium binding. This assumption depends, of course, on the reactivity and lifetime of the CTFE radical. If the CTFE radical lifetime is long with respect to the binding site residence time, diffusion of the photolysis products will tend to spread the photolabeling distribution. Although nothing has been published on CTFE radical lifetimes, preliminary experiments using transient absorption spectroscopy of halothane in various solvents after ultraviolet laser photolysis have identified a transient species with a half-life of approximately 5 μs (Figure 2.5). Based on the known dissociation constants and EC_{50} values of approximately 1 m*M* for volatile anesthetics, calculated target site residence time estimates[3,9]

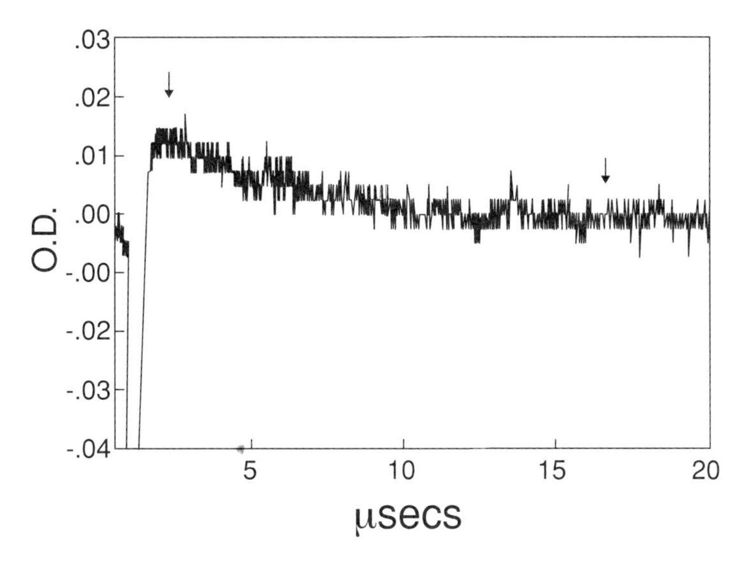

FIGURE 2.5 Halothane photolysis product lifetime. After 266 nm laser flash photolysis of 0.1 *M* halothane solutions, a transient absorbing species with a maximal absorbance at ≈380 nm was noted. The decay of this absorbance was best fit to a single exponential with a half-life of approximately 5 μs, as shown. The same transient species was noted after photolysis in hexane or ethanol, and also after gas phase photolysis in a nitrogen atmosphere. The presence of oxygen had no significant effect on the measured lifetime.

on the order of 100 μs indicate that the photolabeling pattern will indeed reflect the equilibrium distribution of the parent molecule. In addition, the CTFE radical should ideally label different chemical groups on the target equally well (i.e., not show significant chemical discrimination). Although most photoaffinity ligands show some degree of selectivity, lack of chemical discrimination is difficult to establish unambiguously, because selectivity for a specific target group may also be present prior to photolysis. Only by observing the labeling pattern in a wide variety of proteins, under different conditions, can an estimate of selectivity be made. Finally, bound photoaffinity ligand must be observable and measurable in some way. Typically this is accomplished by incorporating a radioisotope such as ^{125}I, ^{14}C, or ^{3}H into the molecule. Photoaffinity ligand binding may also be assessed by monitoring regional protein molecular weight by mass spectrometry.[100-102]

The use of direct photoaffinity labeling should allow mapping of halothane binding sites in soluble proteins, as recently demonstrated in serum albumin.[5,84] Two specific binding sites (Figure 2.6) have been identified in BSA.[84] These correspond to the location of the two tryptophans at positions 134 and 212, in agreement with the fluorescence quenching studies.[6] The anesthetic binds to several amino acids in the IIA domain of BSA (Trp212–Leu217), suggesting (1) weak immobilization within this site (consistent with the K_d of approximately 1 m*M*) and (2) that the CTFE radical binds to several different types of chemical groups, including C-H bonds. The second binding site in BSA also contains tryptophan (Trp134), is more water exposed, and has a tighter interaction with halothane, as suggested by both

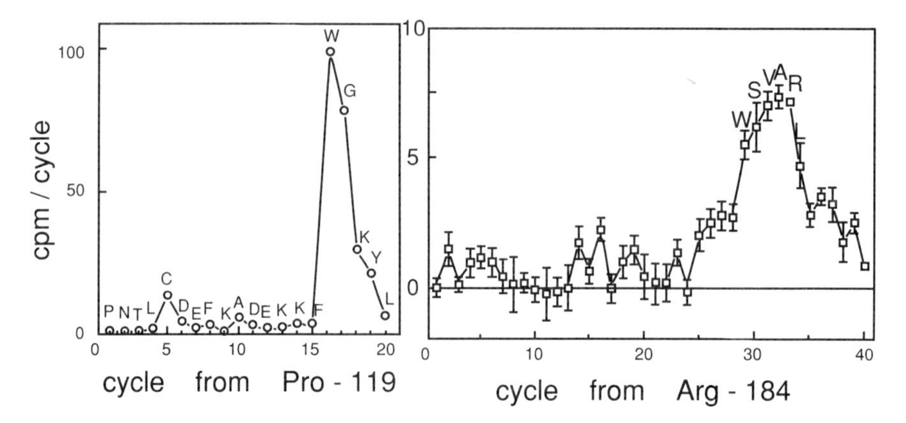

FIGURE 2.6 Halothane-labeled sites in BSA. The two specifically labeled cyanogen bromide fragments of BSA (see Ref. 84) were microsequenced, and the radioactive decay release pattern shown was obtained. In both fragments, a peak is noted in the immediate vicinity of the tryptophan (W) residues. Low release of cpm (≈1%) as observed in this study is common in photoaffinity labeling experiments,[103] and may reflect disruption of the Edman chemistry by the modified (photolabeled) amino acids, or altered solubility of the cleaved phenylthiohydantoin amino acids. (From Eckenhoff, R. G., *J. Biol. Chem.*, 271, 15521–15526, 1996. With permission.)

the photoaffinity results and the fluorescence quenching studies. Of note is that substituting a leucine in this 134 position, as occurs in the case of human serum albumin, completely eliminates halothane binding, hinting at the importance of aromatic residues in providing a favorable interaction for haloalkane anesthetics. Despite competition between fatty acids and anesthetics for BSA binding,[3,5] the photolabeled residues are not part of the presumed long-chain fatty acid binding domains,[61] indicating allosteric communication between the IIA and the fatty acid binding sites.

Direct photoaffinity labeling has allowed halothane binding to be assigned to general domains in a well-characterized membrane protein as well. The halothane binding domains in a ligand-gated channel, the nicotinic acetylcholine receptor (nAChR), was determined in membranes of *Torpedo nobiliana*.[96] The inhaled anesthetics are known to have inhibitory effects on the function of the nAChR.[104-106] Photoaffinity labeling with 0.5 m*M* buffer ^{14}C-halothane concentration revealed that each of the four different subunits of the receptor were specifically labeled to a similar degree, in an agonist-insensitive fashion. After digestion of the α subunit (which is the site of agonist binding) with V8 protease, greater than 90% of the label was associated with the sequences containing the four putative transmembrane sequences. Because three of these sequences were not separated, it remains unclear whether the proposed channel lining helix M2[107] is labeled by halothane. If present, this could indicate a form of inhibition similar to that of local anesthetics. A site-directed mutagenesis study involving the channel-lining residues of the M2 domain[106] indicated that isoflurane may bind to the channel of the nAChR. Although the halothane binding character in this membrane protein is similar to that of soluble proteins in that it is mostly specific (i.e., displaceable with cold halothane), it differs in that membrane protein appears to contain multiple sites, and most seem to be found in the transmembrane region, presumably at the lipid–protein interface. Importantly, the halothane labeling pattern, after digestion of the nAChR protein, differed substantially from that obtained with photoactivating phospholipids,[108] suggesting that the halothane labeling is not the result of simple partitioning into bulk phospholipid. Such data hints at the importance of these functionally crucial transmembrane regions (lipid–protein or protein–protein interhelical interfaces) for anesthetic binding, and possibly action. This is in agreement with other evidence that points toward the lipid–protein interface as a favored binding site for anesthetic-like molecules.[24,109-111]

The principal advantage of photoaffinity labeling is the potential to define binding sites at the amino acid level of resolution. It may also prove possible to orient the anesthetic within protein binding domains, providing details of the underlying bonding interactions. In addition, photoaffinity labeling is the only current approach that allows separation of specific and nonspecific binding for this unique group of low-affinity ligands. The disadvantages are that the short-wavelength ultraviolet light required for halothane photolysis is relatively destructive to biological macromolecules, and sensitivity is limited due to the low affinity of the ligand. Finally, the selection of photolabile inhaled anesthetics is limited, with halothane being the sole example studied to date.

2.8 ELECTRON SPIN RESONANCE SPECTROSCOPY

Electron spin resonance (ESR), or electron paramagnetic resonance (EPR), is a type of absorption spectroscopy that uses microwave frequency radiation to induce energy level transitions between the spin states of unpaired electrons, in a manner analogous to nuclear magnetic resonance spectroscopy.[112,113] The presence of a static magnetic field creates magnetic energy splitting. Unpaired electrons are infrequent in biological specimens but are present in free radicals (such as nitric oxide), various transition metal ions, and triplet electronic states (such as may occur in aromatic amino acids after absorption of ultraviolet radiation).

Nitroxide spin labels are stable, organic free radical reporter groups available in a variety of sizes that can be attached covalently to specific residues (typically amino groups and cysteine thiols) in proteins, and also to phospholipids, via a variety of functional groups. The nitroxide group (>N–O·) is part of a five- or six-member heterocyclic ring, which is connected to a linker of variable size, rigidity, hydrophobicity, and specificity. The linker allows for covalent coupling of the nitroxide spin label to the protein. The ESR spectra obtained are indicative of the environment of the spin label. For example, the local topology, dynamics, solvent exposure, and dielectric features experienced by the probe can be determined.[112,114]

Spin-labeled phospholipids have been used to determine the distribution of anesthetic molecules in lipid bilayers. Trudell and Hubbell[72] used phospholipid vesicles containing nitroxide spin labels at different positions along the acyl chains to probe the location of halothane in the phospholipid bilayer, based on the broadening effect of nitroxide free radicals on the ^{19}F-NMR fluorine doublet, arising from the anesthetic molecule. It was concluded that halothane is rapidly distributed across the entire thickness of the bilayer. This observation is supported by studies with diethylether on the fluidity of sarcoplasmic reticulum membranes spin labeled at various positions along the acyl chains,[109] and by a more recent study that determined the location of halothane and isoflurane in phospholipid bilayers using proton- and deuterium-NMR.[115]

Spin-labeled lipids have also been used to study the effects of alcohols, halothane, methoxyflurane, diethylether, and fluroxene on lipid bilayer order in *Torpedo* electroplaque.[104] These experiments revealed that anesthetic-induced membrane disordering correlates with desensitization of the nAChR. However, whether these two phenomena are directly related remains unclear.

The effect of general anesthetics on the properties of spin-labeled membrane proteins has also received some attention. Thus, the extracellular domain of the anion exchange protein in human erythrocytes was labeled with a nitroxide spin label (*bis*(sulfo-*N*-succinimidyl)doxyl-2-spiro-5′-azelate) and the rotational mobility of the probe was found to be reversibly increased by diethylether.[116] This was interpreted to represent a localized structural change in the protein secondary to diethylether-induced alterations in the surrounding membrane bilayer lipid order. In addition, nitroxide spin labels have been used to examine the effects of general anesthetics on the interaction of the nAChR with neighboring lipid molecules in the bilayer.[117] Isoflurane and hexanol were unable to displace boundary lipids from their association with the receptor. Enhancement of protein side-chain mobility by ethanol, as detected by spectral changes in a maleimide spin label covalently attached to selected cysteine residues on the nAChR has been reported,[118] suggesting that this general anesthetic acts by fluidizing the protein. Using spin labels attached to both bilayer lipids and the skeletal muscle sarcoplasmic reticulum Ca^{2+}-ATPase protein, Bigelow and Thomas[109] reported that diethylether increased the mobility of both the Ca^{2+}-ATPase and the phospholipids adjacent to the protein. Increased Ca^{2+}-ATPase mobility correlated with increased Ca^{2+} pumping activity, indicating that diethylether facilitated the protein conformational changes required for Ca^{2+} transport.

ESR is applicable to low micromolar concentrations of labeled protein and requires only 5 to 10 μl of sample. The principal drawback to the technique is that an extrinsic reporter group is being added to the macromolecule of interest, which could significantly alter its native properties. Although nitroxide spin labels may prove useful for studying the energetics of anesthetic binding to protein targets, cautious interpretation of the results is required, because changes in ESR spectra may result from allosteric effects consequent to anesthetic binding at distant sites, rather than a direct local effect of anesthetic on the nitroxide spin label. Further work with carefully positioned nitroxide spin labels should provide useful information on how general anesthetics alter *local* protein dynamics, and may prove effective for the study of the energetics of anesthetic binding to macromolecular targets.

2.9 CIRCULAR DICHROISM SPECTROSCOPY

2.9.1 Protein Secondary Structure

Circular dichroism (CD) spectroscopy is a variant of absorption spectroscopy that makes use of circularly polarized light to study the three-dimensional conformation of molecules.[119,120] It is one of the basic biophysical tools used to analyze and define protein secondary structure, because the random coil, α-helix, β-sheet, and 3_{10}-helix all display unique CD spectra in the far ultraviolet (UV) region,[120,121] and the contribution of each to the overall spectrum can be quantified. Along with steady-state measurements on the structure of biological molecules, it is also useful for studying kinetic and thermodynamic processes.

Because this form of spectroscopy is sensitive to protein conformation, it is reasonable to find it employed in the examination of the effects that anesthetics might have on protein structure. Thus, Laasberg and Hedley-White[122] reported that halothane decreased the α-helical content of both the β chain of human hemoglobin and high-pH poly(L-lysine) by a small amount. Similarly, Ueda and colleagues[23] reported that anesthetics change α-helix into β-sheet. Because secondary structure is largely formed through intramolecular main-chain hydrogen bonds, these results have been interpreted in terms of the potential hydrogen bond breaking activity of anesthetics.[19-24] More recently, studies on halothane binding to BSA,[6] and to different chain length homopolymers of high-pH α-helical poly(L-lysine),[8] have failed to detect changes in protein secondary structure as monitored by CD spectroscopy, even though anesthetic was shown to be binding to the proteins as assessed by other techniques (fluorescence spectroscopy and direct photoaffinity labeling). Halothane binding to the hydrophobic core of a four-α-helix bundle also has no effect on protein secondary structure.[79] Furthermore, binding of benzene to a three-α-helical coiled-coil peptide does not alter the secondary structure.[80] These recent CD studies therefore suggest that macromolecular hydrogen bond disruption is not a uniform, or even likely, consequence of anesthetic complexation, emphasizing the important role that structural measurements like CD can have in addressing fundamental mechanistic issues.

These results do not mean that inhaled anesthetics are unable to interact with proteins through hydrogen (or electrostatic) bonds. They do, however, suggest that competition with main-chain hydrogen bond donors (amide) or acceptors (carbonyl), sufficient to alter secondary structure, as reflected by CD spectroscopy, does not occur. This is perhaps not surprising because the types of hydrogen bonds that anesthetics are predicted to be capable of forming (Figure 2.7), such as C–H···O and C–X···H (where X is a halogen atom), will be energetically weaker than the C=O···H–N hydrogen bonds responsible for protein secondary structure.[123-128] It is therefore of interest that crystallographic data indicate that between 5 and 24% of the potential hydrogen bonding donors and acceptors in proteins lack a suitable partner.[129,130] Such unfulfilled hydrogen bonding donor or acceptor groups might find suitable partners in an anesthetic molecule, without the need to compete with native protein hydrogen bonding groups. Experimental support for this view is provided by the hydrogen bonding of a water molecule to a backbone carbonyl oxygen (capable of forming two hydrogen bonds) in a hydrophobic cavity containing mutant (Ile76 → Ala) of barnase.[10] The water molecule is, of course, a prototypical hydrogen bond donor and acceptor, yet its interaction with the protein in this case is not characterized by disruption of any main-chain hydrogen bonds.

From an energetic point of view, a protein C=O group would prefer to hydrogen bond with a main-chain N–H group or a neighboring O–H group (on a tyrosine or

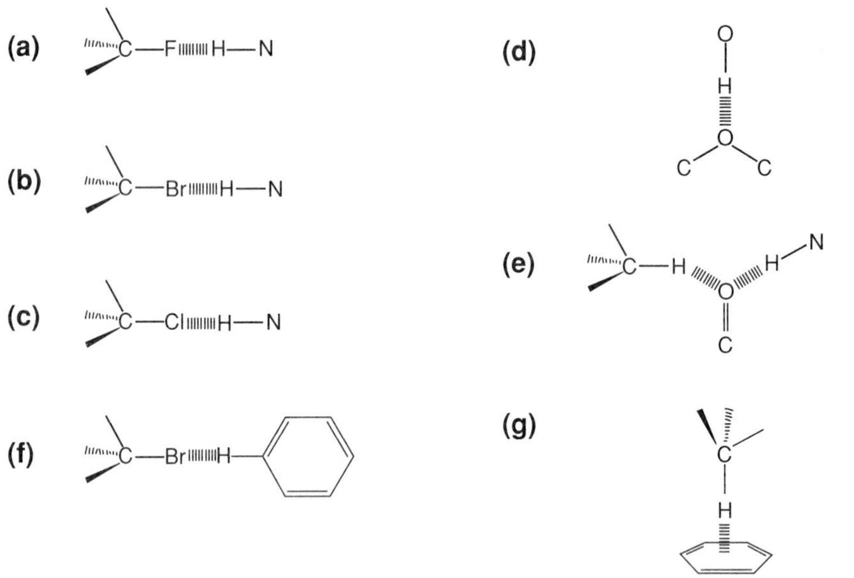

FIGURE 2.7 Selected potential hydrogen bonding interactions for volatile anesthetics and protein groups. Panels a, b, and c show the halogens fluorine, bromine, and chlorine, respectively, acting as the hydrogen bond acceptor. Panel d shows the ether oxygen (on isoflurane, for example) acting as a hydrogen bond acceptor. Panel e shows an anesthetic acting as a hydrogen bond donor to a protein main-chain carbonyl group. Panel f shows an electrostatic interaction between bromine and an aromatic ring hydrogen, and g shows an anesthetic acting as a hydrogen bond donor to the π electrons of an aromatic ring.

serine residue). However, if such groups are not available, the C=O group will satisfy its hydrogen bonding potential by complexing C–H groups,[125] which might exist, for example, on a suitably positioned anesthetic molecule. Similarly, the C–X···H hydrogen bonding potential of the halogen atoms on the inhalational anesthetics may be satisfied by available uncomplexed hydrogen bond donors in proteins, without necessarily disrupting existing protein bonds.

2.9.2 Protein Thermodynamic Stability

CD spectroscopy can also be used to estimate the thermodynamic stability of proteins. Before describing some recent studies relevant to anesthetic mechanisms using this approach, a brief review of stability is required. The folded, biologically active, conformations of proteins are only marginally more stable (5 to 10 kcal/mol) than their unfolded counterparts.[131] Proteins are maintained in their native conformation by hydrogen bonds and the hydrophobic effect. Opposing these organizing forces is the entropic loss associated with folding, which limits the rotational degrees of freedom of the amino acid residues. By raising the temperature of the system, or by adding a chaotropic agent such as guanidinium chloride, it is possible to shift the fold–unfold equilibrium, so that the unfolded or denatured form of the protein is favored. Protein unfolding may be measured by a variety of approaches, including

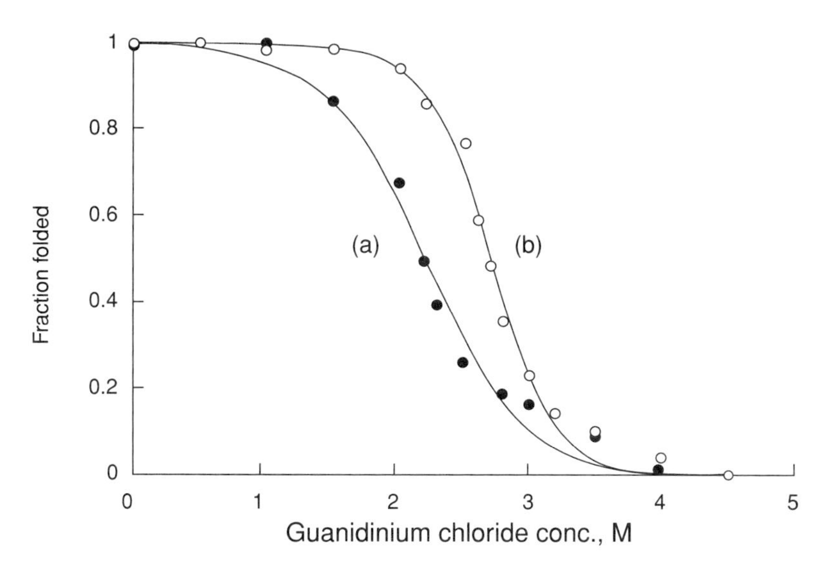

FIGURE 2.8 Chemical denaturation of BSA (5 μ*M*) in the (a) absence and (b) presence of halothane (3.5 m*M*) in 10 m*M* potassium phosphate buffer, pH 7.0. Ellipticity monitored at 222 nm. Curves were fit to the following equation: fraction folded = $(e^{\Delta G_{H_2O}} - m[\text{denaturant}]) / (1 + e^{\Delta G_{H_2O}} - m[\text{denaturant}])$, where ΔG_{H_2O} is the free energy of the transition in the absence of denaturant (an estimate of the conformational stability of the protein), m is the cosolvation term (a measure of the cooperativity of the transition), and [denaturant] is the guanidinium chloride concentration.[138]

ultraviolet difference, fluorescence and CD spectroscopies,[132] and differential scanning calorimetry.[133,134] Although there is some overlap, CD spectroscopy principally detects changes in secondary structure (loss of the main-chain hydrogen bonds as the temperature or the chaotropic agent concentration increases), whereas the other techniques principally measure denaturation of protein tertiary structure.[135] The relationship between the stability of secondary and tertiary structure is an area of much interest currently, and probably is not uniform for proteins of differing size and complexity.

Protein thermal or chemical denaturation as followed by CD spectroscopy[80,132,136,137] can be used to estimate the binding energetics of the anesthetic–protein complex, and also the *global* thermodynamic consequences of anesthetic binding. Figure 2.8 shows a chemical denaturation CD spectroscopy experiment for BSA, and the effect of adding halothane. An increase in the overall structural stability of the protein is observed in the presence of the anesthetic, with the guanidinium chloride concentration required to cause 50% unfolding of the protein increasing approximately 0.5 *M*. This indicates that halothane binds preferentially to the folded protein conformation and that binding of the anesthetic stabilizes the protein, presumably by creating additional favorable van der Waals interactions in the protein interior. The implication of alterations in protein stability may be important for protein activity or function. Because structural flexibility is required for normal protein function,[139,140] this type of anesthetic-induced change in global protein stability is expected to provide much needed quantitation of how anesthetics may ultimately alter protein function. Benzene binding to a T4 lysozyme cavity mutant,[48] and to the hydrophobic core of a three-α-helix coiled-coil protein,[80] likewise stabilizes the overall structures of these folded proteins, as assessed by CD thermal denaturation spectroscopy.

The advantages of far-UV CD spectroscopy (185 to 255 nm) are that it allows a precise description of the relative types of secondary structure present in a protein. In addition, only small quantities of protein are required (300 μl of a 5 to 10 μM solution). Conformational changes in proteins can be detected in some cases following ligand binding or changes in the buffer conditions. The use of near-UV (above 255 nm) CD spectroscopy (requiring higher protein concentrations) to probe the aromatic region of proteins can, in selected cases, provide information on changes in protein tertiary structure.[120] CD spectroscopy can also be used to study the thermodynamic stability of proteins in the absence and presence of ligands, and can be used to estimate the binding energetics. The main disadvantage to using far-UV CD spectroscopy to detect protein structural changes in the presence of anesthetics is that a fairly substantial conformational change is required, involving the alteration of at least 10% of the main-chain hydrogen bonds. More subtle structural changes may be detected by near-UV CD spectroscopy in selected cases.

2.10 INFRARED SPECTROSCOPY

2.10.1 Nitrous Oxide Infrared Spectra

The infrared region of the electromagnetic spectrum includes radiation at wavelengths between 0.7 and 500 μm, or, in wavenumbers (the number of waves in a length of 1 cm), between 14,000 and 20 cm^{-1}. Infrared radiation is absorbed by the bending, stretching, and more complex motions of various functional groups in molecules. The large number of vibrations occurring simultaneously in even a simple molecule results in a complex absorption spectrum, which is characteristic of the functional groups present. The low energy of the absorbed photons in the infrared region is ideal for the detection of low-energy interactions, predicted to be of importance for anesthetic binding to macromolecules.

This methodology has been applied to the study of the interaction of the inhaled anesthetic nitrous oxide (N_2O) with a number of different proteins, including the membrane protein cytochrome *c* oxidase. The linear N_2O molecule exhibits three fundamental absorbance bands in the infrared region. The frequency and bandwidth of the ν_3 antisymmetric stretch of N_2O (Figure 2.9), which occurs near 2230 cm^{-1} in water, is highly dependent on the polarity of the solvent.[141,142] Because of this, the interaction of N_2O with protein targets can be monitored, because the polarity of the buffer environment will differ from that of protein domains. Extensive studies on the behavior of this ν_3 antisymmetric stretch band in a wide range of solvents of differing dielectric properties have allowed further predictions concerning the observed environments in proteins. Such studies suggest that N_2O binds to several, but not all, proteins, and that the binding domains have both aliphatic (valine,

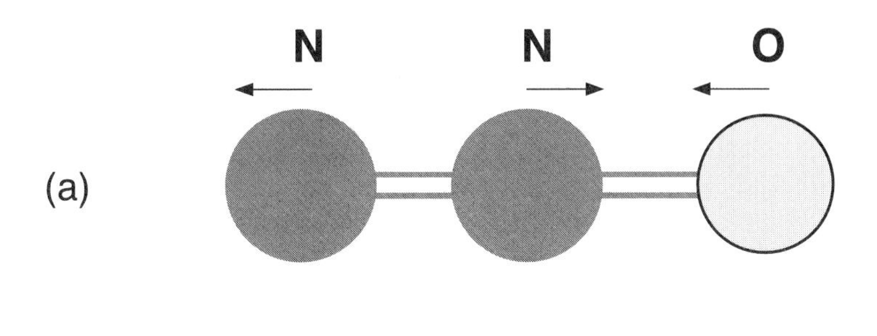

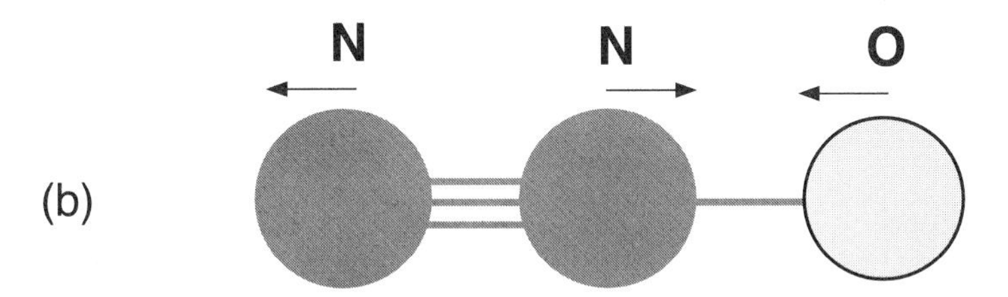

FIGURE 2.9 Antisymmetric stretch vibration, ν_3, of the linear nitrous oxide molecule,[141] which is sensitive to the environment of the anesthetic. The arrows indicate the direction of the atomic vibrations responsible for the antisymmetric stretch. Nitrous oxide resonates between the two structures, a and b.

isoleucine, and leucine) and aromatic (phenylalanine and tyrosine) character. Based on its structure (Figure 2.9), it is anticipated that N_2O should be able to hydrogen bond to suitable donor groups in proteins. However, there is currently no experimental data supporting this prediction.

2.10.2 Protein Infrared Spectra

In an effort to define the structural consequences of binding, and perhaps the location of the anesthetic molecule in the protein matrix, changes in distinct portions of the protein infrared spectrum in the presence of N_2O have been examined. For example, the amide I band (arising from the carbonyl stretch of the main-chain amide groups involved in peptide bond formation) in the region 1600 to 1700 cm^{-1} is used to monitor conformational changes in secondary structure after ligand binding.[143] Accordingly, N_2O binding to oxidized cytochrome *c* oxidase, human serum albumin, cytochrome *c*, myoglobin, and hemoglobin was found to have no effect on the secondary structure of these proteins as monitored by infrared spectroscopy, suggesting that N_2O–protein interactions do not cause major secondary structural changes,[144] as we have also concluded based on CD spectroscopy data for halothane–protein interactions.[6,8,79]

Similarly, the S–H vibration band (of cysteine thiols) infrared spectrum is sensitive to its environment. The bandwidth is indicative of the mobility of the local environment, and the frequency of the absorbance band reflects the strength of the

S–H bond.[145] Thus, although binding of N_2O to human serum albumin was associated with no change in the secondary structure, subtle changes in the S–H band of Cys34 (the only reduced cysteine residue in this protein) at 2563 cm^{-1} suggest tertiary structural changes resulting from anesthetic binding.[144]

The principal advantage of infrared spectroscopy is that no external probe molecules are required, thereby avoiding one possible cause of artifact. The technique provides a snapshot on the time scale of 10^{-12} s, much faster than the resolution of NMR (10^{-5} s). Because of the rapid kinetics displayed by anesthetic molecules, the time scale of infrared spectroscopy avoids the potential for averaging of binding environments. More work with infrared spectroscopy using other anesthetics is likely to provide useful information on protein–anesthetic interactions. The principal disadvantages of infrared spectroscopy are that (1) relatively concentrated protein solutions are required (on the order of 1 to 10 m*M*), (2) water absorbs strongly in spectral regions that may be of interest, and (3) there are a great number of overlapping bands in protein spectra, making interpretation at the single bond level impossible in many cases.

2.11 CONCLUSIONS

The search for mechanisms of general anesthetic action has been ongoing for approximately 100 years. However, only within the past few years has it been possible to study the initial binding step of clinically useful volatile anesthetic molecules to solution protein targets. ^{19}F-NMR spectroscopy allows determination of the energetics and kinetics of fluorinated volatile anesthetic binding, whereas fluorescence spectroscopy and direct photoaffinity labeling permit the investigator to begin to define the characteristics of the binding sites at the molecular level. Site-directed mutagenesis studies on membrane proteins may provide clues as to which parts of the protein are of functional relevance for anesthetic action, but not necessarily the binding site.[2,106] High-resolution X-ray crystallographic or heteronuclear multidimensional NMR studies[62,146] of anesthetic–protein complexes are anticipated to provide the detailed structural information required, and may yield insight as to how anesthetics alter protein activity. Measurement of global and local protein stability and dynamics using differential scanning calorimetry, CD spectroscopy, nitroxide spin labels, and fluorescence and phosphorescence anisotropy is expected to reveal details of how anesthetics alter protein function. Currently, almost nothing is known about how anesthetic binding translates into a change in protein activity, which ultimately must underlie the anesthetic state. Some workers have suggested that anesthetics might disrupt native hydrogen bond networks and cause macromolecular unfolding.[19,20,22,65] However, the majority of the data presented in this chapter indicate that there are only minor protein structural changes following anesthetic binding. Much more work using the varied, and complementary, approaches outlined in this chapter will be needed before an appreciation of the molecular features of anesthetic–protein complexes is achieved. A molecular description of these features will be necessary before it is possible to understand anesthetic mechanisms of action.

ACKNOWLEDGMENTS

Dr. Johansson is the recipient of a Foundation for Anesthesia Education and Research Young Investigator Award, and a grant from the McCabe Foundation. Dr. Eckenhoff is supported by NIH grant GM51595. The authors are indebted to Drs. B. E. Marshall and M. E. Eckenhoff for critically reviewing the manuscript.

REFERENCES

1. Franks, N. P. and Lieb, W. R., Molecular and cellular mechanisms of general anaesthesia, *Nature*, 367, 607, 1994.
2. Harris, R. A., Mihic, S. J., Dildy-Mayfield, J. E., and Machu, T. K., Actions of anesthetics on ligand-gated ion channels: Role of receptor subunit composition, *FASEB J.*, 9, 1454, 1995.
3. Dubois, B. W. and Evers, A. S., ^{19}F-NMR spin-spin relaxation (T_2) method for characterizing anesthetic binding to proteins: Analysis of isoflurane binding to albumin, *Biochemistry*, 31, 7069, 1992.
4. Dubois, B. W., Cherian, S. F., and Evers, A. S., Volatile anesthetics compete for common binding sites on bovine serum albumin: A ^{19}F-NMR study, *Proc. Natl. Acad. Sci. USA*, 90, 6478, 1993.
5. Eckenhoff, R. G. and Shuman, H., Halothane binding to soluble proteins determined by photoaffinity labeling, *Anesthesiology*, 79, 96, 1993.
6. Johansson, J. S., Eckenhoff, R. G., and Dutton, P. L., Binding of halothane to serum albumin demonstrated using tryptophan fluorescence, *Anesthesiology*, 83, 316, 1995.
7. Johansson, J. S., Eckenhoff, R. G., and Dutton, P. L., Reply to Eger and Koblin: Binding of halothane to serum albumin: Relevance to theories of narcosis, *Anesthesiology*, 83, 1385, 1995.
8. Johansson, J. S. and Eckenhoff, R. G., Minimum structural requirement for an inhalational anesthetic binding site on a protein target, *Biochim. Biophys. Acta*, 1290, 63, 1996.
9. Xu, Y., Tang, P., Firestone, L., and Zhang, T. T., ^{19}F nuclear magnetic resonance investigation of stereoselective binding of isoflurane to bovine serum albumin, *Biophys. J.*, 70, 532, 1996.
10. Buckle, A. M., Cramer, P., and Fersht, A. R., Structural and energetic responses to cavity-creating mutations in hydrophobic cores: Observation of a buried water molecule and the hydrophilic nature of such hydrophobic cavities, *Biochemistry*, 35, 4298, 1996.
11. Scapin, G., Young, A. C. M., Kromminga, A., Veerkamp, J. H., Gordon, J. I., and Sacchettini, J. C., High resolution X-ray studies of mammalian and muscle fatty acid-binding proteins provide an opportunity for defining the chemical nature of fatty acid: protein interactions, *Mol. Cell. Biochem.*, 123, 3, 1993.
12. Katz, B. A., Liu, B., and Cass, R., Structure-based design tools: Structural and thermodynamic comparison with biotin of a small molecule that binds to streptavidin with micromolar affinity, *J. Am. Chem. Soc.*, 118, 7914, 1996.
13. Rashin, A. A., Iofin, M., and Honig, B., Internal cavities and buried waters in globular proteins, *Biochemistry*, 25, 3619, 1986.
14. Hubbard, S. J., Gross, K.-H., and Argos, P., Intramolecular cavities in globular proteins, *Protein Eng.*, 7, 613, 1994.

15. Hubbard, S. J. and Argos, P., Cavities and packing at protein interfaces, *Protein Sci.*, 3, 2194, 1994.
16. Hansch, C., Vittoria, A., Silipo, C., and Jow, P. Y. C., Partition coefficients and the structure-activity relationship of the anesthetic gases, *J. Med. Chem.*, 18, 546, 1975.
17. Abraham, M. H., Lieb, W. R., and Franks, N. P., Role of hydrogen bonding in general anesthesia, *J. Pharm. Sci.*, 80, 719, 1991.
18. Taheri, S., Halsey, M. J., Liu, J., Eger, E. I., Koblin, D. D., and Laster, M. J., What solvent best represents the site of action of inhaled anesthetics in humans, rats, and dogs?, *Anesth. Analg.*, 72, 627, 1991.
19. Sandorfy, C., Intermolecular interactions and anesthesia, *Anesthesiology*, 48, 357, 1978.
20. Hobza, P., Mulder, F., and Sandorfy, C., Quantum chemical and statistical thermodynamic investigations of anesthetic activity. 2. The interaction between chloroform, fluoroform, and a N–H···O=C hydrogen bond, *J. Am. Chem. Soc.*, 104, 925, 1982.
21. Brockerhoff, H., Anesthetics may restructure the hydrogen belts of membranes, *Lipids*, 17, 1001, 1982.
22. Brockerhoff, H., Brockerhoff, S., and Box, L. L., Mechanism of anesthesia: The potency of four derivatives of octane corresponds to their hydrogen bonding capacity, *Lipids*, 21, 405, 1986.
23. Shibata, A., Morita, K., Yamashita, T., Kamaya, H., and Ueda, I., Anesthetic-protein interaction: Effects of volatile anesthetics on the secondary structure of poly(L-lysine), *J. Pharm. Sci.*, 80, 1037, 1991.
24. Nakagawa, T., Hamanaka, T., Nishimura, S., Uruga, T., and Kito, Y., The specific binding site of the volatile anesthetic diiodomethane to purple membrane by X-ray diffraction, *J. Mol. Biol.*, 238, 297, 1994.
25. Eisenberg, D. and Hill, C. P., Protein crystallography: More surprises ahead, *Trends Biochem. Sci.*, 14, 260, 1989.
26. Baldwin, J. and Chothia, C., Haemoglobin: The structural changes related to ligand binding and its allosteric mechanism, *J. Mol. Biol.*, 129, 175, 1979.
27. Lewis, M., Chang, G., Horton, N. C., Kercher, M. A., Pace, H. C., Schumacher, M. A., Brennan, R. G., and Lu, P., Crystal structure of the lactose operon repressor and its complexes with DNA and inducer, *Science*, 271, 1247, 1996.
28. Cullen, S. C., Eger, E. I., and Gregory, P., Use of xenon and xenon-halothane in a study of basic mechanisms of anesthesia in man, *Anesthesiology*, 28, 243, 1967.
29. Schoenborn, B. P., Watson, H. C., and Kendrew, J. C., Binding of xenon to sperm whale myoglobin, *Nature*, 207, 28, 1965.
30. Schoenborn, B. P., Binding of cyclopropane to sperm whale myoglobin, *Nature*, 214, 1120, 1967.
31. Nunes, A. C. and Schoenborn, B. P., Dichloromethane and myoglobin function, *Mol. Pharmacol.*, 9, 835, 1973.
32. Tilton, R. F., Kuntz, I. D., and Petsko, G. A., Cavities in proteins: Structure of a metmyoglobin-xenon complex solved to 1.9 Å, *Biochemistry*, 23, 2849, 1984.
33. Miller, K. W., Paton, W. D. M., Smith, E. B., and Smith, R. A., Physicochemical approaches to the mode of action of general anesthetics, *Anesthesiology*, 36, 339, 1972.
34. Schoenborn, B. P., Binding of xenon to horse haemoglobin, *Nature*, 208, 760, 1965.
35. Sachsenheimer, W., Pai, E. F., Schulz, G. E., and Schirmer, R. H., Halothane binds in the adenine-specific niche of crystalline adenylate kinase, *FEBS Lett.*, 79, 310, 1977.

36. Seto, T., Mashimo, T., Yoshiya, I., Kanashiro, M., and Taniguchi, Y., The solubility of volatile anesthetics in water at 25.0 degrees C using ^{19}F-NMR spectroscopy, *J. Pharm. Biomed. Anal.,* 10, 1, 1992.
37. Pai, E. F., Sachsenheimer, W., Schirmer, R. H., and Schulz, G. E., Substrate positions and induced-fit in crystalline adenylate kinase, *J. Mol. Biol.,* 114, 37, 1977.
38. Levitt, M. and Perutz, M. F., Aromatic rings act as hydrogen bond acceptors, *J. Mol. Biol.,* 201, 751, 1988.
39. Burley, S. K. and Petsko, G. A., Weakly polar interactions in proteins, *Adv. Protein Chem.,* 39, 125, 1988.
40. Suzuki, S., Green, P. G., Bumgarner, R. E., Dasgupta, S., Goddard, W. A., and Blake, G. A., Benzene forms hydrogen bonds with water, *Science*, 257, 942, 1992.
41. Dougherty, D. A., Cation-π interactions in chemistry and biology: A new view of benzene, Phe, Tyr, and Trp, *Science*, 271, 163, 1996.
42. Duncum, B. M., *The Development of Inhalational Anaesthesia,* Oxford University Press, London, 1947, 209.
43. Verschueren, K. H. G., Seljée, F., Rozeboom, H. J., Kalk, K. H., and Dijkstra, B. W., Crystallographic analysis of the catalytic mechanism of haloalkane dehalogenase, *Nature,* 363, 693, 1993.
44. Gursky, O., Fontano, E., Bhyravbhatla, B., and Caspar, D. L. D., Stereospecific dihaloalkane binding in a pH-sensitive cavity in cubic insulin crystals, *Proc. Natl. Acad. Sci. USA*, 91, 12388, 1994.
45. Schanstra, J. P., Kingma, J., and Janssen, D. B., Specificity and kinetics of haloalkane dehalogenase, *J. Biol. Chem.,* 271, 14747, 1996.
46. Keuning, S., Janssen, D. B., and Witholt, B., Purification and characterization of hydrolytic haloalkane dehalogenase from *Xanthobacter autotrophicus* GJ10, *J. Bacteriol.,* 163, 635, 1985.
47. Kennes, C., Pries, F., Krooshof, G. H., Bokma, E., Kingma, J., and Janssen, D. B., Replacement of tryptophan residues in haloalkane dehalogenase reduces halide binding and catalytic activity, *Eur. J. Biochem.,* 228, 403, 1995.
48. Eriksson, A. E., Baase, W. A., Wozniak, J. A., and Matthews, B. W., A cavity-containing mutant of T4 lysozyme is stabilized by buried benzene, *Nature*, 355, 371, 1992.
49. Overton, C. E., *Studies of Narcosis,* Chapman and Hall, New York, 1991, 130.
50. Deisenhofer, J. and Michel, H., High-resolution structures of photosynthetic reaction centers, *Annu. Rev. Biophys. Biophys. Chem.,* 20, 247, 1991.
51. Iwata, S., Ostermeier, C., Ludwig, B., and Michel, H., Structure at 2.8 Å resolution of cytochrome *c* oxidase from *Paracoccus denitrificans*, *Nature,* 376, 660, 1995.
52. Tsukihara, T., Aoyama, H., Yamashita, E., Tomizaki, T., Yamaguchi, H., Shinzawa-Itoh, K., Nakashima, R., Yaono, T., and Yoshikawa, S., Structures of metal sites of oxidized bovine heart cytochrome *c* oxidase at 2.8 Å, *Science*, 269, 1069, 1995.
53. Tsukihara, T., Aoyama, H., Yamashita, E., Tomizaki, T., Yamaguchi, H., Shinzawa-Itoh, K., Nakashima, R., Yaono, T., and Yoshikawa, S., The whole structure of the 13-subunit oxidized cytochrome *c* oxidase at 2.8 Å, *Science,* 272, 1139, 1996.
54. Kleywegt, G. J. and Jones, T. A., Where freedom is given, liberties are taken, *Structure,* 3, 535, 1995.
55. Giegé, R., Lorber, B., and Théobald-Dietrich, A., Crystallogenesis of biological macromolecules: Facts and perspectives, *Acta Cryst.,* D50, 339, 1994.
56. Sousa, R., Use of glycerol, polyols and other protein structure stabilizing agents in protein crystallization, *Acta Cryst.,* D51, 271, 1995.

57. Urbanova, M., Dukor, R. K., Pancoska, P., Gupta, V. P., and Keiderling, T. A., Comparison of α-lactalbumin and lysozyme using vibrational circular dichroism. Evidence for a difference in crystal and solution structures, *Biochemistry,* 30, 10479, 1991.
58. Wishnia, A. and Pinder, T., Hydrophobic interactions in proteins: Conformation changes in bovine serum albumin below pH 5, *Biochemistry,* 3, 1377, 1964.
59. Wishnia, A. and Pinder, T. W., Hydrophobic interactions in proteins. The alkane binding site of β-lactoglobulins A and B, *Biochemistry,* 5, 1534, 1966.
60. Dickinson, R., Franks, N. P., and Lieb, W. R., Can the stereoselective effects of the anesthetic isoflurane be accounted for by lipid solubility?, *Biophys. J.,* 66, 2019, 1994.
61. Carter, D. C. and Ho, J. X., Structure of serum albumin, *Adv. Protein Chem.,* 45, 153, 1994.
62. Wüthrich, K., NMR — This other method for protein and nucleic acid structure determination, *Acta Cryst.,* D51, 249, 1995.
63. Rudo, F. G. and Krantz, J. C., Anaesthetic molecules, *Br. J. Anaesth.* 46, 181, 1974.
64. Limbird, L. E., *Cell surface receptors: A short course on theory and methods,* Martinus Nijhoff, Boston, 1986, 58.
65. Ueda, I., Interfacial effects of anesthetics on membrane fluidity, in *Drug and Anesthetic Effects on Membrane Structure and Function,* Aloia, R. C., Curtain, C. C., and Gordon, L. M., Eds., Wiley-Liss, New York, 1991, 15.
66. Eigen, M. and Hammes, G. G., Elementary steps in enzyme reactions, *Adv. Enzymol.,* 25, 1, 1963.
67. Berg, O. G. and von Hippel, P. H., Diffusion-controlled macromolecular interactions, *Annu. Rev. Biophys. Biophys. Chem.,* 14, 131, 1985.
68. Gutfreund, H., Reflections on the kinetics of substrate binding, *Biophys. Chem.,* 26, 117, 1987.
69. Sklar, L. A., Real-time spectroscopic analysis of ligand-receptor dynamics, *Annu. Rev. Biophys. Biophys. Chem.,* 16, 479, 1987.
70. Feher, V. A., Baldwin, E. P., and Dahlquist, F. W., Access of ligands to cavities within the core of a protein is rapid, *Nat. Struct. Biol.,* 3, 516, 1996.
71. Ringe, D., Petsko, G. A., Kerr, D. E., and Ortiz de Montellano, P. R., Reaction of myoglobin with phenylhydrazine: A molecular doorstop, *Biochemistry,* 23, 2, 1984.
72. Trudell, J. R. and Hubbell, W. L., Localization of molecular halothane in phospholipid bilayer model nerve membranes, *Anesthesiology,* 44, 202, 1976.
73. Creed, D., The photophysics and photochemistry of the near-UV absorbing amino acids — II. Tyrosine and its simple derivatives, *Photochem. Photobiol.,* 39, 563, 1984.
74. Sudhakar, K., Wright, W. W., Williams, S. A., Phillips, C. M., and Vanderkooi, J. M., Phenylalanine fluorescence and phosphorescence used as a probe of conformation for cod parvalbumin, *J. Fluorescence,* 3, 57, 1993.
75. He, X. M. and Carter, D. C., Atomic structure and chemistry of human serum albumin, *Nature,* 358, 209, 1992.
76. Cowan, D. O. and Drisko, R. L. E., The photodimerization of acenaphthylene. Heavy-atom solvent effects, *J. Am. Chem. Soc.,* 92, 6281, 1970.
77. Tsao, D. H. H., Casa-Finet, J. R., Maki, A. H., and Chase, J. W., Triplet state properties of tryptophan residues in complexes of mutated *Escherichia coli* single-stranded DNA binding proteins with single-stranded polynucleotides, *Biophys. J.,* 55, 927, 1989.
78. Basu, G., Anglos, D., and Kuki, A., Fluorescence quenching in a strongly helical peptide series: The role of noncovalent pathways in modulating electronic interactions, *Biochemistry,* 32, 3067, 1993.

79. Johansson, J. S., Rabanal, F., and Dutton, P. L., Binding of the volatile anesthetic halothane to the hydrophobic core of a tetra-α-helix bundle protein, *J. Pharmacol. Exp. Ther.*, 279, 56, 1996.
80. Gonzalez, L., Plecs, J. J., and Alber, T., An engineered allosteric switch in leucine-zipper oligomerization, *Nat. Struct. Biol.*, 3, 510, 1996.
81. Richards, F. M. and Kundrot, C. E., Identification of structural motifs from protein coordinate data: Secondary structure and first-level supersecondary structure, *Proteins*, 3, 71, 1988.
82. Postnikova, G. B., Komarov, Y. E., and Yumakova, E. M., Fluorescence study of the conformational properties of myoglobin structure, *Eur. J. Biochem.*, 198, 223, 1991.
83. Tilton, R. F., Singh, U. C., Weiner, S. J., Connolly, M. L., Kuntz, I. D., and Kollman, P. A., Computational studies of the interaction of myoglobin and xenon, *J. Mol. Biol.*, 192, 443, 1986.
84. Eckenhoff, R. G., Amino acid resolution of halothane binding sites in serum albumin, *J. Biol. Chem.*, 271, 15521, 1996.
85. Lopez, M. M. and Kosk-Kosicka, D., How do volatile anesthetics inhibit Ca^{2+}-ATPases?, *J. Biol. Chem.*, 270, 28239, 1995.
86. Beechem, J. M. and Brand, L., Time-resolved fluorescence of proteins, *Annu. Rev. Biochem.*, 54, 43, 1985.
87. Eftink, M. R., Fluorescence quenching: Theory and applications, in *Topics in Fluorescence Spectroscopy, Volume 2: Principles*, Lakowicz, J. R., Ed., Plenum Press, New York, 1991, 53.
88. Jameson, D. M. and Sawyer, W. H., Fluorescence anisotropy applied to biomolecular interactions, *Methods Enzymol.*, 246, 283, 1995.
89. Vanderkooi, J. M., Landesberg, R., Selick, H., and McDonald, G. G., Interaction of general anesthetics with phospholipid vesicles and biological membranes, *Biochim. Biophys. Acta*, 464, 1, 1977.
90. Karon, B. S. and Thomas, D. D., Molecular mechanism of Ca-ATPase activation by halothane in sarcoplasmic reticulum, *Biochemistry*, 32, 7503, 1993.
91. Karon, B. S., Geddis, L. M., Kutchai, H., and Thomas, D. D., Anesthetics alter the physical and functional properties of the Ca-ATPase in cardiac sarcoplasmic reticulum, *Biophys. J.*, 68, 936, 1995.
92. Chowdhry, V. and Westheimer, F. H., Photoaffinity labeling of biological systems, *Annu. Rev. Biochem.*, 48, 293, 1979.
93. Brunner, J., New photolabeling and crosslinking methods, *Annu. Rev. Biochem.*, 62, 483, 1993.
94. Weiland, G. A. and Molinoff, P. B., Quantitative analysis of drug-receptor interactions: I. Determination of kinetic and equilibrium properties, *Life Sci.*, 29, 313, 1981.
95. El-Maghrabi, E. A., Eckenhoff, R. G., and Shuman, H., Saturable binding of halothane to rat brain synaptosomes, *Proc. Natl. Acad. Sci. USA*, 89, 4329, 1992.
96. Eckenhoff, R. G., An inhalational anesthetic binding domain in the nicotinic acetylcholine receptor, *Proc. Natl. Acad. Sci. USA*, 93, 2807, 1996.
97. Eckenhoff, R. G., unpublished data, 1996.
98. Scharf, D., unpublished data, 1995.
99. Eckenhoff, R. G. and Shuman, H., Subcellular distribution of an inhalational anesthetic *in situ*, *Proc. Natl. Acad. Sci. USA*, 87, 454, 1990.
100. Lindeman, J. and Lovins, R. E., The mass spectral analysis of covalently labeled amino acid methylthiohydantoin derivatives derived from affinity-labeled proteins, *Anal. Biochem.*, 75, 682, 1976.

101. Shivanna, B. D., Mejillano, M. R., Williams, T. D., and Himes, R. H., Exchangeable GTP binding site of β-tubulin, *J. Biol. Chem.*, 268, 127, 1993.
102. Grenot, C., Blachère, T., Rolland de Ravel, M., Mappus, E., and Cuilleron, C. Y., Identification of Trp-371 as the main site of specific photoaffinity labeling of corticosteroid binding globulin using Δ^6 derivatives of cortisol, corticosterone, and progesterone as unsubstituted photoreagents, *Biochemistry*, 33, 8969, 1994.
103. Bayley, H., *Photogenerated Reagents in Biochemistry and Molecular Biology*, Elsevier, New York, 1983, 91.
104. Firestone, L. L., Alifimoff, J. K., and Miller, K. W., Does general anesthetic-induced desensitization of the *Torpedo* acetylcholine receptor correlate with lipid disordering?, *Mol. Pharmacol.*, 46, 508, 1994.
105. Raines, D. E., Rankin, S. E., and Miller, K. W., General anesthetics modify the kinetics of nicotinic acetylchloine receptor desensitization at clinically relevant concentrations, *Anesthesiology*, 82, 276, 1995.
106. Forman, S. A., Miller, K. W., and Yellen, G., A discrete site for general anesthetics on a postsynaptic receptor, *Mol. Pharmacol.*, 48, 574, 1995.
107. Unwin, N., Acetylcholine receptor channel imaged in the open state, *Nature*, 373, 37, 1995.
108. Blanton, M. P. and Wang, H. H., Photoaffinity labeling of the *Torpedo californica* nicotinic acetylcholine receptor with an aryl azide derivative of phosphatidylserine, *Biochemistry*, 29, 1186, 1990.
109. Bigelow, D. J. and Thomas, D. D., Rotational dynamics of lipid and the Ca-ATPase in sarcoplasmic reticulum, *J. Biol. Chem.*, 262, 13449, 1987.
110. Fraser, D. M., Louro, S. R. W., Horváth, L. I., Miller, K. W., and Watts, A., A study of the effect of general anesthetics on lipid-protein interactions in acetylcholine receptor enriched membranes from *Torpedo nobiliana* using nitroxide spin-labels, *Biochemistry*, 29, 2664, 1990.
111. Jørgensen, K., Ipsen, J. H., Mouritsen, O. G., and Zuckermann, M. J., The effect of anaesthetics on the dynamic heterogeneity of lipid membranes, *Chem. Phys. Lipids*, 65, 205, 1993.
112. Likhtenshtein, G. I., *Biophysical Labeling Methods in Molecular Biology*, Cambridge University Press, Cambridge, 1993, 1.
113. Brudvig, G. W., Electron paramagnetic resonance spectroscopy, *Methods Enzymol.*, 246, 536, 1995.
114. Millhauser, G. L., Fiori, W. R., and Miick, S. M., Electron spin labels, *Methods Enzymol.*, 246, 589, 1995.
115. Baber, J., Ellena, J. F., and Cafiso, D. S., Distribution of general anesthetics in phospholipid bilayers determined using ^{2}H-NMR and ^{1}H-^{1}H NOE spectroscopy, *Biochemistry*, 24, 6533, 1995.
116. Cobb, C. E., Juliao, S., Balasubramanian, K., Staros, J. V., and Beth, A. H., Effects of diethyl ether on membrane lipid ordering and on rotational dynamics of the anion exchange protein in intact human erythrocytes: Correlations with anion exchange function, *Biochemistry*, 29, 10799, 1990.
117. Abadji, V. C., Raines, D. E., Watts, A., and Miller, K. W., The effect of general anesthetics on the dynamics of phosphatidylcholine-acetylcholine receptor interactions in reconstituted vesicles, *Biochim. Biophys. Acta*, 1147, 143, 1993.
118. Abadji, V. C., Raines, D. E., Dalton, L. A., and Miller, K. W., Lipid-protein interactions and protein dynamics in vesicles containing the nicotinic acetylcholine receptor: A study with ethanol, *Biochim. Biophys. Acta*, 1194, 25, 1994.

119. Johnson, C. W., Secondary structure of proteins through circular dichroism spectroscopy, *Annu. Rev. Biophys. Biophys. Chem.,* 17, 145, 1988.
120. Woody, R. W., Circular dichroism, *Methods Enzymol.,* 246, 34, 1995.
121. Toniolo, C., Polese, A., Formaggio, F., Crisma, M., and Kamphuis, J., Circular dichroism spectrum of a peptide with a 3_{10}-helix, *J. Am. Chem. Soc.,* 118, 2744, 1996.
122. Laasberg, L. H. and Hedley-White, J., Optical rotatory dispersion of hemoglobin and polypeptides. Effect of Halothane, *J. Biol. Chem.,* 246, 4886, 1971.
123. Murray-Rust, P., Stallings, W. C., Monti, C. T., Preston, R. K., and Glusker, J. P., Intermolecular interactions of the C-F bond: The crystallographic environment of fluorinated carboxylic acids and related structures, *J. Am. Chem. Soc.,* 105, 3206, 1983.
124. Desiraju, G. R., The C–H···O hydrogen bond in crystals: What is it?, *Acc. Chem. Res.,* 24, 290, 1991.
125. Steiner, T. and Saenger, W., Role of C–H···O hydrogen bonds in the coordination of water molecules. Analysis of neutron diffraction data, *J. Am. Chem. Soc.,* 115, 4540, 1993.
126. Shimoni, L., Carrell, H. L., Glusker, J. P., and Coombs, M. M., Intermolecular effects in crystals of 11-(trifluoromethyl)-15,16-dihydrocyclopenta[*a*]phenanthren-17-one, *J. Am. Chem. Soc.,* 116, 8162, 1994.
127. Glusker, J. P., Intermolecular interactions around functional groups in crystals: Data for modeling the binding of drugs to biological macromolecules, *Acta Cryst.,* D51, 418, 1995.
128. Karle, I. L., Ranganathan, D., and Haridas, V., A persistent preference for layer motifs in self-assemblies of squarates and hydrogen squarates by hydrogen bonding [X–H···O; X=N, O, or C]: A crystallographic study of five organic salts, *J. Am. Chem. Soc.,* 118, 7128, 1996.
129. Savage, H. F. J., Elliot, C. J., Freeman, C. M., and Finney, J. L., Lost hydrogen bonds and buried surface area: Rationalising stability in globular proteins, *J. Chem. Soc. Faraday Trans.,* 89, 2609, 1993.
130. McDonald, I. K. and Thornton, J. M., Satisfying hydrogen bonding potential in proteins, *J. Mol. Biol.,* 238, 777, 1994.
131. Pace, C. N., Shirley, B. A., McNutt, M., and Gajiwala, K., Forces contributing to the conformational stability of proteins, *FASEB J.,* 10, 75, 1996.
132. Pace, C. N., Shirley, B. A., and Thomson, J. A., Measuring the conformational stability of a protein, in *Protein Structure,* Creighton, T. E., Ed., IRL Press, New York, 1990, 311.
133. Sturtevant, J. M., Biochemical applications of differential scanning calorimetry, *Annu. Rev. Phys. Chem.,* 38, 463, 1987.
134. Freire, E., Thermal denaturation methods in the study of protein unfolding, *Methods Enzymol.,* 259, 144, 1995.
135. Fersht, A. R., Protein folding and stability: The pathway of folding of barnase, *FEBS Lett.,* 325, 5, 1993.
136. Morton, A., Baase, W. A., and Matthews, B. W., Energetic origins of ligand binding in an interior nonpolar cavity of T4 lysozyme, *Biochemistry,* 34, 8564, 1995.
137. Munson, M., Balasubramanian, S., Fleming, K. G., Nagi, A. D., O'Brien, R., Sturtevant, J. M., and Regan, L., What makes a protein a protein? Hydrophobic core designs that specify stability and structural properties, *Protein Sci.,* 5, 1584, 1996.
138. Pace, C. N., Determination and analysis of urea and guanidine hydrochloride denaturation curves, *Methods Enzymol.,* 131, 266, 1986.

139. Shoichet, B. K., Baase, W. A., Kuroki, R., and Matthews, B. W., A relationship between protein stability and protein function, *Proc. Natl. Acad. Sci. USA,* 92, 452, 1995.
140. Broos, J., Visser, A. J. W. G., Engbersen, J. F. J., Verboom, W., van Hoek, A., and Reinhoudt, D. N., Flexibility of enzymes suspended in organic solvents probed by time-resolved fluorescence anisotropy. Evidence that enzyme activity and enantioselectivity are directly related to enzyme flexibility, *J. Am. Chem. Soc.,* 117, 12657, 1995.
141. Vincent-Geisse, J., Soussen-Jacob, J., Tai, N.-T., and Descout, D., Etude des mouvements de molécules linéaires en phase liquide par analyse du profil des bandes d'absorption infrarouge. I. Oxyde nitreux, *Can. J. Chem.,* 48, 3918, 1970.
142. Gorga, J. C., Hazzard, J. H., and Caughey, W. S., Determination of anesthetic molecule environments by infrared spectroscopy, *Arch. Biochem. Biophys.,* 240, 734, 1985.
143. Haris, P. I. and Chapman, D., The conformational analysis of peptides using Fourier transform IR spectroscopy. *Biopolymers,* 37, 251, 1995.
144. Dong, A., Huang, P., Zhao, X. J., Sampath, V., and Caughey, W. S., Characterization of sites occupied by the anesthetic nitrous oxide within proteins by infrared spectroscopy, *J. Biol. Chem.,* 269, 23911, 1994.
145. Dong, A. and Caughey, W. S., Infrared methods for study of hemoglobin reactions and structures, *Methods Enzymol.,* 232, 139, 1994.
146. Clore, G. M. and Gronenborn, A. M., Two-, three-, and four-dimensional NMR methods for obtaining larger and more precise three-dimensional structures of proteins in solution, *Annu. Rev. Biophys. Biophys. Chem.,* 20, 29, 1991.

3 Pressure and Anesthesia

Stephen Daniels

CONTENTS

0-8493-8555-5/01/$0.00+$.50

3.1 INTRODUCTION

Reversible effects of high pressure were first described in marine animals in 1891 by Regnard,[1] who observed deficits in motor activity at pressure. Sporadic reports of experiments on animals continued but the first human experiences of the effects of pressure were not reported until the experiments of Zaltsman[2] in 1961 and the pioneering 1000-ft (3.15-MPa) dive by Keller[3] in 1962. In these reports and in later simulated dives in the laboratory,[4-6] the principal symptoms were dizziness, nausea (with vomiting), and a marked tremor of the hands, arms, and torso, initially known as "helium tremors." These manifestations were originally believed to be the result of a narcotic effect of helium.

That these effects arose as a pure pressure effect was established by an experiment in which (1) identical effects were produced, on Italian great crested newts, by purely hydrostatic compression and by compression using helium–oxygen;[7] (2) liquid-breathing mice, pressurized hydrostatically, exhibited the characteristic tremors at 3 MPa, and at higher pressures (up to 6.7 MPa) a slowly developing generalized contraction of flexor muscles and cessation of breathing;[8] and (3) a subanesthetic concentration of nitrous oxide (0.15 MPa) in mice was not made anesthetic by the addition of 12.25 MPa of helium, as would have been expected if the effects produced by high pressures of helium were essentially narcotic.[7]

The effects of pressure have now been well characterized in both man[9] and animals[10] and are referred to, collectively, as high-pressure neurological syndrome (HPNS). In man the classic signs and symptoms are tremor (especially in the fingers and hands), dizziness, nausea, psychomotor impairment, EEG changes (especially increases in theta activity), and occasional myoclonus. Humans have been exposed to pressures up to 6.61 MPa using an oxygen–helium breathing mixture[11] and breathing a "trimx" of oxygen–helium–hydrogen, utilizing the fact that, under pressure, hydrogen has an anesthetic effect, and anesthetics can ameliorate the effects of pressure (Section 3.1.1.3), to 7.2 MPa.[12] In animals breathing oxygen–helium, tremor is observed, beginning between 3 and 4 MPa at the head and forequarters, and becoming increasingly severe with increasing pressure, until frank convulsions occur between 8 and 9 Mpa. Convulsions are followed by respiratory depression and death, at still higher pressure.[13] In both humans and animals, the signs and symptoms of HPNS appear at lower pressures and are more severe as the rate of compression is increased, but they also remit with time if the pressure is held constant.[10,11]

3.1.1 Pressure Reversal of General Anesthesia

Although the changes in central nervous system (CNS) function brought about by general anesthetics (reduced perception of sensations and unconsciousness) are quite different from those arising from exposure to high pressure, there is nevertheless a remarkable connection, namely the ability of high pressure to reverse general anesthesia[14] and the concomitant amelioration of the effects of pressure by general anesthetics.[15] Early reports of the pressure reversal of general anesthesia followed experiments by Johnson and Flagler,[16] who showed that tadpoles anesthetised using

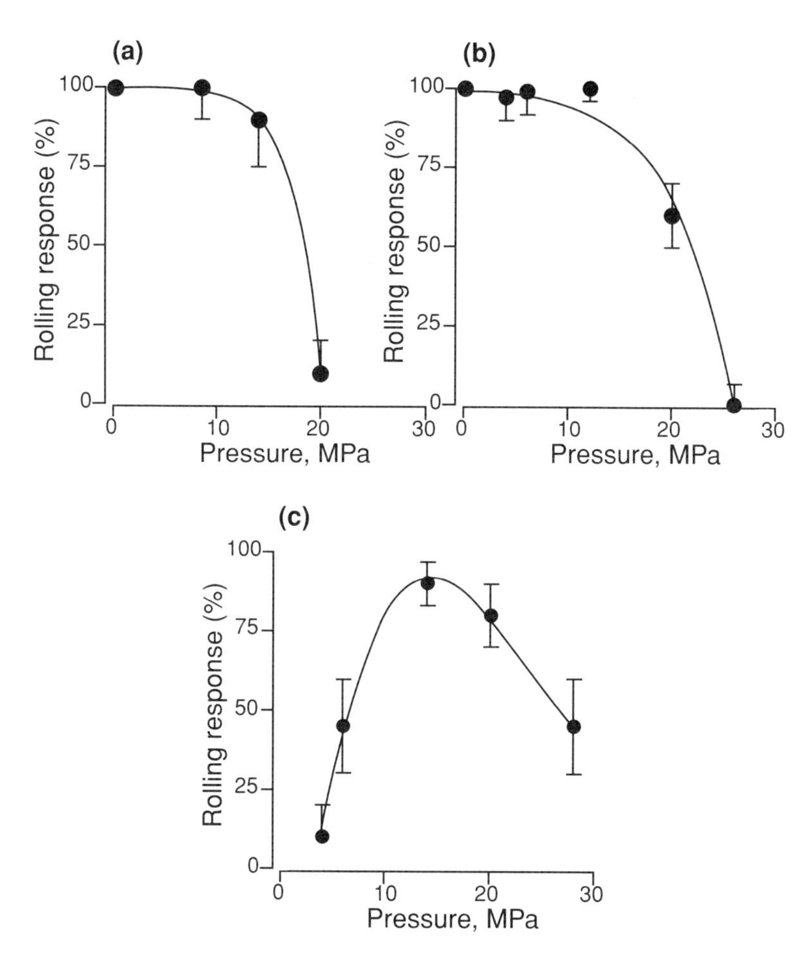

FIGURE 3.1 The rolling response of newts as a function of (a) hydrostatic pressure in water, (b) helium pressure in the presence of 101 kPa oxygen, and (c) helium pressure in the presence of 3.4 MPa nitrogen and 101 kPa oxygen. (Data taken from Lever et al.[32])

ethanol had their swimming ability restored by the application of 10 MPa hydrostatic pressure. This observation languished until it was established that helium acts as a pure transmitter of pressure and the equivalent phenomenon was demonstrated in mammals.[14] A classic example of the pressure reversal of anesthesia is shown in Figure 3.1. It should be noted that Figure 3.1 also shows that the incapacitating effect of pressure is postponed until higher pressures in the presence of anesthetic (3.4 MPa nitrogen).

The spectacular interaction between general anesthetics and pressure leads to speculation as to the nature of the interaction, whether it represents opposing effects at a single site of action or a summation of separate physiologic effects, and whether understanding the interaction will lead to a clearer understanding of the mechanism of action of both general anesthetics and high pressure.

3.1.1.1 Physiochemical Theories

An understanding of the mechanism of action of general anesthetics in a most general sense was sought by applying physicochemical arguments and the interaction between pressure and general anesthetics should, in principal, clarify this analysis.

The most fundamental physicochemical approach was that adopted by Ferguson,[17] who suggested that anesthetic potency was correlated with thermodynamic activity, defined on a Raoult's law basis.[17] This proved to be incorrect, as demonstrated by a number of fluorinated gaseous anesthetics (e.g., SF_6, CF_4) whose thermodynamic activities, at the anesthetic ED_{50} pressure, are approximately 10 times greater than those of other, more conventional, anesthetic gases. This failure has proven vital for the investigation of anesthetic interactions, because otherwise the anesthetic potency of a substance would depend only on the properties of the pure substance and not on any interaction between the substance and its surroundings.

A second hypothesis put forward to explain anesthetic potency was the ability of an anesthetic to order, or otherwise disturb, the aqueous phase of the central nervous system.[18,19] This theory also fails when tested with fluorinated compounds, which do not fit on a correlation between anesthetic potency and hydrate dissociation pressure.[20] Furthermore, many anesthetics, including ether and halothane, do not form hydrates. Finally, this theory cannot account for the simple additivity in potency observed with mixtures of gaseous and volatile anesthetics.[21]

Meyer[22] and Overton[23] observed that anesthetic potency correlates with solubility in fatty substances and suggested that anesthetic potency was related to hydrophobicity. This has been tested using a great many anesthetics and a range of different solvents (including olive oil, octanol, and benzoic acid) and has been found to hold, to within 20%, more than a 10,000-fold range of potency.[24] Variants on the simple hydrophobic theory have been proposed based on the idea that anesthetics, by dissolving at their site of action, lower the chemical potential at that site.[25] This general hypothesis includes those invoking increases in membrane fluidity,[26] although changes in membrane fluidity at clinical concentrations of anesthetic are unlikely.[27]

An alternative theory, developed in part to account for the interaction of high pressure with anaesthesia and the lack of anesthetic effect of helium, sought to relate anesthetic potency to the expansion caused by the incorporation of the anesthetic molecules.[14] Although it was frequently assumed that the anesthetic site of action was the cell membrane, this was not specified and is not a prerequisite.

Finally, a simple binding model (to a hydrophobic site) to explain anesthetic potency has been proposed.[28] In this model anesthetics bind to a site (a neuroactive protein) and inactivate it. This model has much to commend it in preference to the notation that the expansion of the lipid bilayer is causal; namely an explanation for the cut-off in anesthetic potency observed in homologous series of alkanes and alcohols,[29] the observation of stereoselective effects of isoflurane,[30] and NMR spectroscopic measurements showing that in a halothane-anaesthetized rat the halothane was bound, in a saturable manner, and immobile when bound.[31]

3.1.1.2 The Effect of Pressure

The experimentally observed interaction between anesthetics and high pressure can be exploited to resolve which of these hydrophobic theories best fits the available data.

3.1.1.2.1 Meyer–Overton

The Meyer–Overton theory implies that it is the number and density of anesthetic molecules that is important. This in turn suggests that pressure acts by "squeezing out" the dissolved molecules. This is described by the thermodynamic equation

$$\frac{\partial \ln(x / P_a)}{\partial P} = -\frac{\forall_a}{RT} \tag{3.1}$$

where x is the mole fraction solubility, P_a the anesthetic partial pressure, $\forall_a$ the partial molar volume of the gas, R the gas constant, and T the absolute temperature. The Meyer–Overton hypothesis gives

$$P_{50}x_{50} = P_a x_a \tag{3.2}$$

where P_{50} is the anesthetic partial pressure at the ED_{50} and x_{50} is the solubility at this pressure; P_a is the ED_{50} partial pressure at a total pressure P_T and x_a is the solubility at this pressure. Integration of Equation 3.1 between P_T and P_{50}, eliminating x_{50}/x_a, gives

$$\frac{1}{\forall_a}\ln\left(\frac{P_a}{P_{50}}\right) = \frac{1}{2RT}[P_T - P_{50}] \tag{3.3}$$

Thus a graph of the left-hand side of Equation 3.3 against $[P_T - P_{50}]$ should give a straight line. It is clear from Figure 3.2a that, using data from both newts[32] and mice,[33] this is not the case.

3.1.1.2.2 Chemical Potential

In the case of theories of anesthesia based on a lowering of the chemical potential, μ_s, at the site of action by the presence of anesthetic at a mole fraction x_a, then the action of pressure must be to increase the chemical potential. The thermodynamic relationship describing the process is

$$\Delta\mu_s = RT \ln x_s + \forall_s \Delta P \tag{3.4}$$

where x_s is the mole fraction of site material and $\forall_s$ is its molar volume. In dilute solution $x_a << x_s$, thus:

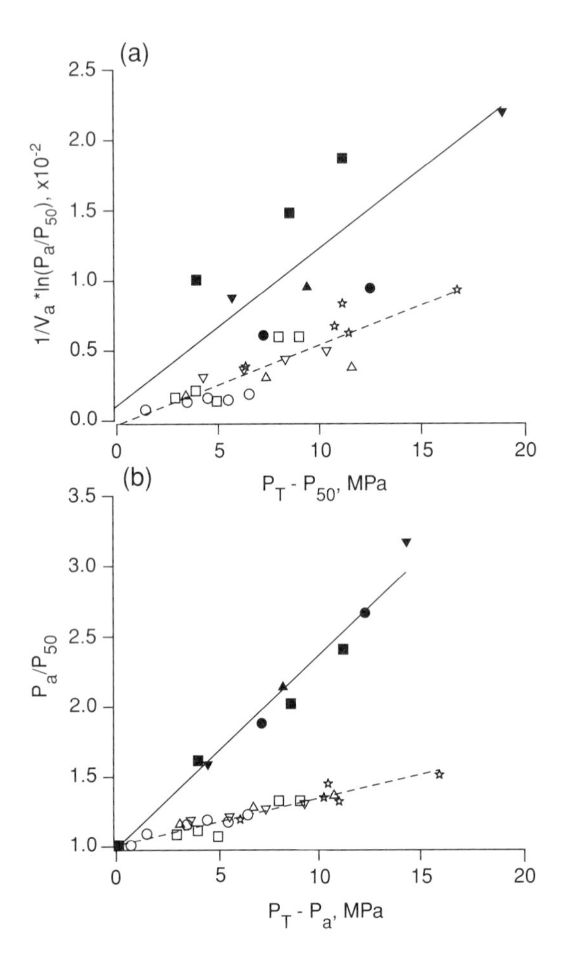

FIGURE 3.2 (a) Test for the Meyer–overton model for pressure reversal of anesthesia. Solid data points, and solid line, are for newts: ■, N_2O; ▼, N_2; ▲, CF_4; and ●, SF_6. Open data points, and dashed line, are for mice: □, N_2O; ▽, N_2; △, CF_4; ○, SF_6; ☆, Ar. (b) Test for the critical volume model. (Data taken from Miller et al.[32] and Miller et al.[33])

$$\Delta\mu_s = RTx_a + \forall_s \Delta P \tag{3.5}$$

If the applied pressure abolishes anesthesia, then $\Delta\mu_s = 0$ and

$$\Delta P = RTx_a / \forall_s \tag{3.6}$$

Available data suggest that the appropriate anesthetic concentration ($x_a/\forall_s$) is 50 m*M*. This implies that pressure reversal of anesthesia would occur with a pressure increase of merely 0.1 MPa, two orders of magnitude less than the experimental value of approximately 10 MPa.

3.1.1.2.3 *Volume Expansion (Critical Volume)*

The critical volume hypothesis suggests that the volume expansion caused by incorporation of the anesthetic molecules at the site of action is opposed by the application of pressure. Thus the fractional expansion is given by

$$\Delta V = \frac{\forall_a x_a P_a}{\forall_s} - \beta P_a = \left[\frac{\forall_a x_a}{\forall_s} - \beta\right] P_a \tag{3.7}$$

where $\forall_s$ is the molar volume of the site and β is the isothermal compressibility. The first term is the expansion due to solution of the anesthetic and the second the compression due to pressure on the site of solution. If the system is compressed using helium, then the fractional compression is given by

$$\Delta V = -\left[\frac{\forall_{He} x_{He}}{\forall_a} - \beta\right][P_T - P_a] \tag{3.8}$$

Thus at the ED_{50} for the anesthetic, at any pressure, Equation 3.7 must equal Equation 3.8, and with rearrangement:

$$\frac{P_a}{P_{50}} = \left[\frac{\beta}{ED_{50}} - \frac{\forall_{He} x_{He}}{ED_{50} \forall_s}\right][P_T - P_a] + 1 \tag{3.9}$$

Figure 3.2b shows a plot of P_a/P_{50} against $(P_T - P_a)$ using data from both newts[32] and mice.[33] In both cases an excellent correlation is obtained. The different slopes are assumed due to the different sites of action for anesthetics in newts and mice. These sites would be expected to have different compressibilities and molar volumes.

3.1.1.2.4 *Protein Binding Model*

In a simple formulation of this model, assuming that anesthetics are acting at the same site as pressure, the equilibrium constant for binding is given as

$$K = \frac{[SA]}{[S][A]} \tag{3.10}$$

where $[SA]$ represents the concentration of bound anesthetic, $[S]$ the concentration of sites, and $[A]$ the concentration of free anesthetic. The effect of pressure on the equilibrium constant is given by the thermodynamic relation

$$\left(\frac{\partial \ln \mathrm{K}}{\partial \mathrm{P}}\right)_{\mathrm{T}} = -\frac{\Delta \mathrm{V}}{\mathrm{RT}} \tag{3.11}$$

If anesthesia is administered at some critical concentration of anesthetic-bound sites, then application of Equation 3.11 leads to

$$\ln\left(\frac{P_a}{P_{50}}\right) = \left(\frac{\Delta V}{RT}\right)[P_T - P_a] \quad (3.12)$$

A plot of $\ln(P_a/P_{50})$ against $[P_T - P_a]$ resembles Figure 3.2b, with an equally good fit of the experimental data to the predicted values. In contrast to the critical volume hypothesis, the slope of the relationship gives the overall volume change accompanying the reaction. In the case of the newt data, the calculated volume change is 195 cm^3 and for mice it is 72 cm^3. Interpretation of this volume change is difficult because it is compounded from the solvation of the anesthetic and conformational changes in the protein following binding (Section 3.2.2.2.2). However, a simple binding model does account for the data and does not conflict with the evidence that the site of action of anesthetics is hydrophobic in nature.

3.1.1.3 Interaction Between Pressure and a Variety of Anesthetics

Although pressure reversal of anesthesia was originally thought to be a universal phenomenon, as implied by the critical volume theory and all other so-called unitary hypotheses, it is now known that this is not the case.

All simple gaseous anesthetics tested in small mammals (Ar, H_2, N_2, N_2O, CF_4, SF_6, and C_2F_6) have been shown to be pressure reversed and to show the symmetric action of ameliorating the effects of pressure.[12,33,34] Furthermore, their pressure protection is in line with their anesthetic potency, although the degree to which they protect against the different components of HPNS (tremor, convulsions, etc.) varies.[34]

In contrast, whereas the volatile anesthetics (halothane, isoflurane) are readily pressure reversed;[35,36] halothane is ineffective against pressure.[35] Such discrepancies are even more marked with the intravenous anesthetics. Thus methohexitone, thiopentone, ketamine, propanidid, and althesin all show pressure reversal, to varying degrees;[37] but whereas ketamine and althesin ameliorate the effects of pressure, thiopentone, methohexitone, and propanidid are ineffective.[38] An inability to ameliorate the effects of pressure is not a universal feature of barbiturates. Pentobarbitone and phenobarbitone are both pressure reversed and both also protect against pressure.[39] Of the steroid anesthetics, 3α-hydroxy-alphaxalone, the major active constituent of althesin, also demonstrates remarkable stereospecificity; it is readily pressure, reversed and very effective against pressure, yet its simple structural analogues, 3β-hydroxy-alphaxalone and Δ-16-alphaxalone, are not anesthetic but are effective against pressure.[37]

These findings indicate that the interaction between anesthetics and pressure is not the simple one envisaged in the physicochemical models, in particular the reciprocity of the interaction is not universal and the degree of pressure reversal for the intravenous anesthetics is not in line with their anesthetic potency. Nor does it seem sensible to take a simple hypothesis, such as the critical volume theory, and elaborate by postulating different sites for anesthetic action and pressure protection and then subdividing those on the basis of the general type of anesthetic agent.[40,41]

Rather, it is more sensible to consider how pressure might be acting before attempting to interpret the interaction between pressure and anesthetics.

3.1.1.4 Species Variation in Pressure Reversal of Anesthesia

Pressure reversal of anesthesia has been demonstrated in many species, including luminous bacteria,[42,43] tadpoles,[16] newts,[32] mice,[33,44] rats,[37] guinea pigs,[45] and humans.[46] The end point by which pressure reversal of anesthesia is, typically, judged in animals is the restoration of motor response[32,33] or response to a painful stimulus.[16,37] In humans, the dose of althesin required to abolish a response to an auditory stimulus was determined at atmospheric pressure and at 3 MPa. It was found that at 3 MPa the dose of althesin had to be increased by approximately 30%. It has also been shown, using rats, that anesthetic-induced changes in cortical-evoked potentials were reversed by 10 MPa pressure, as was the depression of motor responses to tail stimulation.[47] It does appear therefore that, in mammals, pressure will reverse not only the depression of motor response induced by anesthesia but also the sensory response, suggesting that pressure would restore consciousness in humans, and not just the motor component of anesthesia.

There are, however, notable exceptions to species in which pressure reversal of anesthesia can be demonstrated. The pleopod beat frequency of the marine shrimp, *Marinogammarus marinus*, is depressed by halothane and this depression is further enhanced by pressure.[48] Furthermore, the locomotor activity of the brine shrimp,[49] *Artemia salina*, and the freshwater shrimp,[50] *Gammarus pulex*, was depressed by anesthetics and was further depressed by pressure. In all these arthropods the response to anesthetics was similar to that of other species, but they showed no increase in motor activity at pressure, unlike tadpoles or mammals. One difference observed between the arthropods and mammals is that the arthropods do not have the inhibitory, strychnine-sensitive glycine receptor. This may contribute to their failure to show pressure enhancement of activity and pressure reversal of anesthesia. The role of this receptor in the expression of HPNS will be discussed later.

3.2 MECHANISM OF ACTION OF PRESSURE

The effects of increasing pressure, described earlier, may be regarded as increasing hyperexcitability and might be thought analogous to other forms of excitability, such as epilepsy. However, classic antiepileptics, including phenytoin, carbamazepine, and diazepam (in nonsedative doses), are ineffective against pressure.[38,51,52]

Before considering the effects of pressure in detail it is instructive to analyze the site of action of pressure within the CNS. Early experiments showed that mass discharges from subcortical structures were found to precede those from the cortex[53] and that immature animals, in which the cortex is not fully functional, are susceptible to pressure.[54] Classic ablation experiments employing chronically decerebrate and decorticate animals indicated that tremor and convulsions arise subcortically, although a degree of cortical inhibition was apparent.[55] In animals spinally transected (between T_7 and T_{13}), seizure activity was observed caudal to the transection, but in cases where the spinal nerves L_2 to L_6 were also sectioned, the limb remained flaccid.[56] Thus four or five segments of cord provide a sufficient neuronal organization to sustain seizure activity, but the isolated nerve and muscle do not show discharge activity until pressures in excess of 30 MPa are reached.[56]

3.2.1 Pharmacology of Pressure Effects

The range of different interactions observed between general anesthetics and pressure may be taken as an indication that the action of pressure might not be mediated by a nonspecific action on a bulk phase (e.g., compression of a neuronal cell membrane), but may instead arise from a specific action on some aspect of neurotransmission.

3.2.1.1 The Role of Biogenic Amines

A possible role for the biogenic amine group of neurotransmitters was suggested following experiments demonstrating that reserpine-treated mice were very much more susceptible to the effects of pressure than untreated mice.[57] Treatment with reserpine reduces the pressure at which mice exhibit convulsions by approximately 40%, but more dramatically, it removes all compression rate dependence. Normally the pressure at which convulsions occur in CD1 mice is reduced from approximately 10 MPa to 6 MPa by an increase in compression rate from 0.001 to 100 Mpa · h^{-1}. Mice treated with reserpine show convulsions at 6 MPa irrespective of the rate of compression. L-Tryptophan (5-HT precursor), tranylcypromine (nonspecific monoamine oxidase inhibitor), and amphetamine (cytosolic catecholamine release) all produce a partial (40 to 60%) reversal of the effect of reserpine.[57] Studies with more selective agents, such as α-methyl-*p*-tyrosine (tyrosine hydroxylase inhibitor), *p*-chloropheny-lalanine (tryptophan hydroxylase inhibitor), and FLA-63 (dopamine-β-hydroxylase inhibitor), failed to elucidate which of the catecholamine transmitters was causal in mediating the inhibitory control removed by treatment with reserpine.[58,59]

However, intracerebral injection of the neurotoxins 6-hydroxydopamine and 5,6- and 5,7-dihydroxytryptamine (DHT), which physically destroy monoaminergic neurons, substantially reduces the onset pressure for tremor and convulsions in mice, suggesting that it is noradrenergic transmission that is the key factor.[55] This conclusion was reinforced by the finding that desmethylimpramine, which prevents the neurotoxic action of 5,7-DHT on noradrenergic neurons, also abolishes the effect of 5,7-DHT on the onset pressure for tremor and convulsions.[55]

The reduction in the onset pressure for convulsions, caused by reserpine treatment, to one equivalent to that caused by very rapid compression but not to a lower pressure, is similar to the effect of decerebration. This suggests that cortical and midbrain noradrenergic processes act to inhibit the onset of tremor and convulsions without being responsible for originating the seizure activity, in a manner analogous to that in other induced seizure states.[60] This inhibition, however, is clearly limited. In addition to a dependence on compression rate for the appearance of signs and symptoms of HPNS, it is well documented in humans that once HPNS has occurred, if the pressure is kept constant then HPNS remits over the course of hours.[11,61] It is possible that this "adaptation" occurs as noradrenergic processes are upregulated in response to the excitatory stimulus.

3.2.1.2 The Role of Cholinergic Mechanisms

A possible role for cholinergic mechanisms in the expression of HPNS was investigated because of the well-established role of cholinergic processes in arousal and motor control.[62] The effects of atropine (centrally acting competitive antagonist at muscarinic acetylcholine [Ach] receptors), methylatropine (competitive antagonist at mACh receptors that does not cross blood–brain barrier), and mecamylamine (centrally acting nicotinic receptor channel blocker) have been examined at high pressure.[63] None of these drugs altered the pressure at which tremors or convulsions occurred, with the exception of mecamylamine, which slightly reduced the pressure at which tremors appeared. These results appear to rule out any role for cholinergic processes in the genesis of HPNS, especially given that mecamylamine exacerbated rather than ameliorated tremors.

3.2.1.3 GABA-Mediated Inhibitory Processes

GABA (γ-aminobutyric acid) is the principal inhibitory neurotransmitter in the CNS, and both competitive antagonists (e.g., bicuculline) and drugs that block the chloride channel associated with the GABA receptor (e.g., picrotoxin) are convulsant in action. The $GABA_A$ receptor also possesses a recognition site for benzodiazepines (e.g., diazepam, clonazepam), which are powerful sedatives and anticonvulsants for a range of seizure states (e.g., epilepsy). This leads to the idea that GABAergic neurotransmission may have a role in the etiology of HPNS.

The benzodiazepines clonazepam and diazepam have been found to provide some protection against the effects of pressure,[51] although at sedative doses. Furthermore, the protective effect of flurazepam is abolished by the selective benzodiazepine antagonist Ro15-1788,[64] indicating that the protective action of flurazepam arises via a specific interaction at its receptor, which leads to a potentiation of GABAergic inhibition.

Other drugs that act to potentiate the action of GABA — aminooxyacetic acid (a GABA transaminase inhibitor) and 2,4-diaminobutyric acid (a GABA reuptake blocker) — postpone the onset of both tremors and convulsions by approximately 25%. However, at the doses necessary to produce this degree of protection against pressure the convulsions caused by the GABA antagonist bicuculline are completely abolished.[65] Sodium valproate, a weak inhibitor of both GABA transaminase and succinic semialdehyde, has a more pronounced protective effect, increasing the pressure for the onset of tremors and convulsions by 75%, similar to the increase by flurazepam.[65] However, the action of sodium valproate cannot be ascribed simply to its effect on the metabolism of GABA, because it has a number of other actions that may be more important with respect to its anticonvulsive properties; namely, a reduction in membrane excitability by blocking Na^+ and Ca^{2+} channels.[66]

Muscimol, a GABA agonist at $GABA_A$ receptors, is without effect when given systemically intraperitoneal [IP] or intracerebroventricularly. Similarly, baclofen, a GABA agonist at $GABA_B$ receptors, is without effect.[55] The converse to testing the effect of GABA agonists is to test the additivity of action of the GABA antagonists with pressure. The actions of both picrotoxin[67] and bicuculline[68] have been tested in combination with pressure by establishing dose–response curves for both at fixed

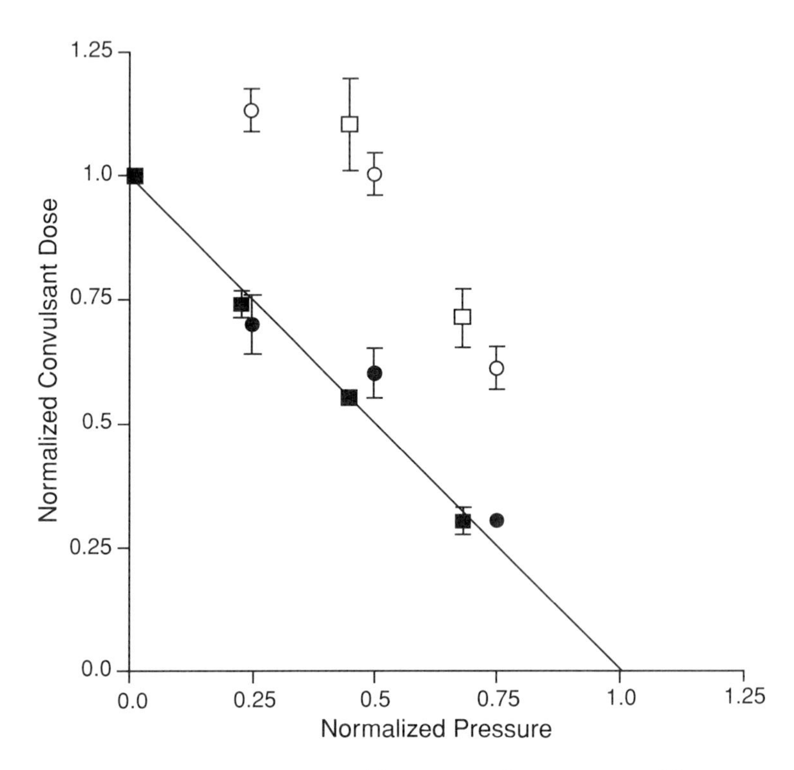

FIGURE 3.3 ED_{50} values for strychnine (■), NMDA (●), bicuculline (□), and picrotoxin (○), normalized relative to their respective ED_{50} values at atmospheric pressure, plotted against the pressure at which they were determined, and normalized to the ED_{50} for pressure seizures in the absence of drug. The line of unit slope represents simple additivity. (Data taken from Bowser-Riley et al.,[67] strychnine and picrotoxin; Bowser-Riley et al.[68] bicuculline; unpublished data, NMDA.)

pressures: atmospheric pressure and pressures equivalent to 25, 50, and 75% of the pressure that elicits convulsions in the absence of convulsant drug. The convulsant ED_{50} values, normalized with respect to the ED_{50} at atmospheric pressure, are shown in Figure 3.3 plotted against the pressure at which they were determined and normalized with respect to the pressure that elicits convulsions in the absence of drug. Bicuculline and picrotoxin show marked divergence from the additivity of action expected if the pressure effect was through some action on the $GABA_A$ receptor.

Pressure up to 13 MPa had no effect on the response of the rat superior cervical ganglion to exogenous GABA, as recorded extracellularly, nor did this pressure reverse the potentiation of the GABA response produced by either ketamine or pentobarbitone.[65] Furthermore, pressure up to 10 MPa had no effect on either spontaneous or electrically evoked release of GABA from the frog hemisected spinal cord unless the cord was electrically stimulated prior to pressurization, in which case an increase in GABA release was observed.[65] An effect of pressure on GABA release cannot, therefore, be postulated to underlie the observed increase in excitability seen at high pressure.

3.2.1.4 Glycine-Mediated Inhibitory Processes

Glycine is now recognized as one of the principal inhibitory neurotransmitters, particularly in subcortical structures.[69] A possible link between glycine-mediated inhibitory neurotransmission and the mechanism of action of pressure arose from studies into the interaction between pressure and a group of centrally acting muscle relaxants based on mephenesin.[55] These compounds, although not normally regarded as anticonvulsant, reveal crucial differences in action. The aromatic compounds, mephenesin and methocarbamol, are effective anticonvulsants against strychnine, but not against bicuculline or picrotoxin (see Structure 1). In contrast, the aliphatic compounds, meprobamate, carisoprodol, and the related compound orphenadrine (see Structure 2), which are much more potent centrally acting muscle relaxants, are effective anticonvulsants against bicuculline and picrotoxin but not against strychnine.

This distinction in the action of these two groups of compounds suggests that there may be a link between the mechanism of action of strychnine (the classic competitive antagonist for glycine) and that of pressure. Strong support for this hypothesis comes from experiments in which the ED_{50} values for strychnine were determined at fixed pressures (Figure 3.3). Strict additivity of action, in eliciting seizures, between strychnine and pressure is observed.[67]

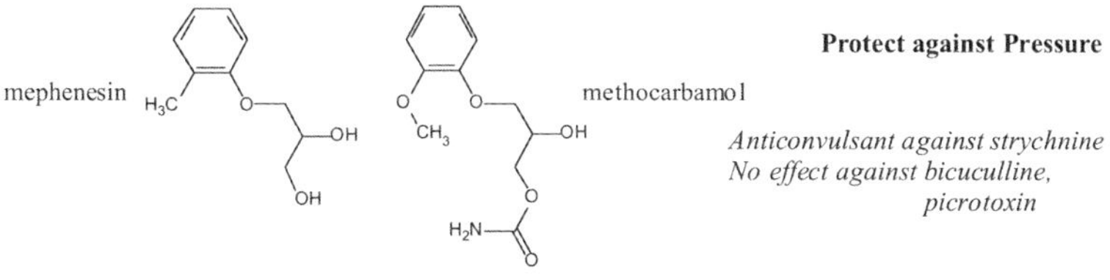

mephenesin — methocarbamol

Protect against Pressure

Anticonvulsant against strychnine
No effect against bicuculline, picrotoxin

Structure 1

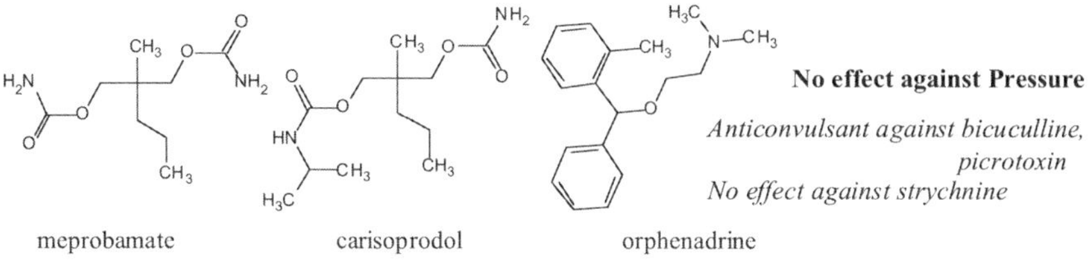

meprobamate — carisoprodol — orphenadrine

No effect against Pressure

Anticonvulsant against bicuculline, picrotoxin
No effect against strychnine

Structure 2

That the pressure protection afforded by mephenesin and related aromatic propandiols does not arise simply from their muscle relaxant action is emphasized by experiments with a series of closely related compounds. The mephenesin analogue, 3[2-methylphenoxy]propan-2-ol is six times more potent than mephenesin, 3[2-methylphenoxy] propan-1,2-diol, as a muscle relaxant, yet it is half as potent against pressure (or strychnine).[70] These experiments also revealed that mephenesin was the most potent of this range of propandiols against both pressure and strychnine.

It was also shown that if the terminal hydroxy group was exchanged for an amine group, then the activity of the compound changed from a pressure-protecting compound to a proconvulsant that acts additively with pressure (see Structure 3). Molecular modeling of these compounds, and of strychnine, suggested that the lowest energy form placed the terminal nitrogen at a point in space similar, relative to the aromatic ring, to that occupied by the amino nitrogen in strychnine (see Structure 4). Confirmation of the importance of this aspect of the structure was provided by experiments with a series of compounds with structures based on strychnine.[70]

Gramine, like strychnine, acts in an additive manner with pressure. However, as predicted, the compound 3-methoxymethylindole protected against pressure, with a potency approximately 50% that of mephenesin, and the compound 3-methylindole-2-butyrolactam was without effect against pressure; it acted as a competitive antagonist to strychnine.[70]

Finally, a separate group of compounds, related to benzimidazole, that are also centrally acting muscle relaxants and anticonvulsants against strychnine but not against bicuculline or picrotoxin are also effective against pressure.[71] The structure–activity relationship of this group of compounds against pressure matched that against strychnine, and their potency against pressure (or strychnine) was unrelated to their muscle relaxant potency.

Taken together, these findings provide a compelling body of evidence in support of a role for glycine-mediated inhibitory neurotransmission in the etiology of HPNS.

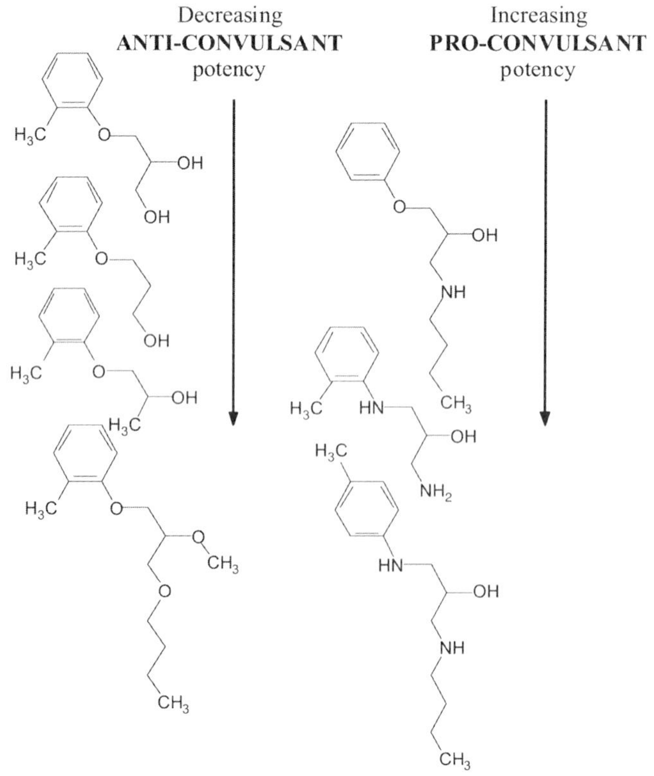

Structure 3

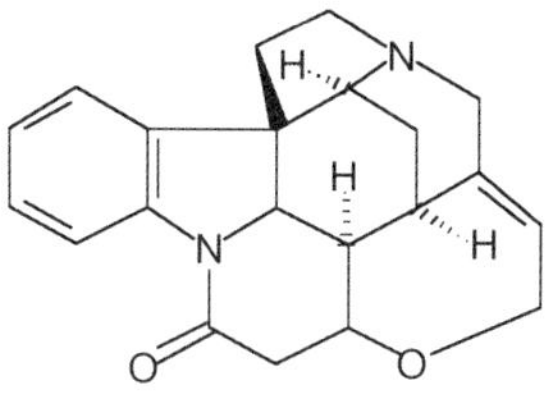

strychnine

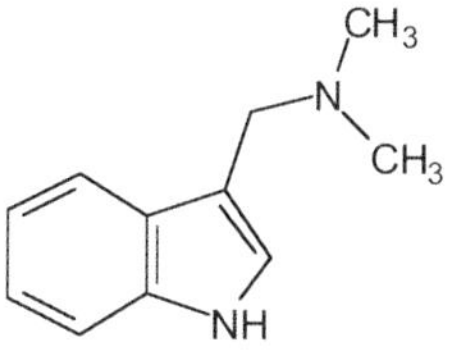

gramine

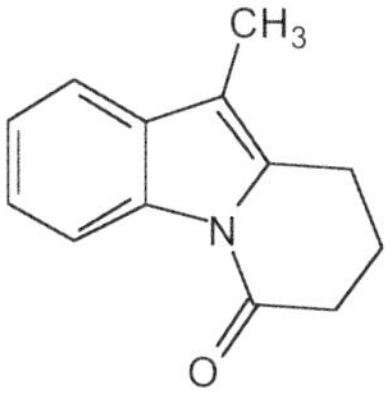

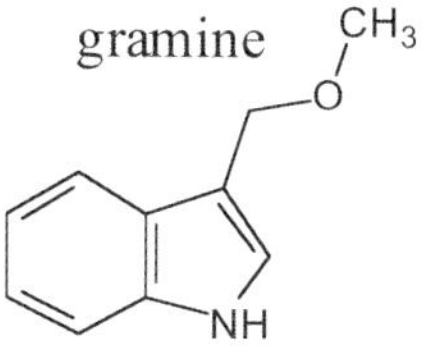

3-methoxymethylindole

3-methylindole-2-butyrolactam

Structure 4

3.2.1.5 Glutamate-Mediated Excitatory Processes

A role for glutamate-mediated excitatory neurotransmission was postulated following experiments that showed effective protection against pressure was provided by the *N*-methyl-D-aspartate (NMDA) antagonist 2-amino-7-phosphonoheptanoic acid (AP7).[72]

It has recently become clear that competitive antagonists at the NMDA-type glutamate receptor generally provide protection against both tremor and seizure phases of HPNS, although there is considerable variation in potency. Thus, D-AP7 is the most potent competitive NMDA antagonist tested to date, with 3-((+/–)-2-carboxypiperazine-4-yl)-propyl-1-phosphonate (CCP), DL-(E)-2-amino-4-methyl-5-phosphono-3-pentanoic ethylester (CGP 39551), and β-D-aspartylaminomethyl phosphonate being progressively less potent.[73-76]

In the case of noncompetitive NMDA antagonists, widely different effects have been observed. In the rat, ketamine is effective against all aspects of HPNS[77] but MK801 ((+)-5-methyl-10,11-dihydro-5H-dibenzo[a,d]cyclohepten-5,10-imine) has no effect.[78] However, in the baboon MK801 gives protection.[79] Whether this discrepancy arises because different ionotropic NMDA receptor subtypes exist in the rat compared with baboon is not known. Finally, (+/–)-cis-2,3-piperidinedicarboxylic acid, a partial agonist at NMDA receptors, provides moderate protection against both phases of HPNS.[73]

The action of pressure at non-NMDA-type glutamate receptors is even less clear. The kainate/AMPA selective antagonist γ-D-glutamylaminomethylsulfonic acid (GAMS) is effective in the rat against the tremor phase of HPNS but not against the seizures,[76] whereas LY 293558 is effective against both phases.[80] The noncompetitive amino-3-hydroxy-5-methyl-4-isoxazole proprionic acid (AMPA) antagonist L-glutamic acid diethyl ester (GDEE) has no effect on HPNS in the rat.[73] However, in the baboon the noncompetitive AMPA antagonist 1-(4-aminophenyl)-4-methyl-7,8-methylenedioxy-5H-2,3-benzodiazepine (GYKI-52466) is, like GAMS in the rat, effective against the tremor phase.[80]

NMDA acts additively with pressure in mice (Figure 3.3), suggesting, by analogy with the additivity observed between pressure and strychnine, that pressure may have a direct effect on glutamate as well as glycine receptors. However, the finding that NMDA and strychnine also are additive (unpublished results) suggests, despite the lack of additivity observed between pressure and bicuculline or picrotoxin or between strychnine and bicuculline, the alternative possibility that the actions of NMDA, strychnine, and pressure might be mechanistically independent but interact via a physiologic summation.

3.2.2 Cellular and Molecular Effects of Pressure

Experimental evidence from *in vitro* studies using both invertebrate- and vertebrate-derived preparations indicates that pressure prolongs action potential duration but has little effect on either amplitude or conduction velocity.[81] In addition, a propensity for repetitive firing is induced.[81] The effects of pressure on axonal conduction are relatively small and frequently are manifested only at pressures above 20 MPa, twice that at which substantial effects (seizures) are seen *in vivo*. These findings suggest that the hyperexcitability observed at pressure must arise from some effect on synaptic transmission.

3.2.2.1 Effect of Pressure on Peripheral Synaptic Transmission

Pressure has a pronounced, although paradoxical, effect on excitatory synaptic transmission. Pressures as low as 3.5 MPa depress excitatory synaptic transmission in rat superior cervical ganglia.[82] Such a depression has also been observed in a variety of invertebrate preparations.[81]

3.2.2.1.1 Presynaptic Effects

Evidence suggests that this decrease in fast excitatory synaptic transmission arises from a presynaptic action of pressure that leads to a reduction in the spontaneous release of transmitter. Thus at the frog neuromuscular junction the frequency of miniature end-plate currents (MEPC) was reduced by as much as 75% at 10 MPa.[83] An explanation for the reduction in transmitter release is suggested by experiments using a lobster abdominal muscle preparation. These studies indicate that high pressure (6 to 10 MPa) decreases Ca^{2+} influx into presynaptic nerve terminals.[84-86]

3.2.2.1.2 Postsynaptic Effects

The effects of pressure on postsynaptic events are less marked than the presynaptic effects. Pressures of 10 MPa did not alter the response to iontophoretically applied ACh of *Aplysia* neurons,[87] although it has been reported that pressures of 15 MPa reduce the amplitude and slow the decay phase of MEPC.[83] These latter changes might reflect an increase in channel open time. The binding of ACh to its receptor is reduced by approximately 25% at 30 Mpa.[88]

3.2.2.2 Effect of Pressure on Central Synaptic Transmission

The effects of pressure on peripheral synaptic transmission are generally associated with a decrease in excitability, and they are observed at somewhat higher pressures than those that will stimulate hyperexcitability *in vivo*. The effects of pressure on peripheral cholinergic synapses may be responsible for the motor paralysis observed *in vivo* at high pressures and which precedes death at pressure, but are not important at the lower pressures (below 10 MPa) at which hyperexcitability is observed. Evidence that central synaptic transmission is affected differently by pressure than that described for peripheral and invertebrate synapses was provided by experiments showing that, in isolated neonatal rat spinal cord, monosynaptic reflexes were unaffected by 10 MPa pressure but polysynaptic reflexes were potentiated.[89]

3.2.2.2.1 Presynaptic Effects

Pressure (8 MPa) has been reported to decrease depolarization-induced Ca^{2+} uptake into brain synaptosomes,[90] and for both chromaffin and mast cells, exocytosis is more sensitive to pressure than the voltage-gated Na^+ and K^+ channels.[91,92]

In vivo voltametry has revealed that extracellular dopamine concentrations in the nucleus accumbens and striatum were increased by pressures of 8 Mpa.[93,94] Moreover, although drugs that affect dopaminergic transmission produce little effect when given systemically,[58,59] selective dopamine antagonists given intracerebroventricularly (D_1, e.g., 7-chloro-2,3,4,5-tetrahydro-3-methyl-5-phenyl-1H-3-benzazepine-7-ol; D_2, e.g., sulpiride, haloperidol) ameliorate the effects of pressure. The D_2 antagonists are effective against all aspects of HPNS and the D_1 antagonists are effective only against the early tremor.[95] It is possible that the observed increase in dopamine release could be brought about by stimulation of 5-HT1b receptors[96] or by pressure on membrane Na^+/Ca^{2+} exchange processes, possibly an inhibition of the activity of membrane-bound Na^+,K^+-ATPase,[97] or a combination of both effects. Activation of dopaminergic transmission in the nigrostriatum is associated with stereotypical motor activity and in the mesolimbic system with a general increase in motor activity. Potentiation by pressure of dopaminergic transmission through an effect of pressure on presynaptic processes may well contribute to the general hyperexcitability observed at high pressure.

3.2.2.2.2 Effects of Pressure on Isolated Postsynaptic Receptors

That neurons do exhibit excitability at pressure was demonstrated by Fagni and colleagues,[98] who showed that the postsynaptic excitability of hippocampal CA_1 cells was increased at 8 MPa, even though afferent input was reduced. Intracellular recordings at pressure from CA_1 cells revealed bursts of transient depolarizations accompanied by repetitive action potentials and increased excitability.[99]

The possibility that pressure might have a selective action on postsynaptic receptors has been investigated by isolating mammalian receptors using heterologous expression in *Xenopus* oocytes.[101-105] The effects of pressure on the dose–response relationships for kainate, NMDA, GABA, and glycine of oocytes microinjected with mRNA extracted from rat brain are shown in Figure 3.4. The effects of pressure on these ionotropic receptors show an interesting symmetry. For the receptors mediating excitatory transmission, NMDA-sensitive glutamate receptors show a marked potentiation (128%) at 10 Mpa,[105] whereas kainate-sensitive receptors are relatively unaffected, showing little potentiation of the maximum response (<14%) and no change in the EC_{50}.[101,102] For the receptors mediating inhibitory transmission, glycine

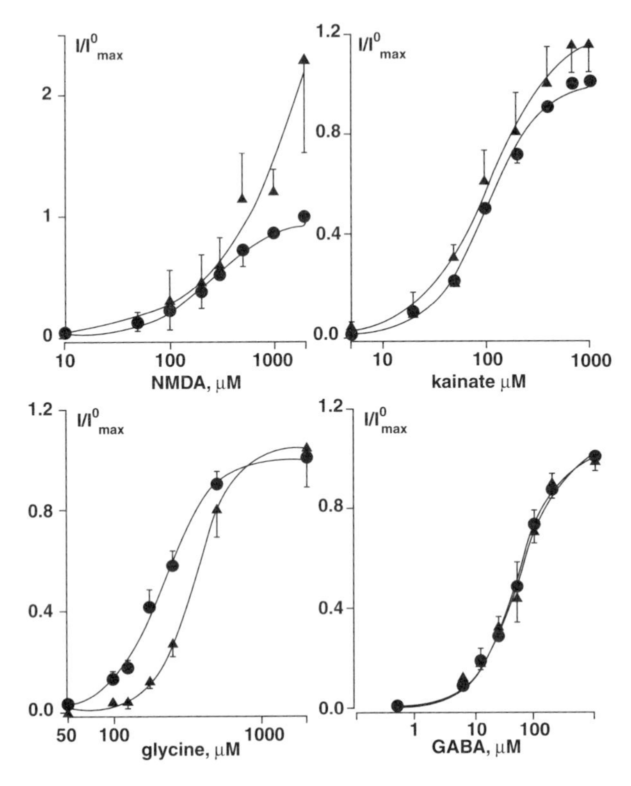

FIGURE 3.4 Dose–response curves representing the normalized current, relative to the maximum current, elicited from *Xenopus* oocytes injected with mRNA extracted from rat brain by NMDA, kainate, glycine, and GABA. In all cases the control dose–response data are given by ● and data at 10 MPa are given by ▲. The curves, with the exception of that for NMDA at 10 MPa which is drawn by eye, represent the best fit to a model of the general form $I/I^0_{max} = 1/[1 + (x/x_{50})^{-k}]$, where x is the agonist concentration, x_{50} is the EC_{50} value, and k is the Hill coefficient. The bars represent individual standard errors. (Data taken from Williams et al.,[105] NMDA; Shelton et al.,[102] kainate and glycine; Shelton et al.,[104] GABA.)

receptors show no change in the maximum response but a significant increase in the EC_{50} (60%) at 10 MPa,[102,103] whereas pressure has no effect on the $GABA_A$ receptor.[104] The effects of pressure on both the NMDA and glycine receptor become progressively greater at higher pressures.

Pressure affects the two receptors that possess glycine binding sites. At the glycine receptor, glycine binding activates the ion channel (Cl^-), whereas at the NMDA receptor, glycine acts as a modulator, not activating the ion channel (Na^+, K^+, Ca^{2+}) but promoting the binding of NMDA, which does act as a modulator.[106,107]

In the case of the simpler glycine receptor, a classic interpretation of the dose–response data would be that pressure is acting, pharmacologically, as a simple noncompetitive antagonist. This action could be manifested either as a reduction in the efficacy of glycine binding or in the transduction of binding to channel opening. A direct effect on channel opening or lifetime seems unlikely because the closely related $GABA_A$ receptor is unaffected. The apparent volume change in glycine binding was estimated as 110 $cm^3 \cdot mol^{-1}$, which is very large for simple chemical equilibria and indicates that considerable conformational changes occur when glycine binds.[102] The effect of pressure may therefore be envisaged as opposing the conformational change that follows glycine binding, thereby reducing the effective inhibitory transmission available. Although the effect is not as large as that on the NMDA receptor, if inhibitory synapses *in vivo* operate at submaximal agonist concentrations then a profound effect *in vivo* would be observed.

The picture with the NMDA receptor is not as simple. In this case the apparent volume change associated with glycine binding must be negative so that the interaction would be augmented by pressure. Pressure must favor the existence of the glycine-bound form of the receptor. In both cases a more detailed understanding of the effect of pressure on glycine binding at these two receptors must await experiments on single channels.

3.3 SUMMARY

Despite the early success of the generalized physicochemical theories in providing a framework within which the mechanism of action of both general anesthetics and high pressure could be understood, it is now clear that both exert their effects through more specific actions. Although neither mechanisms are understood in detail at a molecular or cellular level, there is consensus that both exert their effects primarily on central synaptic transmission.

In this context the major actions of pressure appear to be to enhance excitatory transmission mediated by the NMDA subtype of the glutamate receptor and to diminish postsynaptic inhibition mediated by glycine. In addition, a generalized increase in excitability in CNS motor areas may arise through presynaptic actions that increase the availability of dopamine.

The question, therefore, is how pressure and general anesthetics might interact. There is little direct evidence of anesthetic–pressure interactions at a molecular or cellular level. The lack of pressure reversal of anesthesia in invertebrate species that do not use glycine as an inhibitory transmitter is suggestive but as yet is not confirmed by any direct study, and it may be a reflection of an atypical response to anesthetics rather than to pressure. The increase in extracellular dopamine reported at pressure (Section 3.2.2.2.1) has a corollary in that increased nitrogen partial pressures produces a fall in extracellular dopamine.[108] However, there is also no evidence to link these effects directly, and although dopaminergic mechanisms may be involved in motor disorders there is little reason to believe that they are implicated in the loss of consciousness. Finally, there is no evidence for a direct link between GABAergic or non-NMDA glutaminergic mechanisms and pressure reversal in that 9 MPa pressure did not reverse the pentobarbitone-induced inhibition of the kainate-sensitive receptor[101] or the potentiation of the $GABA_A$ response.[104]

Although there are as yet no direct studies on the interaction between anesthetics and pressure at either the glycine or the NMDA receptor, it appears that the most likely explanation of the interaction is that *in vivo* there is a physiologic summation between the overall excitatory effects of pressure and the depressant effects of general anesthetics.

REFERENCES

1. Regnard, P., *Recherches Experimentals sur les Conditions Physique de la Vie dans les Eaux*, Masson, Paris, 1891.
2. Zaltsman, G. L., *Physiological principals of a sojourn of a human in conditions of raised pressure of the gaseous medium*, [English translation] Foreign Technology Division, Wright-Patterson Air Force Base, Ohio, AD 655 360, 1967.
3. Keller, H. & Buhlmann, A. A., Deep diving and short decompression by breathing mixed gases, *Journal of Applied Physiology*, 20, 1267, 1965.
4. Bennett, P. B., *Psychometric impairment in men breathing oxygen-helium at increased pressures*, Medical Research Council, RN Personnel Research Committee, Underwater Physiology Sub-committee, Report Number 251, 1965.
5. Bennett, P. B., Performance impairment in deep diving due to nitrogen, helium, neon and oxygen, in *Proceedings of the 3rd Symposium on Underwater Physiology*, Lambertsen, C. J., Ed., Williams & Wilkins, Baltimore, 1967, 327.
6. Bennett, P. B. & Dossett, A. N., *Undesirable effects of oxygen-helium breathing at great depths*, Medical Research Council, RN Personnel Research Committee, Underwater Physiology Sub-committee, Report Number 260, 1967.
7. Miller, K. W., Paton, W. D. M., Street, W. B. & Smith, E. B., Animals at very high pressures of helium and neon, *Science*, 157, 97, 1967.
8. Kylstra, J. A., Nantz, R., Crowe, J., Wagner, W. & Saltzman, H. A., Hydraulic compression of mice to 166 atmospheres, *Science,* 158, 793, 1967.
9. Bennett, P. B., The high pressure nervous syndrome in man, in *The physiology and medicine of diving*, 3rd edition, Bennett, P. B. & Elliott, D. H., Eds., Balliere Tindall, London, 1982, 262.
10. Brauer, R. W., The high pressure nervous syndrome: Animals, in *The physiology and medicine of diving*, 1st edition, Bennett, P. B. & Elliott, D. H., Eds., Balliere Tindall, London, 1982, 231.
11. Hempleman, H. V., Florio, J. T., Garrard, M. P., Harris, D. J., Hayes, P.A., Hennessy, T. R., Nichols, G., Török, Z. & Winsborough, M. M., U.K. deep diving trials, *Philosophical Transactions of the Royal Society of London*, B304, 119, 1984.
12. Lafay, V., Barthelemy, P., Comet, B., Frances, Y. & James, Y., EEG-changes during the experimental human dive Hydra-10 (71-atm 7,200kPa), *Undersea & Hyperbaric Medicine*, 22, 51, 1995.
13. Bowser-Riley, F., Lashbrook, N. M., Paton, W. D. M. & Smith, E. B., An evaluation of mephenesin and related drugs in controlling the effects of high pressure on animals, in *Proceedings of the VIII symposium on underwater physiology*, Bachrach, A. J. & Matzen, M. M., Eds., Undersea Medical Society Inc., Bethesda, MD, 1984, 557.
14. Lever, M. J., Miller, K. W., Paton, W. D. M. & Smith, E. B., Pressure reversal of anaesthesia, *Nature*, 231, 368, 1971.
15. Miller, K. W., Inert gas narcosis, the high pressure neurological syndrome and the critical volume hypothesis, *Science*, 185, 867, 1974.
16. Johnson, F. H. & Flagler, E. A., Hydrostatic pressure reversal of narcosis in tadpoles, *Science*, 112, 91, 1950.
17. Ferguson, J., The use of chemical potentials as indices of toxicity, *Proceedings of the Royal Society of London*, B127, 387, 1939.

18. Pauling, L., A molecular theory of general anaesthesia, *Science*, 134, 15, 1961.
19. Miller, S. L., A theory of gaseous anaesthesia, *Proceedings of the National Academy of Sciences of USA*, 75, 4906, 1961.
20. Miller, K. W., Paton, W. D. M. & Smith, E. B., The anaesthetic pressures of certain fluorine-containing gases, *British Journal of Anaesthesia*, 39, 910, 1967.
21. Eger, E. I., Lungren, C., Miller, S. L. & Stevens, W. C., Anesthetic potencies of SF_6, CF_4, chloroform and ethane in dogs, *Anesthesiology*, 30, 129, 1969.
22. Meyer, H. H., Theories of narcosis, in *Harvey Lectures*, Lippincott, Philadelphia, 1906, 11.
23. Overton, E., *Stüdien uber die Narkose, zugleich ein Beitrag zur allgemeiner Pharmakologie*, Gustav Fischer, Jena, 1901.
24. Miller, K. W., Paton, W. D. M., Smith, E. B. & Smith, R. A., Physiochemical approaches to the mode of action of general anaesthetics, *Anesthesiology*, 36, 339, 1972.
25. Hill, M. W., The Gibbs free energy hypothesis in general anaesthesia, in *Molecular mechanisms in general anaesthesia*, Halsey, M. J., Millar, R. A. & Sutton, J. A., Eds., Churchill & Livingstone, Edinburgh, 1974, 510.
26. Trudell, J. R., Biophysical concepts in molecular mechanisms of anaesthesia, in *Molecular mechanisms of anaesthesia (Progress in Anesthesiology, Vol. 2),* Fink, B. R., Ed., Raven Press, New York, 1980, 261.
27. Lieb, W. R., Kovalyesik, M. & Mendelsohn, R., Do clinical levels of general anaesthetic affect lipid bilayers? Evidence from Raman scattering, *Biochimica et Biophysica Acta*, 688, 388, 1982.
28. Franks, N. P. & Lieb, W. R., Molecular mechanisms of general anaesthesia, *Nature*, 300, 487, 1982.
29. Franks, N. P. & Lieb, W. R., Mapping of general anaesthetic target sites provides a molecular basis for cutoff effects, *Nature*, 316, 349, 1985.
30. Franks, N. P. & Lieb, W. R., Stereospecific effects of volatile general anaesthetic optical isomers on nerve ion channels, *Science*, 254, 427, 1991.
31. Evers A. S., Berkowitz, B. A. & d'Avignon, D. A., Correlation between the anesthetic effect of halothane and saturable binding in brain, *Nature,* 328, 157, 1987.
32. Miller, K. W., Paton, W. D. M., Smith, R. A. & Smith, E. B., The pressure reversal of general anaesthesia and the critical volume hypothesis, *Molecular Pharmacology*, 9, 131 , 1973.
33. Miller, K. W., Wilson, M. W. & Smith, R. A., Pressure resolves two sites of action of inert gases, *Molecular Pharmacology*, 14, 950, 1978.
34. Smith, R. A., Dodson, B. A. & Miller, K. W., The interactions between pressure and anaesthetics, *Philosophical Transactions of the Royal Society of London*, B304, 69, 1984.
35. Halsey, M. J. & Wardley-Smith, B., Pressure reversal of narcosis produced by anaesthetics, narcotics and tranquillisers, *Nature*, 257, 811, 1975.
36. Kent, D. W., Halsey, M. J., Eger, E. I. & Kent, B., Isoflurane anaesthesia and pressure antagonism in mice, *Anaesthesia & Analgesia*, 56, 97, 1977.
37. Halsey, M. J., Green, C. J. & Wardley-Smith, B., Renaissance of nonunitary molecular mechanisms of general anaesthesia, in *Molecular Mechanisms of Anesthesia (Progress in Anesthesiology, Vol. 2)*, Fink, B. R., Ed., Raven Press, New York, 1980, 273.
38. Wardley-Smith, B. & Halsey, M. J., Pressure reversal of narcosis: Possible separate molecular sites for anesthetics and pressure, in *Molecular Mechanisms of Anesthesia (Progress in Anesthesiology, Vol. 2)*, Fink, B. R., Ed., Raven Press, New York, 1980, 489.
39. Beaver, R. W., Brauer, R. W. & Lahser, S., Interaction of central nervous system effects of high pressures with barbiturates, *Journal of Applied Physiology*, 43, 221, 1977.
40. Halsey, M. J., Wardley-Smith, B. & Green, C. J., The pressure reversal of anaesthesia — a multisite expansion hypothesis, *British Journal of Anaesthesia*, 50, 1091, 1978.

41. Richards, C. D., Martin, K., Gregory, S., Keightley, C. A., Hesketh, T. R., Smith, G. A., Warren, G. B. & Metcalfe, J. C., Degenerate perturbations of protein structure as the mechanism of anaesthetic action, *Nature*, 276, 775, 1978.
42. Middleton, A. J. & Smith, E. B., General anaesthetics and bacterial luminescence. I: The effect of diethyl ether on the in vivo light emission of Vibrio fischeri, *Proceedings of the Royal Society of London Series*, B193, 159, 1976.
43. Middleton, A. J. & Smith, E. B., General anaesthetics and bacterial luminescence. II: The effect of diethyl ether on the in vitro light emission of Vibrio fischeri, *Proceedings of the Royal Society of London Series*, B193, 173, 1976.
44. Winter, P. M., Smith, R. A., Smith, M. & Eger, E. I., Pressure antagonism of barbiturate anaesthesia, *Anesthesiology*, 44, 416, 1976.
45. Tobey, R. E., McCracken, L. E., Small, A. & Homer, L. D., Effect of hyperbaric helium on anaesthetic action of thiopental, in *Proceedings of Sixth Symposium on Underwater Physiology*, Shilling, C. S. & Beckett, M. W., Eds., FASEB, Bethesda, MD, 1978, 267.
46. Dundas, C. R., Alphaxalone/alphadolone in diving-chamber anaesthesia, *Lancet*, 1, 378, 1979.
47. Angel, A., Gratton, D. A., Halsey, M. J. & Wardley-Smith, B., Pressure reversal of the effect of urethane on the evoked somatosensory cortical response in the rat, *British Journal of Pharmacology*, 70, 147, 1980.
48. Youngson, A. F. & Macdonald, A. G., Interaction between halothane and hydrostatic pressure, *British Journal of Anaesthesia*, 42, 801, 1970.
49. Simon, S. A., Parmentier, J. L. & Bennett, P. B., Anesthetic antagonism of the effects of high hydrostatic pressure on locomotory activity of the brine shrimp, Artemia, *Comparative Biochemistry & Physiology*, 75A, 193, 1983.
50. Smith, E. B., Bowser-Riley, F., Daniels, S., Dunbar, I. T., Harrison, C. B. & Paton, W. D. M., Species variation and the mechanism of pressure-anaesthetic interactions, *Nature*, 311, 56, 1984.
51. Halsey, M. J. & Wardley-Smith, B., The high pressure neurological syndrome: Do anticonvulsants prevent it?, *British Journal of Pharmacology*, 72, 502, 1981.
52. Wardley-Smith, B., Dore, C., Hudson, S. & Wann, K. T., Effects of four common anticonvulsants on the high pressure nervous syndrome in the rat, *Undersea Biomedical Research*, 19, 13, 1992.
53. Brauer, R. W., Mansfield, W. M., Beaver, R. W. & Gillen, H. W., Stages in development of high pressure neurological syndrome in the mouse, *Journal of Applied Physiology*, 46, 756, 1979.
54. Mansfield, W. M., Brauer, R. W. & Gillen, H. W., Age dependence of seizure patterns in mice exposed to high pressure heliox atmospheres, *Federation Proceedings of the Federation of American Societies for Experimental Biology*, 33, 350, 1974.
55. Bowser-Riley, F., Mechanistic studies on the high pressure neurological syndrome, *Philosophical Transactions of the Royal Society of London*, B304, 31, 1984.
56. Kaufmann, P. G., Bennett, P. B. & Farmer, J. C., Jr., Differential effects of pressure on the mammalian central nervous system, in *Proceeding of the VII Symposium on Underwater Physiology*, Bachrach, A. J. & Matzen, M. M., Eds., Undersea Medical Society Inc., Bethesda, MD, 1981, 371.
57. Brauer, R. W., Beaver, R. W. & Sheehan, M. E., Role of monoamine neurotransmitters in the compression rate dependence of HPNS convulsions, in *Proceedings of Sixth Symposium on Underwater Physiology*, Shilling, C. S. & Beckett, M. W., Eds., FASEB, Bethesda, MD, 1978, 49.

58. Koblin, D. D., Little, H. J., Green, A. R., Daniels, S., Smith, E. B. & Paton, W. D. M., Brain monoamines and the high pressure neurological syndrome, *Neuropharmacology*, 19, 1031, 1980.
59. Daniels, S., Green, A. R., Koblin, D. D., Lister, R. G., Little, H. J., Paton, W. D. M., Bowser-Riley, F., Shaw, S. G. & Smith, E. B., Phamacological investigation of the high pressure neurological syndrome: Brain monoamine concentrations, in *Proceedings of the VIIth Symposium on Underwater Physiology*, Bachrach, A. J. & Matzen, M. M., Eds., Undersea Medical Society Inc., Bethesda, MD, 1981, 329.
60. Mason, S. J. & Corcoran, M. E., Catecholamines and convulsions, *Brain Research*, 170, 497, 1979.
61. Bennett, P. B. & McLeod, M., Probing the limits of human diving, *Philosophical Transactions of the Royal Society of London*, B304, 105, 1984.
62. Pepeu, G., Brain acetylcholine: An inventory of our knowledge on the 50th anniversary of its discovery, *Trends in Pharmacological Science*, 4, 416, 1983.
63. Wardley-Smith, B., Angel, A., Halsey, M. J. & Rostain, J.-C., Neurochemical basis for the high pressure neurological syndrome: Are cholinergic mechanisms involved?, in *Proceedings of the VIIIth Symposium on Underwater Physiology*, Bachrach, A. J. & Matzen, M. M., Eds., Undersea Medical Society Inc., Bethesda, MD, 1984, 621.
64. Bichard, A. R. & Little, H. J., The benzodiazepine antagonist, RO 15-1788 prevents the effects of flurazepam on the high pressure neurological syndrome, *Neuropharmacology*, 21, 877, 1982.
65. Bichard, A. R. & Little, H. J., γ-aminobutyric acid transmission and the high pressure neurological syndrome, in *Proceedings of the VIII symposium on underwater physiology*, Bachrach, A. J. & Matzen, M. M., Eds., Undersea Medical Society Inc., Bethesda, MD, 1984, 545.
66. Chapman, A. G., Keane, P. E., Meldrum, B. S., Simiand, J. & Vernieres, J. C., Mechanism of action of valproate, *Progress in Neurobiology*, 19, 315, 1982.
67. Bowser-Riley, F., Daniels, S. & Smith, E. B., Investigations into the origin of the high pressure neurological syndrome: The interaction between pressure, strychnine and 1,2-propandiols in the mouse, *British Journal of Pharmacology*, 94, 1069, 1988.
68. Bowser-Riley, F., Daniels, S., Hill, W. A. G., Learner, T. S. & Smith, E. B., The additive effects of pressure and chemical convulsants in mice, *Journal of Physiology*, 409, 36P, 1988.
69. Kuhse, J., Betz, H. & Kirsch, J., The inhibitory glycine receptor: architecture, synaptic localisation and molecular pathology of a postsynaptic ion-channel complex, *Current Opinion in Neurobiology*, 5, 318, 1995.
70. Bowser-Riley, F., Daniels, S., Hill, W. A. G. & Smith, E. B., An evaluation of the structure-activity relationships of a series of analogues of mephenesin and strychnine on the responses to pressure in mice, *British Journal of Pharmacology*, 96, 789, 1989.
71. Bowser-Riley, F., Daniels, S. & Smith, E. B., The effect of benzazole-related centrally acting muscle relaxants on HPNS, *Undersea Biomedical Research*, 15, 331, 1988.
72. Meldrum, B., Wardley-Smith, B., Halsey, M. J. & Rostain, J.-C., 2-Amino-phosphonoheptanoic acid protects against the high pressure neurological syndrome, *European Journal of Pharmacology*, 87, 501, 1983.
73. Wardley-Smith, B. & Meldrum, B., Effect of excitatory amino acid antagonists on the high pressure neurological syndrome in rats, *European Journal of Pharmacology*, 105, 351, 1984.

74. Pearce, P. C., Halsey, M. J., Maclean, C. J., Ward, E. M., Webster, M. T., Luff, N. P., Pearson, J., Charlett, A. & Meldrum, B. S., The effects of the competitive NMDA receptor antagonist CCP on the high pressure neurological syndrome in a primate model, *Neuropharmacology*, 30, 787, 1991.
75. Pearce, P. C., Halsey, M. J., Maclean, C. J., Ward, E. M., Pearson, J., Henley, M. & Meldrum, B. S., The orally active NMDA receptor antagonist CGP 39551 ameliorates the high pressure neurological syndrome in *Papio anubis*, *Brain Research*, 622, 177, 1993.
76. Wardley-Smith, B., Meldrum, B. S. & Halsey, M. J., The effect of two novel dipeptide antagonists of excitatory amino acid neurotransmission on the high pressure neurological syndrome in the rat, *European Journal of Pharmacology*, 138, 417, 1987.
77. Angel, A., Halsey, M. J., Little, H. J., Meldrum, H. J., Ross, J. A. S., Rostain, J.-C. & Wardley-Smith, B., Specific effects of drugs at pressure: animal investigations, *Philosophical Transactions of the Royal Society of London*, B304, 85, 1984.
78. Wardley-Smith, B. & Wann, K. T., The effect of non-competitive NMDA receptor antagonists on rats exposed to hyperbaric pressure, *European Journal of Pharmacology*, 165, 107, 1989.
79. Pearce, P. C., Dore, C. J., Halsey, M. J., Luff, N. P., Maclean, C. J. & Meldrum B. S., The effects of MK801 on the high pressure neurological syndrome in the baboon, *Neuropharmacology*, 29, 931, 1990.
80. Pearce, P. C., Maclean, C. J., Shergill, H. K., Ward, E. M., Halsey, M. J., Tindley, G., Pearson, J. & Meldrum, B. S., Protection from high pressure induced hyperexcitability by the AMPA Kainate receptor antagonists GYKI-52466 and LY-293558, *Neuropharmacology*, 33, 605, 1994.
81. Wann, K. T. & Macdonald, A. G., Actions and interactions of high pressure and general anaesthetics, *Progress in Neurobiology*, 30, 271, 1988.
82. Kendig, J. J., Trudell, J. R. & Cohen, E. N., Effects of pressure and anaesthetics on conduction and synaptic transmission, *Journal of Pharmacology and Experimental Therapeutics*, 195, 216, 1975.
83. Ashford, M. L. J., Macdonald, A. G. & Wann, K. T., The effects of hydrostatic pressure on the spontaneous release of transmitter at the frog neuromuscular junction, *Journal of Physiology*, 333, 531, 1982.
84. Grossman, Y., Colton, J. S. & Gilman, S. C., Interaction of Ca-channel blockers and high-pressure at the crustacean neuromuscular-junction, *Neuroscience Letters*, 125, 53, 1991.
85. Golan, H. & Grossman, Y., Synaptic transmission at high-pressure — effects of $[Ca^{2+}]_o$, *Comparative Biochemistry and Physiology A — Comparative Physiology*, 103, 113, 1992.
86. Golan, H., Moore, H. J. & Grossman, Y., Quantal analysis of presynaptic inhibition, low $[CA^{2+}]_0$, and high-pressure interactions at crustacean excitatory synapses, *Synapse*, 18, 328, 1994.
87. Parmentier, J. L., Shrivastav, B. B. & Bennett, P. B., Hydrostatic pressure reduces synaptic efficiency by inhibiting transmitter release, *Undersea Biomedical Research*, 8, 175, 1981.
88. Sauter, J. F., Braswell, P., Wankowicz, P. & Miller, K. W., The effects of high pressures of inert gases on cholinergic receptor binding and function, in *Proceedings of the VII symposium on underwater physiology*, Bachrach, A. J. & Matzen, M. M., Eds., Undersea Medical Society Inc., Bethesda, MD, 1981, 629.

89. Tarasiuk, A., Schleifsteinattias, D. & Grossman, Y., High-pressure effects on reflexes in isolated spinal-cords of newborn rats, *Undersea Biomedical Research*, 19, 331, 1992.
90. Gilman, S. C., Kumaroo, K. K. & Hallenbeck, J. M., Effect of pressure on uptake and release of calcium by brain synaptosomes, *Journal of Applied Physiology*, 60, 1446, 1986.
91. Heinemann, S. H., Conti, F., Stuhmer, W. & Neher, E., Effects of hydrostatic-pressure on membrane processes — sodium-channels, calcium channels, and exocytosis, *Journal of General Physiology*, 90, 765, 1987.
92. Heinemann, S. H., Stuhmer, W. & Conti, F., Single acetylcholine-receptor channel currents recorded at high hydrostatic pressures, *Proceedings of the National Academy of Science, USA*, 84, 3229, 1987.
93. Abraini, J. H. & Rostain, J-.C., Effects of the administration of α-methyl-para-tyrosine on the striatal dopamine increase and the behavioral motor disturbances in rats exposed to high-pressure, *Pharmacology Biochemistry and Behavior*, 40, 305, 1991.
94. Abraini, J. H. & Rostain, J-.C., Dopamine increase in the nucleus-accumbens of rats exposed to high-pressure, *Neuroreport*, 2, 233, 1991.
95. Abraini, J. H., Tomei, C. & Rostain, J-. C., Role of dopamine-receptors in the occurrence of the behavioral motor disturbances in rats exposed to high-pressure, *Pharmacology Biochemistry and Behavior*, 39, 773, 1991.
96. Kriem, B., Abraini, J. H. & Rostain, J.-C., Role of 5-HT1b receptor in the pressure-induced behavioral and neurochemical disorders in rats, *Pharmacology Biochemistry and Behavior*, 53, 257, 1996.
97. Paul, M. L. & Philp, R. B., Effect of pressure on the release of endogenous dopamine from rat striatum and the role of sodium-calcium exchange, *Undersea Biomedical Research*, 19, 1, 1992.
98. Fagni, L., Zinebi, F. & Hugon, M., Evoked-potential changes in rat hippocampal slices under helium pressure, *Experimental Brain Research*, 65, 513, 1987.
99. Wann, K. T. & Southan, A. P., The action of anaesthetics and high pressure on neuronal discharge patterns, *General Pharmacology*, 23, 993, 1992.
100. Wann, K. T. & Southan, A. P., Pressure modifies the excitability of pyramidal neurones in the rat hippocampus, *Biophysical Journal*, 70, W-P140, 1996.
101. Daniels, S., Zhao, D. M., Inman, N., Price, D. J., Shelton, C. J. & Smith, E. B., Effects of general anaesthetics and pressure on mammalian excitatory receptors expressed in *Xenopus* oocytes, in *Molecular and Cellular Mechanisms of Alcohol and Anesthetics*, Roth, S. H. & Miller, K. W., Eds., Annals of the New York Academy of Sciences, 625, 108, 1991.
102. Shelton, C. J., Doyle, M. G., Price, D. J., Daniels, S. & Smith, E. B., The effect of high pressure on glycine and kainate sensitive receptor channnels expressed in *Xenopus* oocytes, *Proceedings of the Royal Society of London Series B*, 254, 131, 1993.
103. Roberts, R. J., Shelton, C. J., Daniels, S. & Smith, E. B., Glycine activation of human homomeric α1 glycine receptors is sensitive to pressure in the range of the high pressure nervous syndrome, *Neuroscience Letters*, 208, 125, 1996.
104. Shelton, C. J., Daniels, S. & Smith, E. B., Rat brain $GABA_A$ receptors expressed in *Xenopus* oocytes are insensitive to high pressure, *Pharmacology Communications*, 7, 215, 1996.
105. Williams, N., Roberts, R. J. & Daniels, S., A study into pressure sensitivity of ionotropic receptors, *Progress in Biophysics and Molecular Biology*, (in press).

106. Betz, H., Structure and function of inhibitory glycin receptors, *Quarterly Review of Biophysics*, 25, 381, 1992.
107. Watkins, J. C., Krogsgaard, L. P. & Honore, T., Structure-activity relationships in the development of excitatory aminoacid receptor agonists and competitive antagonists, *Trends in Pharmacological Science*, 11, 25, 1990.
108. Barthelemy-Requin, M., Semelin, P. & Risso., J. J., Effect of nitrogen narcosis on extracellular levels of dopamine and its metabolites in the rat striatum, using intracerebral microdialysis, *Brain Research*, 667, 1, 1994.

4 Genetics and Anesthetic Mechanism

Philip G. Morgan and Margaret M. Sedensky

CONTENTS

4.1 INTRODUCTION

Volatile agents have been used in operating rooms for nearly 150 years, with little understanding of how they produce their effects. Each clinically used agent carries with it a low margin of safety and unwanted side effects. Despite a great deal of research, no studies have yielded a molecular picture of the site of action of volatile anesthetics. A genetic approach can identify the components of an anesthetic site of action, because it can connect changes in behavior to changes in molecular structure. For example, recent articles have described the cloning of genes responsible for cystic fibrosis, malignant hyperthermia, and premature aging. These studies used sophisticated techniques of molecular biology in addition to the traditional methods of classical genetics to characterize the molecular basis of these clinical entities. We

0-8493-8555-5/01/$0.00+$.50

will outline the use of molecular genetics, along with basic techniques of classical genetics, to study the molecular nature of the site of action of volatile anesthetics.

Over the years, a great number of sophisticated techniques have been recruited to solve the mystery of how volatile anesthetics work. Why then the current interest in the use of genetics to understand the mechanism of volatile anesthetic action? There is a two-part answer to this question. First, our genes contain the blueprint for every molecular component of the anesthetic site of action. Therefore, regardless of the exact chemical nature of an anesthetic site (i.e., lipid or protein or both), its structure is dictated by an invariant material contained within virtually all our cells. Second, a genetic approach is capable of tracing behavior back to a discreet set of molecular changes. Reducing a behavior such as reversible loss of consciousness to the molecules that interact to cause the behavior is the key step in any genetic approach. Both of these aspects of a genetic approach avoid presuppositions about the molecular nature of an anesthetic site. Other experimental designs make a "best guess" as to the nature of molecules that directly interact with volatile anesthetics, and then measure the effect of these agents on the suspect molecule. Even though these "guesses" are the end product of well-researched data, the results of such studies often leave doubts as to whether the effects represent phenomena superfluous or incidental to the state of anesthesia. That so many systems can be perturbed by these highly lipid-soluble compounds has certainly contributed to the current lack of useful information as to how volatile anesthetics work at a molecular level. In contrast, if one can identify genes that code for molecules that directly control an animal's response to volatile anesthetics, it is possible to understand the molecular nature of these molecules and their function. Thus, at its best, genetics can give information about molecules that are indisputably part of an anesthetic site.

Clearly, finding genes that control the whole animal response to volatile anesthetics is the crucial first step in a genetic approach. Generally, in order to understand a gene's function, one has to first change a gene (i.e., make a mutation in that gene) to see the effect on the system. Thus, in the case of volatile anesthetics, one tries to change aspects of an animal's behavior by mutating the animal. Classical genetics pinpoints that change to a single mutation within a gene. Molecular genetics then analyzes the nature of the mutation, relating it to the normal function of the normal gene product.

4.2 BACKGROUND

Anesthesiologists are quite comfortable with research in which a condition is changed and the resultant effect on an experimental animal or system is measured. However, genetics uses what may seem to most of us like a backward approach to answering a question. In classical genetics one changes the experimental animal, then looks for the altered condition (a mutated gene). Although one does not necessarily know what molecules caused that trait to change, it is clear a permanent change has occurred in the gene that controls that trait. As previously mentioned, the beauty of this approach is its lack of preconceptions as to the molecular nature of a given trait. On the other hand, if a complicated pathway or cascade of events leads to the ultimate expression of a given trait, one may have mutated any of a

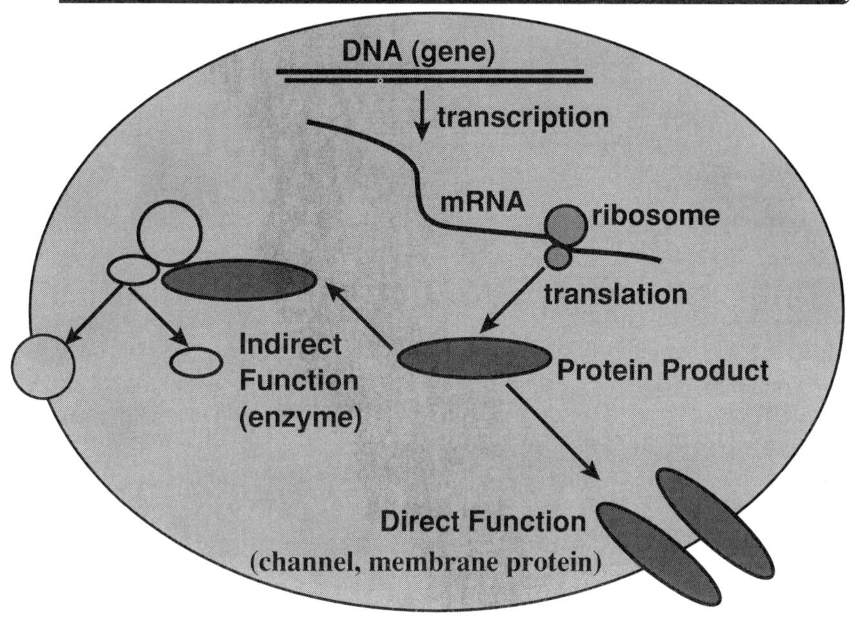

FIGURE 4.1 Genetic control of anesthetic sensitivity can occur at a number of possible positions in a cascade of events synthesizing components of a site of action. Changes can occur in the transcription or translation of genes that primarily affect anesthetic response. Mutations in protein products of various types could directly affect the site of action. These include ion channels or other membrane proteins that interact with anesthetics or affect nearby ion channels. In addition, enzymes that modify channels or membrane proteins or are involved in synthesizing other components of the membrane (i.e., lipids) could be affected. Finally, gene products that affect uptake or metabolism of anesthetics could be altered.

great number of processes that culminates in that particular trait. For example, mutations that change sensitivity to volatile anesthetics could arise from changes in molecules at the anesthetic site of action, from elimination of enzymes that make these molecules, or from molecules that change elimination or availability of the anesthetic itself (Figure 4.1). The possibilities are many. Data are sorted out in part by collecting multiple mutations that affect anesthetic sensitivity and studying their interactions and relative importance. Finally, one must usually identify, using molecular techniques, the protein products of these genes.

Classical genetics can define the specific chromosomal region in which a gene is located. These techniques can also position genes relative to each other, but cannot place genes on specific physical regions of DNA. Molecular biology, however, can be used to pinpoint a gene to a specific physical fragment of DNA.[1,2,3] It is also possible to manipulate pieces of DNA such that that they can be moved into novel environments. For example, a normal gene can be introduced into a defective or mutant animal, giving rise to a normal animal. This is the first step in treating genetic diseases (i.e., the beginning of genetic therapeutics). A second use is to introduce a defective gene into a previously normal organism. This can "genetically engineer" desired mutations into organisms to study a gene's function. Often it is referred to as generating a "targeted mutation." However, perhaps the most current common application of genetic engineering is to introduce eukaryotic genes (e.g., genes from mammals) into bacteria. These genes can be grown in large quantities in such systems, leading to isolation of large amounts of a gene. This technique has led to large-scale production of human insulin or interferon.

4.3 THE CENTRAL DOGMA

In all eukaryotes, the genetic material contained in the chromosomes is deoxyribonucleic acid (DNA). (In some viruses the genetic material is ribonucleic acid [RNA].)[4] The blueprint for making an organism comes entirely from the information contained in certain regions, termed genes, of this DNA. DNA consists of a phosphate–sugar backbone (deoxyribose) bonded to a series of purines and pyrimidines,[5] either adenine, thymine, guanine, or cystine. These four molecules are termed bases and are abbreviated as A, T, G, and C. Genes encode information because they contain a particular order of As, Ts, Gs, and Cs that is unique to that gene. These stretches of coding information can go on for many thousands of bases; the order is termed the DNA sequence of a gene. The unique sequence of a gene is copied (transcribed) to another species of nucleic acid, known as RNA. RNA is comprised of the ribose sugars (not deoxyribose) of A, G, C, and uracil, a pyrimidine that replaces thymine. When a gene is copied into RNA, the new form is termed messenger RNA (mRNA). It carries the information from the chromosomes to the part of a cell that constructs protein.

Once mRNA leaves the nucleus of a cell, the information is used by two other types of RNA to sequentially piece together amino acids into a protein. These RNAs are ribosomal RNA (rRNA) and transfer RNA (tRNA).[6] They convert consecutive sequences of three bases each into amino acids. This is termed translation. It is as if each of the bases represents a letter in the genetic code, forming a series of three-letter words.[7] These groups of three bases are called codons. For example, CUG specifies the amino acid leucine, and UGA specifies serine. Certain codons signal a stop and start to the translation of mRNA into protein.[7] Thus, reading the sequence of an mRNA is analogous to looking for a capital letter to start a sentence, and a period to stop it, with a series of words, comprised of three letters each, in between.[8] Although this may sound too simple to generate the complexity and diversity of life, keep in mind that each sentence can be thousands of words long, and that there are many thousands of sentences in most organisms.

A change in a single base in a gene can change an amino acid within its encoded protein. This is termed a point mutation. For example, when GAG is changed to GUG, valine is inserted into a protein instead of glutamic acid. If this happens in the sixth amino acid of the β chain of the hemoglobin molecule, the result is HgS, the basis of sickle cell disease.[9] The deletion or insertion of a base can have more global effects on a protein by shifting the alignment, the "frame" of all subsequent codons. Such a mutation is termed a frame shift mutation and can give rise to severe alterations. Obviously many complicated types of mutations can occur, with consequences of varying severity. The important point to remember is, whether it is a structural protein such as hemoglobin or an enzyme such as pseudocholinesterase,[10] the sequence of DNA within the gene dictates the particular amino acids within that molecule. The structures of other molecules that are not proteins, such as lipids or sugars, are likewise determined by the ability of proteins (enzymes, the original gene products) to correctly synthesize a myriad of compounds. All aspects of our physiology are ultimately dependent on this genetic material, and can be perturbed by changes in it. These molecular mechanisms are essentially identical (i.e., conserved) across the animal and plant kingdoms. In many ways different-appearing organisms are very similar at the DNA level, and simple organisms often can be good initial models for molecular processes of more complex ones. This is particularly true if the physiologic process being studied appears to be conserved across species. In fact, most of our detailed knowledge of the functioning of mammalian cellular machinery is derived directly from the brilliant studies of bacteria and viruses during the past 40 years.[7,8,11,12] During an era of intense study of the human genome, it is important to remember that much useful information can still come from simple model systems.

4.4 CLASSICAL GENETICS

Traditionally, geneticists have tackled questions of inheritance by observing the passage of clearly identifiable traits from one generation to the next.[13] As described above, an organism's observed features, its phenotype, must arise from the content of its genetic material, its genotype. Classical genetics relies on differences in phenotype (e.g., blue eyes vs. brown eyes) as the probes for identifying different genes and locating their relative positions. These differences are ascribed to individual genes and positioned in the genome (all the DNA of an organism) only after genetic manipulations, usually matings, have been carried out.

In classic genetic animals such as fruit flies, one can usually mate an animal with a new mutation to an animal that carries strategically placed genetic markers. These markers are characteristics that are easy to see (such as eye color or wing configuration) that are known to be coded by DNA of specific regions of specific chromosomes. These markers are already "mapped." The number of offspring that show the new mutation in combination with these mapped genes eventually pinpoints the mutation to a specific region of a specific chromosome. If two mutations are in the same chromosome, the distance between them is specified in terms of "map units." The further apart two mutations are, the greater the number of map units between them. Two mutations that are one map unit apart are genetically very close to each other; they have about one chance in 100 of separating from each other in meiosis.

A second, equally important aspect of classical genetics is to determine the interaction between genes. One may determine the effect of one mutation (A) on a second mutation (B) by constructing an animal containing both mutations. If such an animal has elements of both phenotypes, then A and B may function independently. If such an animal has the phenotype of only one mutation, that mutation is said to be *epistatic* to the other.[14] Such information is useful in determining functional pathways involving many gene products (Figure 4.2). In general, the mutation that is *epistatic* (controls the phenotype of an animal containing more than one mutation) is

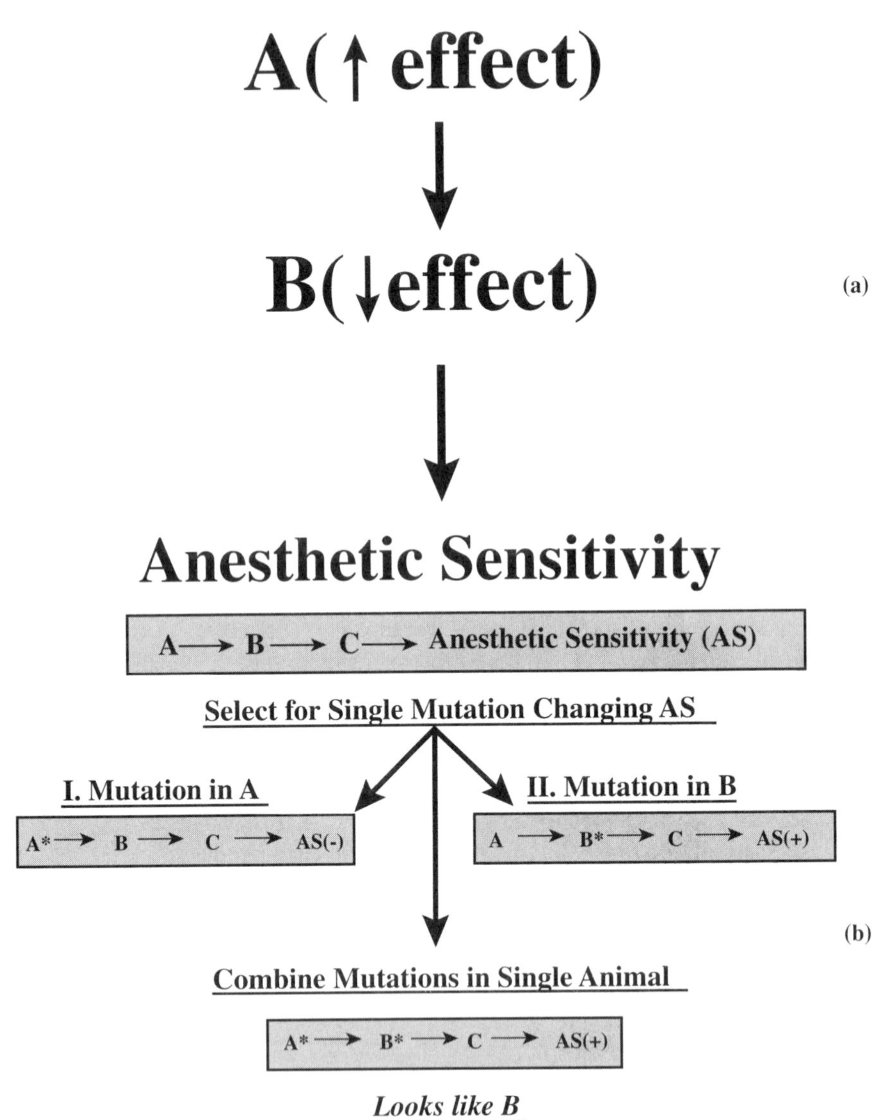

FIGURE 4.2 (a) A simplified genetic pathway involving two genes, A and B, controlling sensitivity to anesthetics. In this pathway genes A and B have opposite effects on sensitivity. However, without knowing the effects of mutations in A and B on each other, it is not possible to order them in a pathway. (b) A model pathway similar to that in (a). AS(+) refers to increased anesthetic sensitivity, whereas AS(–) refers to decreased sensitivity (resistance). Because the product of gene A works through the product of gene B, mutations in B will override mutations in A. Thus an animal with mutations in both A and B will look like B. This is known as *epistasis*. If the pathway is expanded to include gene C, it follows that mutations in C will override mutations in both A and B. By constructing double and triple mutants and observing the resulting phenotypes, one can determine the structure of a genetic pathway. In turn, this identifies crucial elements of control of anesthetic response.

downstream of the other mutation. By "downstream" we mean that one gene product (say A) exerts its effects by functioning through a second one (in this case B) (see legend for Figure 4.2). In such a case, when an animal has the B phenotype, it does not matter what the state of A is because A's effects are dependent on a functioning B product. A mutation resulting in loss of the B product will always produce the B phenotype.[14,15] As described in Figure 4.2, other types of interaction are also possible.

4.5 MOLECULAR GENETICS

Molecular biology has been applied with great success to classical genetic systems. This has led to the emergence of a new field, molecular genetics, which can identify the molecular basis of macroscopic changes characterized in classical genetics. Genes and their protein products are routinely characterized, as well as the molecular changes associated with mutations and disease states. The most important discoveries that led to the development of molecular genetics are discussed below.

Several groups of researchers independently discovered a class of enzymes (restriction endonucleases) that cut DNA at very specific spots, called restriction sites.[16] For example, one enzyme, EcoR I, cuts between the G and A in the sequence GAATTC,[17] and another enzyme, Bgl II, cuts between the first A and G in AGATCT.[17] In any one animal the DNA is identical in all cells. If that animal's DNA is cut by a specific endonuclease, a reproducible array of DNA fragments is generated. Mutations often lead to very clear differences between the digests of DNA of different individuals if the correct endonucleases are used to digest the DNA.[18] These are detected by digesting an individual's DNA with specific enzymes and running the resulting array of fragments on an electrophoretic gel which separates the fragments by size. Mutant DNA is compared to an identical treatment of nonmutated (i.e., normal) DNA. The resulting differences in restriction fragments are known as restriction fragment length polymorphisms (RFLPs).

To study a DNA fragment, usually one needs to generate large amounts of these pieces of DNA generated by restriction endonucleases. This is done by incorporating them into naturally occurring viruses or synthetic molecules that can accommodate foreign DNA. These carrier DNA molecules often are called vectors.[2,16] Such vectors have been engineered to contain series of DNA sequences that can be cut by a variety of restriction enzymes. If the vector is cut by the same endonuclease that generated the foreign fragment, this new DNA can be inserted into the carrier and the insertion patched into the vector by another enzyme. Thus, one physically cuts an RFLP out of a gel, mixes it with vectors and appropriate enzymes, and creates a new molecule of DNA that is a chimera for the original vector and the DNA of interest to the investigator. This new molecule then infects a bacterial culture, where it uses the bacterium's cellular machinery to reproduce itself every 20 min or so. For example, in the case of the gene for human factor VIII, one can commercially grow this gene in huge amounts, relatively inexpensively and free from other human cell contaminants. This isolation of a gene and its introduction into a bacterial cell is called cloning, and is used to generate enough copies of a gene to study the DNA in detail. These cloned fragments can then be used for sorting out the order of its bases (sequencing the DNA), for introducing new mutations in a gene, or for synthesis of the protein product of a gene.

Once the coding sequence of a gene is known one can begin to understand the function of the protein product. Genes coding for similar proteins in different organisms often have a similar order of bases (i.e., of As, Ts, Gs, and Cs). Proteins with similar functions usually have regions of similar amino acids. By comparing these similarities, called homologies, one can often assign a function to a newly identified protein. A truly immense amount of sequence data has accumulated over the past 10 years.[19-21] These data include sequences from organisms as simple as viruses and as complex as humans. By comparing newly identified sequences with other known genes, one can often identify homologies. If no such homology exists, all is not lost. Certain aspects of a protein's structure and function (such as a membrane spanning region) give characteristic recognizable patterns of amino acids. Thus much information can be gained from the sequence even in the absence of known homologies. The structures of proteins can lead to conclusions about protein function. In addition, the cloned gene often allows one to synthesize the protein product and study it directly in a variety of other surroundings. Thus, we have gone from a gross change in the whole animal to a specific gene and protein product, which can be isolated and studied independently of the organism.

This very powerful approach is now being applied to the question of how volatile anesthetics work. The basic plan is to find mutants in which the site of action of volatile anesthetics is changed. These mutants can then be analyzed at the level of the DNA of the animals. The changed region then serves as a probe to isolate, clone, and sequence the normal gene. One then knows the amino acid sequence of that gene's protein and can attempt to deduce its function. Whether this protein is a structural subunit of the $GABA_A$ receptor or an enzyme in a pathway of lipid biosynthesis, one has identified a molecule that has changed the whole animal response to volatile anesthetics. Eventually one can determine all the molecular components of the anesthetic site of action by studying a collection of specific mutants and/or the interaction of various mutants with one another.

Clearly, a tractable model system is necessary to implement these techniques to their maximum effect.[22] What should the characteristics of such a model be? It should have a simple genetic system that is amenable to manipulation. It should be inexpensive to study and easy and quick to grow. The organism must have behaviors that can be easily identified and disrupted by anesthetics. These behaviors should correspond to behaviors in more complex organisms. Finally, tools of molecular biology should be easy to apply to the model system. When picking a model system, one clearly trades off simplicity for evolutionary similarity to humans. Currently the genetic determinants of volatile anesthetics are being studied in four very different organisms: the yeast, *Saccharomyces cerevisiae*; the nematode, *Caenorhabditis elegans*; the fruit fly, *Drosophila melanogaster*; and the mouse, *Mus musculus*. Below, we detail these studies, from most simple to most complex animal model.

4.5.1 *Saccharomyces cerevisiae*

Keil and colleagues[23] have studied the effects of volatile anesthetics on the yeast, *S. cerevisiae*. As the authors note, yeast has both advantages and disadvantages compared with more complex eukaryotes. Yeast offers superb, extremely powerful genetics that control a complicated life cycle. It possesses a very short generation time and a great number of mapped genes. In addition, it recently became the first eukaryote to have its entire genome sequenced. Thus, for a molecular genetic approach to many questions of interest, it is unparalleled in its advantages.

Clearly to its disadvantage for our purposes, yeast does not have a nervous system. The precise mechanism of action of volatile anesthetics will certainly differ between yeast and higher eukaryotes. However, as Keil and colleagues state, any mutations isolated must "reflect molecular effects of the anesthetics." Fundamental interactions between volatile anesthetics and target molecules can be relatively easily sorted out. Such an approach has been useful in determining the mechanism of action of other drugs in mammals. The authors found that all volatile anesthetics inhibit growth in yeast, although at quite high concentrations. However, the inhibiting concentrations followed the Meyer–Overton relationship as noted for inhibition of the nervous system in more complex organisms. In addition, this inhibition was reversible even after 24 h of exposure to the anesthetic.

The authors screened for mutations conferring resistant to the inhibitory effect of volatile anesthetics. They isolated a single mutation, termed zzz4, which conferred resistance to 13% isoflurane (EC50 for isoflurane is 12% for wild type) (Figure 4.3).

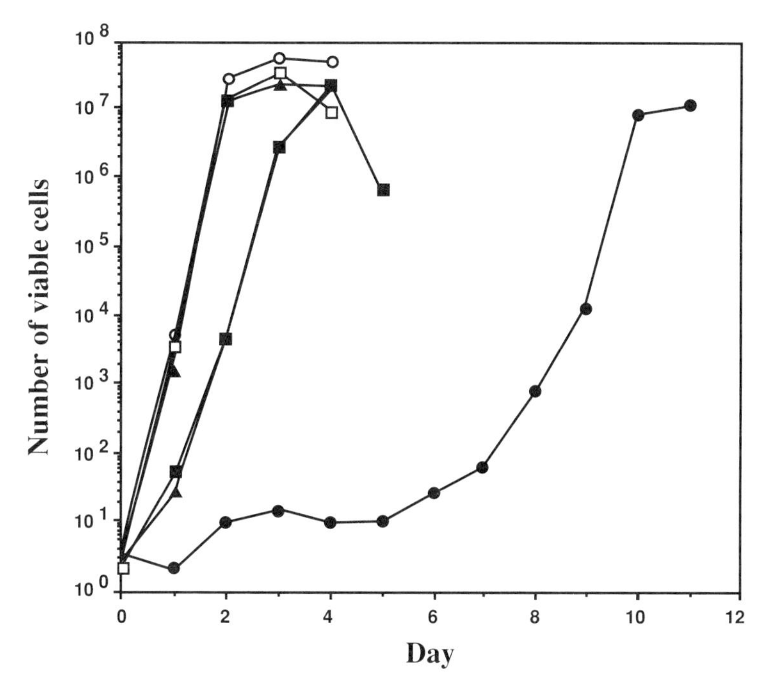

FIGURE 4.3 Effect of isoflurane on growth of wild-type and zzz4 cells in liquid culture. Wild-type cells and zzz4 mutants were grown in the presence and absence of 13% isoflurane. Circles, wild type; squares, zzz4-1 (a mutation conferring partial loss of gene function); triangles, zzz4::LEU2 (a null or complete loss of gene function mutation); open symbols, 0% isoflurane; closed symbols, 13% isoflurane. (Reproduced with permission of authors.)

zzz4 was cloned using transposon mutagenesis (see *unc-80* discussion in *Caenorhabditis elegans* section for a description of transposon mutagenesis). Keil and colleagues have identified this gene as a membrane protein whose amino acid structure is similar (homologous) to a rodent phospholipase A2-activating protein (PLAP). PLAP itself is homologous to G proteins, and so falls into one class of candidates proposed to be affected by volatile anesthetics.[24] It remains to be seen whether homologues of this class of protein also affect anesthetic sensitivities in organisms with nervous systems. Keil notes that one such homologue has been found in the genome of *C. elegans*, although no mutation has yet been isolated. The authors isolated mutations in six other genes that also confer resistance to isoflurane, and mutations in another two genes that suppress the isoflurane resistance of zzz4 mutations. The molecular nature of these other genes and the genetic pathway governing their interactions await further study.

4.5.2 *Caenorhabditis elegans*

Although not as simple as yeast, the nematode, *C. elegans* is also a very simple genetic model. In addition, it possesses a well-defined nervous system. Thus, one can use behavioral changes as anesthetic end points. The adult is about 1 mm long and consists basically of a cuticle, nerves, muscles, gut, and gonad.[25] The nonmutated worm, called N2, always consists of exactly 959 somatic cells in the hermaphrodite. The embryonic ancestry of every adult cell is known back to the first cell division of the fertilized egg.[26,27] Of these 959 adult cells, 302 (in the hermaphrodite) are neurons. Furthermore, this animal has been serially sectioned, from head to tail, and examined via electron microscopy, such that every synapse of these 302 neurons is known.[28] Although very simple, the animal has complex behaviors and shares at least four neurotransmitters with more complex animals (acetylcholine [Ach], GABA, serotonin, and dopamine).[29] Its total DNA content (genome) is 100,000,000 bases, only about 20 times that of *Escherichia coli*.[29] The *C. elegans* genome currently is being physically mapped. This, simply speaking, means that the DNA of the organism has been enzymatically digested to yield small fragments and these fragments ordered relative to each other and to the genetic map of each chromosome.[29] This is the first step in spelling out the DNA sequence of the entire organism and is instrumental in cloning genes of interest. In addition, the *C. elegans* genome is being sequenced (recently it was announced the first 50,000,000 were finished) and should be finished in approximately 2 to 3 years. In these ways *C. elegans* is similar to yeast, the fruit fly, and the mouse, because similar genome projects are being undertaken in all four organisms.

C. elegans is in many ways an ideal organism for classical genetic studies. It has a short generation time and exists as a self-fertilizing hermaphrodite. Males do exist and can be used for genetic crosses. More than 100 mutants that have known defects in neuromuscular function have already been isolated and genetically mapped.[29] It is relatively easy to identify new mutations in *C. elegans* and to study already identified mutations under new conditions (i.e., exposure to anesthetics).

It may appear at first glance that *C. elegans* is a far-fetched model for anything of clinical relevance to anesthesiologists. It is interesting to note that when first introduced by Sydney Brenner some 20 years ago,[25] *C. elegans* was proposed as *the* animal in which to make inroads on the question of genetic control of behavior. It is simple enough to yield to a detailed molecular study, but complicated enough to possess fundamental behavioral responses. Important for a study of volatile anesthetics, *C. elegans* responds to anesthetics in a manner similar to all other animals.[22] The "normal worm," N2, usually moves in a constant sinuous motion across an agar plate. Upon placement in a volatile anesthetic, N2 increases its motion, as if avoiding a noxious stimulus. This progresses to an increasingly uncoordinated pattern of movement, until the animal is finally immobilized in a state of flaccid paralysis. In addition, it no longer moves backward when tapped on the snout. These animals quickly become mobile again when removed from the anesthetic, with no apparent residual effect. This reversible immobility can be used as the end point for "anesthesia" and the construction of dose–response curves. However, other end points for anesthetics have also been used in *C. elegans*. Crowder[30] and others have argued that alternative behaviors better reflect loss of consciousness in mammals than does immobility. These studies are discussed later in this section.

A log–log plot of EC_{50} values for N2 (the effective dose at which 50% of worms are immobilized) versus oil/gas partition coefficient yields a straight line ($R^2 = .99$), with a slope of –.96. Because this behavior follows the Meyer–Overton relationship,[31,32] *C. elegans* represents a reasonable model in which to use genetics to study the site of action of volatile anesthetics. The first step is to isolate animals that have mutations at the anesthetic site of action. Recognition of an interesting mutant animal is obviously of key importance.

The first interesting mutant was identified because it is very sensitive to halothane, with an EC_{50} approximately one third that of N2.[22,33] This animal, when not in an anesthetic, moves in a motion described as "fainting" — that is, bursts of normal motion of a few body lengths are followed by several seconds of no motion at all. This worm, by classic genetic mapping, was found to be a new form, a new allele, of a gene already identified in *C. elegans* called *unc-79*. It was suggested that another fainter mutant animal, *unc-80*, be examined for similar changes (Jim Lewis, personal communication). *unc-80* is also more sensitive to halothane, but its EC_{50} is between that of *unc-79* and N2.[34] Although very similar in appearance and behavior in anesthetics, *unc-79* and *unc-80* have different mutations that lie on completely different chromosomes. An animal containing both mutations, written *unc-79;unc-80,* behaves largely like *unc-79* in volatile anesthetics, suggesting that *unc-79* functions closer to final control of normal anesthetic sensitivity than *unc-80*.[34,35]

Although very sensitive to halothane, it was clearly possible that these two similar mutants were simply sluggish animals that were somehow paralyzed more easily by halothane for some trivial reason, unrelated to the anesthetic site of action. However, *unc-79* and *unc-80* were not sensitive to all anesthetics, but extremely sensitive only to the most lipid-soluble anesthetics (e.g., methoxyflurane and halothane)[34] and slightly resistant or unchanged to others (e.g., enflurane and isoflurane, Figure 4.4). If these mutants were simply sluggish, we would expect them to show an increased sensitivity to all the anesthetics. We were also concerned that these changes in sensitivity may reflect altered uptake or metabolism of the anesthetics.

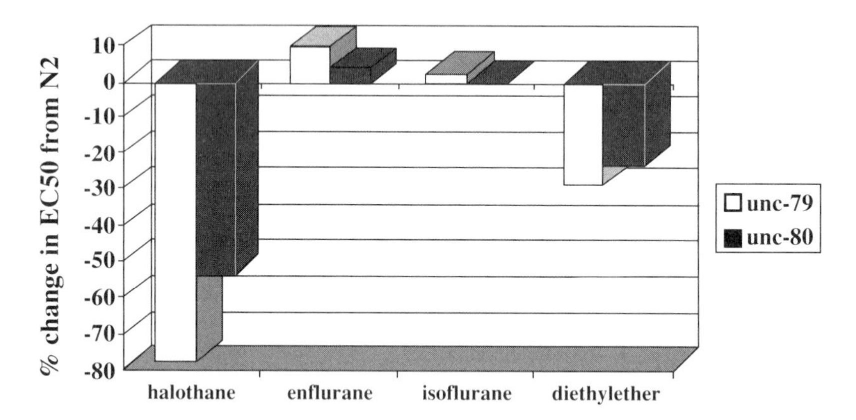

FIGURE 4.4 Changes in the EC50 values of *unc-79* and *unc-80* compared to those of N2 (the wild-type strain) in four anesthetics. The EC50 values of N2 are represented as the zero (0) line, and the EC50 values of the mutants are represented as % changes from the EC50 values of N2. A negative change indicates an increase in sensitivity; a positive change indicates a decrease in sensitivity (or resistance). Anesthetics are indicated on the horizontal axis. Percentage changes are indicated on the vertical axis.

However, such changes would change the time course of an anesthetic's effect, but not the steady state data response as represented here.

Obviously, *unc-79* and *unc-80* appeared quite interesting for an immediate molecular analysis of their function. They seemed fairly unique, in that a screen of about 100 other uncoordinated-type mutant animals in high and low doses of both halothane and enflurane yielded no other mutants with an altered anesthetic response.[35] However, in addition to beginning a molecular analysis of these two genes, we screened animals for mutations changing sensitivities to other anesthetics. To identify genes that may produce other components of the anesthetic site of action, we sought mutations that altered sensitivity to enflurane, isoflurane, or diethylether. The interactions between isolated genes were then analyzed. These studies have led to the characterization of a pathway involving several genes controlling sensitivity to volatile anesthetics in *C. elegans* (Figure 4.5).[35]

unc-79 and *unc-80* represent the first step in this pathway; *unc-1*, *unc-7*, and *unc-9* are of interest since the normal genes are necessary for the increase in sensitivity to halothane in *unc-79* animals. *unc-1* is unique in that its function is required continuously throughout the life of the nematode; thus it is a candidate as a target for halothane. The molecular cloning of these three genes (*unc-79*, *unc-80*, and *unc-1*) is currently underway, and described below.

What conclusions can be drawn from this accumulation of data based on classical genetic techniques? The simplest explanation of the data is that, in *C. elegans*, multiple sites of action of volatile anesthetics exist. Previous neurophysiologic

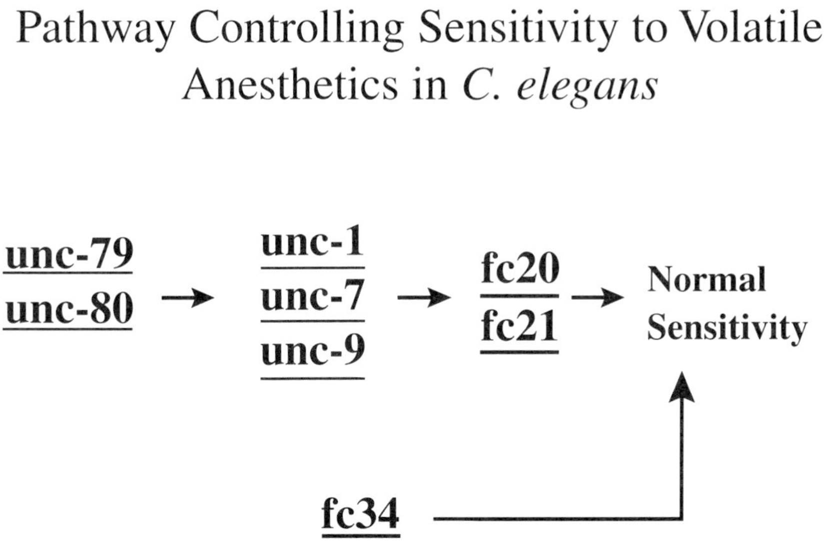

Pathway Controlling Sensitivity to Volatile Anesthetics in *C. elegans*

unc-79
unc-80 → unc-1
unc-7
unc-9 → fc20
fc21 → Normal Sensitivity

fc34 → Normal Sensitivity

FIGURE 4.5 A genetic pathway controlling sensitivity to volatile anesthetics in *C. elegans*. The genes *unc-79*, *unc-80*, *unc-1*, *unc-7*, and *unc-9* are discussed in text. The mutations *fc20*, *fc21*, and *fc34* not discussed in Ref. 59. They are mutations that change sensitivity to all volatile anesthetics and are epistatic to the previously discussed mutations. The presence of normal copies of all these genes is necessary for normal sensitivity to volatile anesthetics in *C. elegans*.

studies in mammalian systems by MacIver and colleagues[36,37] have independently supported multiple sites of action of anesthetics. They found different neurophysiologic effects in rat hippocampus by different volatile anesthetics, a finding inconsistent with a unitary mechanism of action.

4.5.2.1 Molecular Characterization of *unc-80*

unc-80 is particularly suited to a method of gene cloning that begins with transposon tagging. Transposons are special small pieces of DNA, studied in depth in many organisms, that can hop in or out of the genes of an organism.[38] In *C. elegans*, special strains of worm exist in which these transposons are unusually mobile ("high hoppers"). Thus for *unc-80* one begins with the normal gene, as present in N2, denoted *unc-80*$^+$. One screens the "high hopper" strains for a mutant *unc-80* animal. Such an animal presumably has the special piece of DNA, the transposon, somewhere within the gene. The transposon insertion disrupts the normal *unc-80* function, producing a mutant *unc-80*. One grows the offspring of this particular animal in very large quantities and extracts its DNA. The DNA is cut with a restriction endonuclease, as is a separate sample of DNA from a wild-type animal. Running each digest in an electrophoretic gel gives a spread of various sized pieces of DNA that collectively represent the entire genome of the animal. The fragment from the spontaneous mutant will have a transposon in it and will be larger than its counterpart from the wild-type animal.

The above approach was used to identify a 10-kb fragment containing most of the *unc-80* gene. This gene contains a 55% homology at the amino acid level to a lysomal α-1,4-glucosidase in humans. This homology extends over the entire length of the gene. The human homologue for the glucosidase has been localized to the 17q.23 region. Mutations in this gene give rise to Pompe's disease.[39] Limited clinical experience with this disease indicates that these patients are extremely sensitive to halothane; normal doses of halothane have led to intraoperative death.[40] Interesting, at least one variant of a human disease, malignant hyperthermia (MH), has also been localized to this general region, from 17q.2 to 17q.24.[41] It would be interesting if dominant mutations in an α-glucosidase conferred the hypermetabolic response to volatile anesthetics seen in some variants of MH.

4.5.2.2 Molecular Characterization of *unc-79*

unc-79 is the gene that when mutated causes *C. elegans* to be hypersensitive to the most liquid-soluble volatile anesthetic. The approach to cloning *unc-79* was first to localize the approximate position of the gene on the physical map. *unc-79* was mapped to a narrow region of chromosome III and then mapped relative to RFLPs between the *unc-79* genome and the genome of a related strain of *C. elegans*, Bergerac. These data place *unc-79* in the region of one cloned piece of *C. elegans* DNA, E03A3. Injections of the DNA of E03A3 into *unc-79* animals can restore the offspring of the mutant animal to normal. Using this technique, called mutant rescue, the position of *unc-79* is now narrowed to one gene, E03A3.6.

The theoretical gene product E03A3.6 is a protein with homology to several transmembrane proteins. These include the gene *pecanex* (a neurogenic protein from *Drosophila),*[42,43] the α-1 subunit of calcium channels, the CD1D glycoprotein from humans,[44] and the protein *period* in *Drosophila.*[45] All of these gene products are glycoproteins and most function in the nervous system. Although they are diverse in exact function, it is exciting that *unc-79* falls within a set of genes with nervous system activity. They are all transmembrane proteins and thus may be available to interact with gaseous anesthetics. Because they are glycoproteins, it is possible the function of *unc-79* may be affected by changes in glycosylation. This could explain the effects of *unc-80* on sensitivity to volatile anesthetics.

Eckenhoff[46] has taken a unique approach with the mutants *unc-79* and *unc-80* to understand how the volatile anesthetics work. He can anatomically localize the distribution of halothane at the subcellular level. Using a technique called EPMA (electron probe microanalysis), he can detect binding of halothane to various fractions of cells (e.g., cell membranes, mitochondria, cell nuclei, etc.). After an animal's exposure to halothane, it is rapidly frozen to –160°C, fixing the halothane in place. The sample is then exposed to a high-voltage electron beam, which lyses halothane but leaves a fragment containing the bromine atom in place. This bromine represents the original distribution of halothane molecules. Unexpectedly, bromine concentrations were decreased in the mitochondria and plasma membrane of *unc-79* and *unc-80*. (For example, in mitochondria, *unc-79* had 75% of normal binding, whereas *unc-80* had 58% compared to N2.) These data are consistent with decreased halothane binding in certain subcellular fractions of halothane-sensitive animals and may indicate that *unc-79* and *unc-80* are directly related to halothane binding in *C. elegans*.

4.5.2.3 Molecular Characterization of the Suppressor Genes

unc-1 is a gene that restores to normal the behavior of an *unc-79* or *unc-80* animal. For example, the animal that carries mutations in both *unc-79* and *unc-1* is like N2 in sensitivity to halothane or enflurane. *unc-7* and *unc-9* are very similar to *unc-1* in their effects, whereas *unc-24* is a partial suppressor of *unc-79* and *unc-80*. *unc-1* has been localized to a region of the genome containing a neurocalcin gene.[47,48] Neurocalcins are thought to function as a ball valve, inhibiting the flow of calcium through transmembrane channels. Loss of such a protein would therefore increase intracellular calcium. Neurocalcins are also tethered to the membrane, presumably to keep the protein in correct proximity to calcium channels. The nature of the tethering sites in the membrane is not known. In addition, neurocalcin associates with other proteins that presumably modulate its action.[49]

Starich and colleagues[50] have cloned and identified the *unc-7* gene. They find it to be a member of a group of transmembrane proteins, with unknown functions. Homologous proteins are found in flies and mice and are candidates for parts of gap junctions between neurons.[51] In addition, Barnes and colleagues[52] have molecularly characterized *unc-24*. This gene codes for a stomatin-like protein and probably serves as part of a transmembrane cation channel.

Taken together, these data can be used to construct a model. *unc-1* may be necessary to maintain calcium homeostasis, a factor thought to be important in the effects of volatile anesthetics.[53] *unc-79* may be involved in the correct tethering (i.e., it is a modulator of neurocalcin). Loss of *unc-79* led to increased blocking of transmembrane calcium flux. This would lower intracellular calcium and may lead to the fainting phenotype as well as increased sensitivity to some anesthetics. In turn, because *unc-79* is homologous to a variety of membrane-associated glycoproteins, *unc-80* may be necessary for the correct glycosylation of *unc-79*. It must be remembered that these results are preliminary, and further proof of their identity is necessary. The roles of *unc-7* and *unc-24* are also unclear, but they may function with *unc-1* in determining calcium flux or be part of the downstream functions affected by the changes in intracellular calcium. However, it is clear that a genetic approach can be used to determine the molecular nature of pathways affected by volatile anesthetics.

4.5.2.4 Other End Points for Anesthesia in Nematodes

Crowder[30] and colleagues have argued that the high EC_{50} values for immobility in nematodes raises the question of the appropriateness of this end point as a model for loss of consciousness in mammals. They have identified several other behaviors in nematodes that are altered at concentrations of 0.5 to 2% halothane. These end points include normal motion, mating, chemotaxis, and egg laying. In general, these behaviors also follow the Meyer–Overton relationship and are quickly reversible.

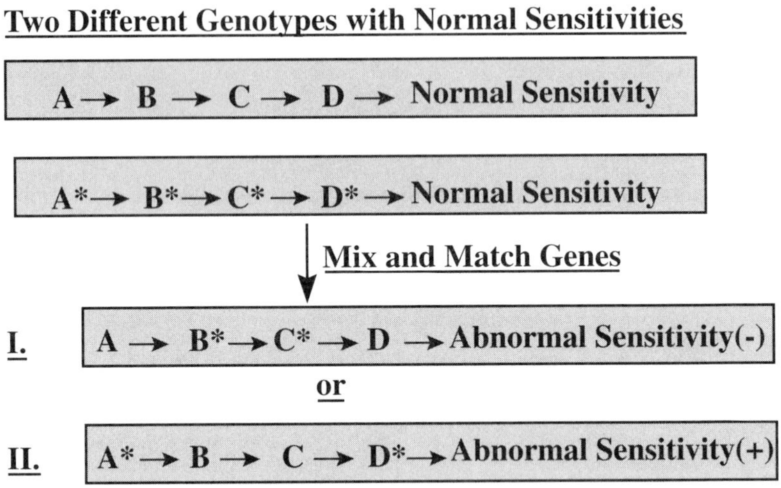

Two Different Genotypes with Normal Sensitivities

A → B → C → D → Normal Sensitivity

A* → B* → C* → D* → Normal Sensitivity

Mix and Match Genes

I. A → B* → C* → D → Abnormal Sensitivity(-)

or

II. A* → B → C → D* → Abnormal Sensitivity(+)

FIGURE 4.6 The general plan for obtaining alterations in sensitivity to anesthetics using recombinant inbred strains. Two genetically different strains, both with normal phenotypes, are mated to produce hybrid offspring. Because the normal phenotypes in the parents are the result of interaction of many genes, all of which may have undergone small genetic changes, the mixing of the gene products may lead to wide variation in function of the pathway involving the genes. As a result, one may get a pathway with multiple changes in it and observe large changes in sensitivity. Such changes may not be seen in animals with single mutations. The problem then is to determine which changes are most important in determining the phenotype (i.e., anesthetic sensitivity).

At present, no single mutations have been isolated that specifically alter sensitivities to volatile anesthetics using these end points.

Crowder, et al. has taken a different approach to isolating genes affecting these more sensitive endpoints. They have mated worms from two different wild-type strains of *C. elegans* (Bristol and Bergerac) and isolated hybrid strains from the offspring. These hybrid lines are inbred and termed recombinant inbred strains. Some of the resulting strains are significantly altered in their sensitivity to anesthetics (Figure 4.6). As an example they have identified strains varying by a factor of 5 in sensitivity to halothane when using normal mating as an end point. In addition, they have isolated strains resistant to 5% halothane using mobility as an end point, an isolation that was not possible in screens of single mutations by Morgan and Sedensky (Crowder, personal communication). These strains are interesting in that they contain combinations of entirely normal sets of genes, some from one strain and some from the other. In theory no genes have been mutated. Rather, it is the interaction of, and changes between, groups of genes (and not one specific gene) that causes the varied response. By examining a large number of hybrid strains with altered sensitivities they are now mapping the regions of the genome responsible for the changes. It will be of great interest to correlate these two approaches as the characterizations progress.

In an evolutionary sense, *C. elegans* is far removed from *Homo sapiens*. The phenomenon of "anesthesia," however, appears to be a highly conserved phenomenon shared by very disparate phyla. The excellent adherence of so many different species, including *C. elegans*, to the Meyer–Overton relationship, speaks to this point. Clearly, volatile anesthetics disrupt some of the most fundamental processes of neuronal function, ones that exist in many species. As in *C. elegans*, there must exist in humans a group of genes that specifies the molecular components of the anesthetic site of action. It is likely that these genes may bear some degree of resemblance to those of *C. elegans* (certainly other *C. elegans* genes have their counterpart in mammalian species). Many genes of basic importance to neuronal function, such as the $GABA_A$ or ACh receptor, are very similar throughout much of their sequences across very different species, attesting to phylogenetic conservation of some basic neuronal components.[53] Genes isolated from *C. elegans* can be used to probe other species for DNA of similar structure in hopes of finding genes of similar function.

4.5.3 *Drosophila melanogaster*

A similar approach to that used in *C. elegans* is being undertaken in the fruit fly, *Drosophila melanogaster,* which has many well-studied mutants with easily seen morphologies. However, *Drosophila* is more complicated than *C. elegans*.[54] For example, it contains 100 times as many neurons with *C. elegans*. This complexity can certainly be viewed as an advantage compared with *C. elegans*, in that the nervous system may more closely model that of mammals.[55,56] Although more complex than *C. elegans* it is very amenable to techniques of cloning and sequencing genes.[57,58] As noted by Nash and colleagues, the EC_{50} values for immobility in *Drosophila* more closely match those of mammals than do those of the nematode. (These difference have been discussed in Refs. 56 and 59.)

More than 15 years ago, Gamo et al.[60,61] isolated a variant of the normal fruit fly, *D. melanogaster*, that was resistant to certain anesthetics. This variant, termed Eth-29, was a spontaneously arising mutant, derived from a strain that carried several other mutations, with an unknown number of changes from the wild type. Eth-29 was resistant to anesthetizing doses of diethylether, chloroform, and halothane, compared with the strain from which it was derived.[61] These strains differed from each other by approximately 50 to 100% in their respective EC_{50} values. However, each characteristic resistance was not controlled by the same gene. For example, the mutation that conferred halothane resistance was genetically mapped as an X-linked recessive mutation (a mutation on the X or sex chromosome), whereas the resistance to chloroform was an incompletely dominant gene on an autosome (not on the sex chromosome). Thus, although both are among the most lipid-soluble volatile anesthetics, and both are halogenated hydrocarbons, at least two different genes control their function separately in *Drosophila*.

Whole flies of the Eth-29 strain were analyzed for changes in lipid composition compared with the sensitive strain. No difference was found in total lipids or phosphorylation of certain lipids between the two strains. However, a difference was found in the number of various molecular forms of phosphatidylethanolamine (PE); the resistant strain contained a relative decrease of unsaturated fatty acids.[61] This particular property of the resistant strain was related to genes on both the X and the third chromosome.

In addition, Eth-29 has many other interesting properties. It is also resistant to killing doses of diethylether, chloroform, and halothane, although very sensitive to X-rays at both lethal and nonlethal ("knock-down," a dyskinetic result of X-rays felt to be related to neuronal damage) doses.[62] All of these properties map to different chromosomes, with major and minor effects from different genes. Clearly, Eth-29 is complicated, containing multiple mutations that affect many different processes, which may interact with one another in complex and as yet undefined ways. It is illustrative of the importance of isolating single gene changes that control anesthetic response. More recent work by Gamo et al.[63] describes attempts to identify the molecular changes in the ether-resistant mutants.

Krishnan and Nash[55] have isolated mutants of *D. melanogaster* that have altered responses to halothane. These authors mutagenized flies and performed column fractionation of the flies in an "inebriometer." Individuals sensitive to halothane fell out of the column while those that were resistant settled on baffles within the column. Flies not falling out of the column were collected and used to establish mutant strains. The authors mapped these mutations to the X chromosome. Four halothane-resistant mutants were isolated in this fashion. They were termed *har* (*ha*lothane *r*esistant) *38, har 56, har 63,* and *har 85*. Dose–response curves of these strains exposed to halothane are shown in Figure 4.7.

Genetic characterization of these mutants shows that har 38 and har 85 are recessive mutations, and har 63 is dominant. The authors then mapped the two recessive mutations to a narrow region on the X chromosome, so close to each other that they may be mutations in the same gene. To test whether har 38 and har 85 are two alleles of the same gene, the authors constructed females carrying one X chromosome for each strain. These animals, called heterozygotes, behaved similarly to an animal with both X chromosomes from either har 38 or har 85. This is called partial complementation and suggests that har 38 and har 85 are alleles of the same gene, or somehow interact with each other to produce resistance to anesthetics.

The behavior of unanesthetized har 38 and har 85 is also quite interesting. Krishnan and Nash describe their motion as walking by fits and starts, taking a few steps and then stopping before starting again. This phenotype is very reminiscent of the "fainting" phenotype of the *unc-79* and *unc-80* mutants in *C. elegans*, which show altered sensitivity to halothane. In addition, Campbell and Nash[64] have found that har 38 and har 85 exhibit altered large changes in sensitivities to methoxyflurane, chloroform, and halothane, but less striking changes in sensitivities to enflurane, isoflurane, and diethylether. Nash and colleagues have concluded that multiple sites of action exist for volatile anesthetics in *Drosophila*. A similar pattern is seen in the *C. elegans* mutants *unc-79* and *unc-80*. It is fascinating that these phenotypic changes are similar across a wide phylogenetic gap. A highly conserved neurophysiologic function may be common to these very different species. It is clear that these mutants in *Drosophila* offer an exciting opportunity to learn more about an anesthetic site of action. As was discussed earlier, these mutations offer a mechanism to exploit molecular genetics to determine the protein product of these genes.

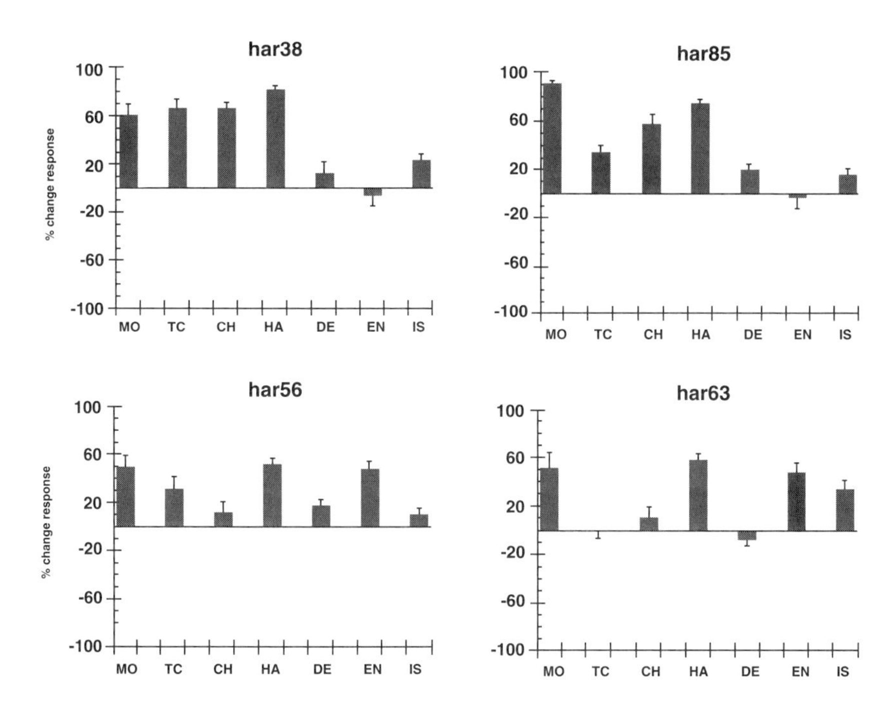

FIGURE 4.7 Change in anesthetic response in *Drosophila melanogaster* as described by Nash and colleagues.[64] The figure shows the responses of four mutants isolated by this group when exposed to a variety of volatile anesthetics. Note that for *har38* and *har85*, there are large changes in sensitivity to halothane, methoxyflurane, and chloroform, but smaller changes to ether, enflurane, and isoflurane. These results are reminiscent of the changes seen in *unc-79* and *unc-80* in *Caenorhabditis elegans*, although the changes are in the opposite direction. The authors also noted that the pattern of sensitivities was dependent on the end point used to determine anesthesia. (Reproduced with the permission of the authors.)

Nash and colleagues[65] have also studied specific mutations, isolated in other screens, that alter known ion channels. They have found that these mutations also alter sensitivity to volatile anesthetics. They conclude that, in the complicated picture emerging from these genetic studies, ion channels have roles in the pathways governing sensitivity to volatile anesthetics. A separate approach has been used by Dapkus et al.[66] These authors studied halothane sensitivity in two different wild-type strains of *Drosophila* known as 91R and Canton-S. They found that the EC_{50} of 91R was 69% greater than that for Canton-S in females. (The difference was 48% in males.) By measuring the sensitivities of offspring between these strains they showed that this difference mapped to the second chromosome and acted as a dominant gene.

In summary, multiple studies have identified genetic components of sensitivity to volatile anesthetics in *D. melanogaster*. In addition, multiple genes are clearly involved; altered sensitivity to one volatile anesthetic does not imply alterations in sensitivities to all. Again, as in the nematode (and mouse, see below), multiple sites of action seems to be the simplest explanation.

4.5.4 *Mus musculus*

In studies involving the mouse, *Mus musculus*, one deals with the same tradeoffs described for flies and nematodes. One sacrifices ease of genetics and a simple nervous system in mice, but one gains a nervous system that more closely approximates that of man.[67] In addition, the advancement of molecular genetics has extended to this organism as well, such that many genes have been cloned in mice.[68] Eventually many of the advantages of the more simple organisms will be present in the mouse as well.

Whereas Eth-29 was a spontaneous mutant in flies, Koblin and colleagues[69] purposely bred mice to generate differences in sensitivity to N_2O. Mice were picked that were either resistant or sensitive to N_2O. By mating these mice to each other over several generations, strains were developed that, at the extremes of behavior, were different from each other in sensitivity to N_2O by nearly a factor of 2. Presumably Koblin et al. isolated naturally occurring variants of multiple genes controlling sensitivity to N_2O. Membranes from neural synapses were extracted from the brains of these mouse strains. No difference was found in fatty acid composition of any of the five phospholipids analyzed in the study.

Other genetic studies of anesthetic action in mice center on the susceptibility of certain strains of mice to alcohol. (For excellent reviews, see Phillips et al.[70] and Harris and Allan[71].) Like the volatile anesthetics, the straight-chain alcohols and alkanes follow the Meyer–Overton relationship, using the oil/water partition coefficient of these aqueous solutions plotted against ED_{50}. This fact led to the assumption (now under question) that these agents worked at the same site of action as did the volatile anesthetics. One candidate for a site of ethanol action in the mouse has been the $GABA_A$ receptor, a transmembrane channel that mediates flow of Cl^- ions.[70] In general, ethanol seems to increase chloride flux through such channels. However, these studies were initially *in vitro* experiments. A genetic approach was necessary to extrapolate these results to the *in vivo* situation.

These genetic studies take advantage of selected genetic lines, lines started from breeding pairs with large differences in sensitivity to certain drugs (similar to Koblin's approach described previously).[72,73] It is crucial to remember that these lines probably differ in multiple genes. In this case most studies have centered on strains with an increased sensitivity (called long sleep [LS]) or decreased sensitivity (called short sleep [SS]) to ethanol.[74-77] However, other lines have also been developed with differences in sensitivity to diazepam (diazepam sensitive [DS] or diazepam resistant [DR]),[78] or differences in acute sensitivity to ethanol (high acute sensitivity [HAS] or low acute sensitivity [LAS]).[79] The sensitivities of each of these lines is then forced to deviate further from normal by constant selection and breeding of those individuals with the greatest differences in sensitivities. These strains are then analyzed for molecular changes that may account for these differences. Ideally, once this selection has been carried out for many generations, only those genetic differences that lead to the phenotypic differences in question (altered sensitivity to ethanol) should persist. The real world is seldom ideal, and these studies are no exception. For example, LS and SS mice have been shown to differ in a number of responses other than sensitivity to ethanol.

In general the strains with increased sensitivity to ethanol (LS) have increased sensitivity to GABA agonists and drugs that potentiate GABA.[76,80-83] All of the ethanol-sensitive strains show increased ethanol enhancement of $GABA_A$-stimulated Cl^- flux. These studies and others strongly implicate the flux of chloride through $GABA_A$ channels as a site of action for ethanol intoxication. However, as was pointed out by Dudek and Abbott,[84] the LS and SS strains differ in at least nine genes related to ethanol sensitivity. Therefore, multiple molecular changes may be affecting ethanol sensitivity in these mice. The effect of ethanol on other sites (voltage-gated calcium channels, sodium channels, potassium channels) is unclear at present. Thus, these lines of mice present the additional problem of sorting out which molecular changes are of importance. Mice, although clearly more similar to humans, also possess enough complexity that the molecular analysis of genetic changes is still equivocal. Such a problem necessitates a multifactorial analysis of the altered sensitivity. However, any results will probably be easily applied to a polygenic control of anesthetic response in humans.

The relationship of sites of action for ethanol with those for volatile anesthetics is unclear. Erwin et al.[85] showed that sensitivity to diethylether did not increase in LS mice. Baker et al.[86] found that LS mice did not have an increased sensitivity to halothane. These studies led to the conclusion that volatile anesthetics probably did not work at the same sites of action as ethanol. However, in 1981, Koblin and Deady[87] showed that LS mice did have increased sensitivities to the chemical isomers enflurane and isoflurane. These studies may have indicated as early as 1981 that not all volatile anesthetics work by identical mechanisms. In addition, it was the first indication that sensitivity to ethanol may be similar in mechanism to that of enflurane or isoflurane. Related studies in rats, however, indicate that some genetic changes can simultaneously alter sensitivity to ethanol, isoflurane, and halothane.[88] It is quite possible that ethanol and enflurane may preferentially affect the same site, whereas halothane and diethylether may affect other targets (a result observed also in nematodes and flies).

4.5.5 *Homo sapiens*

If genetic differences in sensitivities exist in various model organisms, why do we not see them in man? The answer is that it is very difficult to look for them. The initial determinations of minimum alveolar concentration (MAC) for the various anesthetics were done in healthy individuals and in a small sample size (by genetic standards). Thus, although these studies determine the mean for a population quite well, they are not intended to identify outlying groups of responders. To do so one would have to look at a much larger sample size and, of course, be able to study patients with other neurologic changes.

One could expect that such individuals, with abnormal responses to volatile anesthetics, would appear in the operating room occasionally. Their altered sensitivity should be noted and appear as a case report. The problem is almost none of our patients receive just one volatile anesthetic for their anesthetic technique. The addition of sedatives, opiates, and muscle relaxants makes it very difficult, if not impossible, to determine an altered sensitivity to volatile anesthetics. If one adds to this the complication of possible altered sensitivity to one volatile anesthetic and not others, recognition becomes very difficult indeed. Probably all of us in practice have cared for patients who seemed very sensitive to our anesthetics as a whole. The usual response in the clinical world is to modify our usual techniques, changing doses of several medications. It becomes impossible to identify one causative critical component.

However, the studies in model organisms should eventually aid us in identifying populations of humans at risk for altered sensitivities. A great number of mutations in these animals have corresponding diseases in humans (see, for example, Pompe's disease). In the end it may be of great value to be forewarned about potential increased sensitivities to specific volatile anesthetics in specific inherited conditions.

4.5.6 Future Work

It is clear that molecular genetics is poised to make great contributions to the understanding of how volatile anesthetics work. This will initially require cloning and molecular identification of the genes that control anesthetic responses. In *C. elegans* and *S. cerevisiae* some of the involved genes have been identified, and their interactions with volatile anesthetics are now being characterized. Work is underway to characterize the genes involved in *D. melanogaster.* It will be of interest to determine whether similar changes in more complicated organisms (i.e., mammals) also confer altered sensitivities to volatile anesthetics.

In mice (and flies) the ability exists to obtain targeted mutations, a technique in which one introduces a mutation in a specific gene in order to observe the resulting effect on the whole organism. An initial possibility will be to determine if genes exist in mice that are homologous to those isolated in yeast or worms or flies. If so, these genes (or their homologous parts) can be cloned into plasmids. These sequences can then be altered in desired ways while in the plasmid, and the altered sequences can be reintroduced into mouse embryos. At a low rate these sequences may be taken up into the homologous gene in the mouse, thus introducing a mutation into a targeted position. This allows one to return to the more usual approach of altering a specific condition (in this case a gene) and observing the effect on the entire animal (see the first paragraph in Section 4.2). Such a technique can then be used to identify if mutations in similar genes alter sensitivities in mice.

The use of targeted mutations can be used in other experiments as well. First, one can make educated guesses as to what might be an anesthetic target (say a GABA channel subunit). Mutations may be targeted in this gene and the resulting mutant animals studied for altered sensitivity to anesthetics. Once a particular gene is implicated small mutations can be introduced in specific regions of the protein to identify important structure–function relationships (i.e., to analyze which part of the protein is important for the action of volatile anesthetics).

In yeast, nematodes, and flies, one can also collect multiple alleles of genes affecting sensitivity to volatile anesthetics and directly sequence the alleles. By determining where in the protein the changes occur, one can find out which regions of a protein are important in determining anesthetic sensitivities. This approach attempts to gather data similar to those discussed above via targeted mutations.

Obviously, one must be able to collect many alleles of a gene to make this approach work. Thus, this approach is most useful in yeast, nematodes, and flies, but not as useful in mice.

4.6 SUMMARY

The genetic data from these model organisms can be grouped to form an interesting picture. For example, in each organism genetic control of sensitivity to halothane differs from that of enflurane, indicating that the volatile anesthetics may disrupt different specific sites. These separate sites may be highly conserved across species as are, for example, the family of chemically gated cation channels.[53] If so, one should be able to use the organisms described here to probe the human response to volatile anesthetics. It is difficult to predict exactly how this work may affect clinical practice. Certainly synthesis of a "perfect" anesthetic, or induction of endogenous anesthetics, awaits some basic knowledge of the mechanism of action of volatile anesthetics.

To date, no studies have yielded a molecular picture of the site of action of volatile anesthetics. However, a genetic approach to understanding how the volatile anesthetics work is becoming increasingly attractive as it becomes increasingly easy to understand at the molecular level how a gene works. This rapidly evolving field offers the exciting possibility that in the very near future we will know something of the molecular nature of the site of action of volatile anesthetics, and thus something of their mechanism of action.

REFERENCES

1. Prokop A, Bajpai RK, eds. In recombinant DNA technology. *Ann. N.Y. Acad. Sci.* 1991; 646: 1–385.
2. Sanbrook J, Frich EF, and Maniatis T, eds. *Molecular Cloning: A Laboratory Manual,* 2nd ed. Cold Spring Harbor, NY. Cold Spring Harbor Laboratory (Publ) 1989, 1.1–1.105.
3. Sanger F et al. Cloning in single-stranded bacteriophage as an aid to rapid DNA sequencing. *J. Mol. Biol.* 1980; 143: 161–178.
4. Avery OT, MacLeod CM and McCarty M. Studies on the chemical nature of the substance inducing transformation of pneumococcal types. *J. Exp. Med.* 1944; 79: 137–157.
5. Watson JD, Crick FHC. Molecular atructure of nucleic acids: A structure for deoxyribose nucleic acid. *Nature* 1953; 171: 737–738.
6. Clark BFC, Marcker KA. The Role of N-formyl-methyl sRNA in protein biosynthesis. *J. Mol. Biol.* 1966; 17: 394–406.
7. Stent GS, Calendar R. RNA translation. In Stent GS, Calendar R eds. *Molecular Genetics, An Introductory Narrative*, 2nd ed. San Francisco, CA. WH Freeman and Co. 1979, 501–539.
8. Crick FHC, Barrett L, Brenner S, and Watts-Tobin RJ. General nature of the genetic code for proteins. *Nature* 1961; 192: 1227–1232.

9. Luban NLC, Epstein BS, and Watson SP. Sickle cell disease and anesthesia. *Adv. Anesth.* 1984; 1: 289–336.
10. Kalow W, Genest K. A method for the detection of atypical forms of human serum cholinesterases. Determination of dibucaine numbers. *Can. J. Biochem.* 1957; 35: 339–346.
11. Chapeville F et al. On the role of soluble ribonucleic acid in coding for amino acids. *Proc. Nat. Acad. Sci. USA* 1962; 48: 1086–1092.
12. Gold L et al. Translational initiation of prokaryotes. *Annu. Rev. Microbiol.* 1981; 35: 365–403.
13. In Ayala FJ, Kiger JA, Jr. eds. *Modern Genetics*, 2nd ed. Menlo Park, CA. Benjamin/Cummings Publishing Company Inc. 1984; 28–44.
14. Ayala FJ, Kiger JA, Jr. Modifier genes and epsistatic genes in modern genetics. In Ayala FJ Kiger JA, Jr, eds. *Modern Genetics*, 2nd Ed. Menlo Park, CA. Benjamin/Cummings Publishing Company Inc. 1984; 628–631.
15. Ayala FJ, Kiger JA, Jr. Genetic effects on translation of the code. In Ayala FJ, Kiger JA, Jr, eds. *Modern Genetics*, 2nd ed. Menlo Park, CA. Benjamin/Cummings Publishing Company Inc. 1984; 384–388.
16. Murray K, Murray NE. Phage lambda receptor chromosomes for DNA fragments made with restriction endonuclease III of *Haemophilus Influenza* and restriction endonuclease I of *Escherichia coli*. *J. Mol. Biol.* 1975; 98: 551–564.
17. Ayala FJ, Kiger JA, Jr. DNA restrictor and modification enzymes. In Ayala FJ, Kiger JA, Jr, eds. *Modern Genetics*, 2nd ed. Menlo Park, CA. Benjamin/Cummings Publishing Company Inc. 1984; 267–275.
18. Nathans D, Smith HO. Restriction endonucleases in the analysis and restructuring of DNA molecules. *Annu. Rev. Biochem.* 1975; 44: 273–293.
19. Pearson WR, Lupman DJ. Improved tools for biological sequence comparison, *Proc. Natl. Acad. Sci. USA* 1988; 85: 2444–2448.
20. Gish W, States DI. Identification of protein coding regions by database similarity search. *Nat. Genet.* 1993; 13: 266–272.
21. Altschul SF, et al. Basic local alignment search tool. *J. Mol. Biol.* 1990; 215: 403–410.
22. Morgan PG, Cascorbi HF. Effect of anesthetics and a convulsant on normal and mutant *Caenorhabditis elegans*. *Anesthesiology* 1985; 62: 738–744.
23. Keil RL, Wolfe D, Reiner T, Peterson CJ and Riley JL. Molecular genetic analysis of volatile anesthetic action. *Mol. Cell. Biol.* 1996, 16: 3446–3453.
24. Franks NP, Lieb WR. Molecular and cellular mechanisms of general anaesthesia. *Nature* 1994; 367: 607–614.
25. Brenner S. The genetics of *Caenorhabditis elegans*. *Genetics* 1974; 77: 71–94.
26. Sulston JE, Horvitz HR. Post-embryonic cell lineages of the nematode *Caenorhabditis elegans*. *Dev. Biol.* 1977; 56: 110–156.
27. Sulston JE, Schierenberg E, White JG, and Thomson JN. The embryonic cell lineages of the nematode *Caenorhabditis elegans*. *Dev. Biol.* 1983, 100: 64–119.
28. White JG, Southgate E, Thomson JN, and Brenner S. The structure of the ventral nerve cord of *Caenorhabditis elegans*. *Philos. Trans. Roy. Soc. London* 1986; 314B: 1–340.
29. Wood WB, ed. *The Nematode, Caenorhabditis elegans*. Cold Spring Harbor, NY. Cold Spring Harbor Laboratory (Publ) 1989.
30. Crowder CM. A genetic screen in *Caenorhabditis elegans* for mutations conferring resistance to halothane. *Anesthesiology* 1995; 83 (3A): A762.
31. Meyer HH. Zur Theorie der alkoholnarkose. I. Mitt. Welche eigenschaft der anasthetika bedingt ihre narkotische wirkung. *Arch. Exp. Pathol. Pharmak.* 1989; 42: 109–119.

32. Overton E. Studien uber die narkose. Jena, Verlag von Guston Fischer, 1901.
33. Sedensky MM, Meneely PM. Genetic analysis of halothane response in *Caenorhabditis elegans. Science* 1987; 236: 952–954.
34. Morgan PG, Sedensky MM, and Meneely PM. Multiple sites of action of volatile anesthetics in *Caenorhabditis elegans. Proc. Nat. Acad. Sci. USA* 1990; 87: 2965–2969.
35. Sedensky MM, Morgan PG. Genetics of response to volatile anesthetics in *Caenorhabditis elegans. Ann. N.Y. Acad. Sci.; Molecular and Cellar Mechanisms of Alcohol and Anesthetics* 1991; 625: 524–531.
36. MacIver MB, Roth SH. Inhalation anaesthetics exhibit pathway-specific and differential actions on hippocampal synaptic responses *in vitro. Br. J. Anaesth.* 1988; 60: 680–691.
37. MacIver MB, Tauck DL, and Kendig JJ. General anaesthetic modification of synaptic facilitation and long-term potentiation in hippocampus. *Br. J. Anaesth.* 1989; 62: 301–310.
38. Emmons SW. *Transposable Elements in the Nematode, Caenorhabditis elegans.* Cold Spring Harbor, NY. Cold Spring Harbor Laboratory (Publ) 1989; 66–72.
39. Lin CY, Shieh JJ. Identification of a de novo point mutation resulting in an infantile form of Pompe's disease. *Biochem. Biophys. Res. Commun.* 1995; 208: 886–893.
40. McFarlane HJ, Soni N. Pompe's disease and anaesthesia. *Anaesthesia* 1986; 41: 1219–1224.
41. Levitt, RC et al. Evidence for genetic heterogeneity in malignant hyperthermia susceptibility. *Genomics* 1991; 11: 543–547.
42. LaBonne SG, Sumantha I, and Mahowald AP. Molecular genetic of *pecanex*, a maternal effect locus of *Drosophila Melanogaster* that potentially encodes a large transmembrane region. *Dev. Biol.* 1989; 136: 1–16.
43. Hillier L, Clark N, Dubuque T et al. Genbank Accession #H 16865.
44. Balk SP, Blercher PA, and Terhosrst C. Isolation and characterization of a cDNA and gene coding for a fourth CD1 molecule. *Proc. Nat. Acad. Sci. USA* 1989; 86, 252–256.
45. Reddy P, Jacquier AC, Abovich N, Petersen G, and Rosbash M. The *period* clock locus of *D. melanogaster* codes for a proteoglycan. *Cell* 1986; 46: 53–61.
46. Eckenhoff RG. Subcellular distribution of halothane in anesthetic sensitive mutants of *C. elegans. FASEB J.* 1991; 5: A868.
47. Faurobert E, Chen CK, Hurley JB, and Teng DH. *Drosophila* neurocalcin, a fatty acylated, Ca^{2+}-binding protein that associates with membranes and inhibits *in vitro* phosphorylation of bovine rhodopsin. *J. Biol. Chem.* 1996; 271: 10256–10262.
48. Teng DH, Chen CK, and Hurley JB. A highly conserved homologue of bovine neurocalcin in *Drosophila melanogaster* is a Ca^{2+}-binding protein expressed in neuronal tissues. *J. Biol. Chem.* 1994; 269: 31900–31907.
49. Okazaki K, Obata NH, Inoue S, and Hidaka H. S100 beta is a target protein of neurocalcin delta, an abundant isoform in glial cells. *Biochem. J.* 1995; 306: 551–555.
50. Starich TA, Herman RK, and Shaw JE. Molecular and genetic analysis of *unc-7*, a *Caenorhabditis elegans* gene required for coordinated locomotion. *Genetics* 1993; 133: 527–541.
51. Starich TA, Lee RYN, Panzarella C, Avery L, and Shaw JE. *eat-5* and *unc-7* represent a mutigene family in *Caenorhabditis elegans* involved in cell-cell coupling. *J. Cell Biol.* 1996; 134: 537–548.
52. Barnes TM, Jin Y, Horvitz HR, Ruvkin G, and Hekimi S. The *Caenorhabditis elegans* behavioral gene *unc-24* encodes a novel bipartite protein similar to bothe erythrocyte band 7.2 stomatin and nonspecific lipid transfer protein. *J. Neurochem.* 1996; 67: 46–57.

53. Schofield PR, Darlison MG, Fujita N, et al. Sequence and functional expression of the $GABA_A$ receptor shows a ligand-gated receptor super-family. *Nature* 1987; 328: 221–227.
54. Wilkins AS. *Drosophila melanogaster* in *Genetic Analysis of Animal Development*. New York, John Wiley and Sons. 1986; 78–146.
55. Krishnan KS, Nash HA: A genetic study of the anesthetic response: Mutants of *Drosophila melanogaster* altered in sensitivity to halothane. *Proc. Natl. Acad. Sci. USA* 1990; 87: 8632–8636.
56. Allada R, Nash HA: *Drosophila melanogaster* as a model for study of general anesthesia: The quantitative response to clinical anesthetics and alkanes. *Anesth. Analg.* 1993; 77: 19–26.
57. Papazian DM, Schwartz TL, Tempel BL, Jan YN, and Jan LY. Cloning of genomic and complementary DNA from *Shaker*, a putative potassium channel gene from *Drosophila. Science* 1987: 237; 749–753.
58. Royden CS, Pirrotta V, and Jan LY. The *tko* locus, site of a behavioral mutation in *D. Melanogaster*, codes for a protein homologous to prokaryotic ribosomal protein S12. *Cell* 1987: 51; 165–173.
59. Morgan PG, Sedensky MM. Mutations conferring new patterns of sensitivity to volatile anesthetics in *C. elegans*. *Anesthesiology* 1994; 81: 888–898.
60. Gamo S, Ogaki M, and Nakashima-Tanaka E. Strain differences in minimum anesthetic concentrations in *Drosophila melanogaster. Anesthesiology* 1981; 54: 289–293.
61. Gamo S, Nakashima-Tanaka E, and Ogaki M. Alteration in molecular species of phosphatidylethanolamine between anesthetic resistant and sensitive strains of *Drosophila melanogaster. Life Sci.* 1982; 30: 401–408.
62. Gamo S, Megumi T, and Nakashima-Tanaka E. Sensitivity to ether anesthesia and to gamma-rays in mutagen-sensitive strains of *Drosophila melanogaster. Mut. Res.* 1990; 236: 9–13.
63. Gamo S, Morioka K, Dodo K, Taniguchi F and Tanaka Y. Studies on ether anesthesia mutations in *Drosophila melanogaster. Prog. Anesth. Mech.* 1995; 3 (Special Issue): 120–131.
64. Campbell DB, Nash HA. Use of *Drosophila* mutants to distinguish among volatile general anesthetics. *Proc. Natl. Acad. Sci. USA* 1994; 91: 2135–2139.
65. Leibovitch BA, Campbell DB, Krishnan KS and Nash NA. Mutations that affect ion channels change the sensitivity of *Drosophila melanogaster* to volatile anesthetics. *J. Neurogenet.* 1995; 10: 1–13.
66. Dapkus D, Ramirez S, and Murray MJ. Halothane resistance in *Drosophila melanogaster*: Development of a model and gene localization techniques. *Anesth. Analg.* 1996; 83: 147–155.
67. Plomin R, DeFries JC, and McClearn GE. *Behavioral Genetics: A Primer.* San Francisco, CA. Freeman 1980.
68. Hill RE, Jones PF, Rees AR et al. A new family of mouse homeo box-containing genes: Molecular structure, chromosomal localization and developmental expression of Hox-7.1. *Genes Dev.* 1989; 3: 26–37.
69. Koblin DD, Dong DE, Deady JE, and Eger EI II: Selective breeding alters murine resistance to nitrous oxide without alteration in synaptic membrane lipid composition. *Anesth.* 1980; 52: 401–407.
70. Phillips TJ, Feller DJ, and Crabbe JC. Selected mouse lines, alcohol and behavior. *Experientia* 1989; 45: 805–826.
71. Harris RA, Allan AM. Alcohol intoxication: Ion channels and genetics. *FASEB J.* 1989; 3: 1689–1695.

72. Eleftheriou BE, Elias PK. Recombinant inbred strains: A novel approach for psychopharmacogeneticists. Eleftheriou, BE, ed. *Psychopharmacogenetics*. New York, Plenum 1975.
73. McClearn GE, Kakihana R. Selective breeding for ethanol sensitivity in mice. *Behav. Genet.* 1973; 3: 409–410.
74. Marley RJ, Miner LL, Wehner JM, and Collins AC. Differential effects of central nervous system depressants in long-sleep and short-sleep mice. *J. Pharm. Exp. Ther.* 1986; 238: 1028–1033.
75. Goldstein B, Kakihana R. Alcohol withdrawal reactions in mouse strains selectively bred for long and short sleep times. *Life Sci.* 1975; 17: 981–986.
76. Allan AM, Harris RA. Gamma-aminobutyric acid and alcohol actions: Neurochemical studies of long sleep and short sleep mice. *Life Sci.* 1986; 39: 2005–2015.
77. Defries JC, Wilson JR, Erwin VG, and Petersen DR. LS X SS recombinant inbred strains of mice: Initial characterization. *Alcoholism: Clin. Exp. Res.* 1989; 13(2): 196–200.
78. Allan AM, Gallagher EJ, Gionet SE, and Harris RA. Genetic selection for benzodiazepine ataxia produces functional changes in the gamma-aminobutyric acid receptor chloride channel complex. *Brain Res.* 1988; 452: 118–126.
79. Baker RC, Smolen T, Smolen A, and Deitrich RA. Relationship between acute ethanol related responses in long sleep and short sleep mice. *Alcoholism: Clin. Exp. Res.* 1987; 11: 574–578.
80. Suzdak PD, Schwartz RD, and Paul SM. Ethanol stimulates gamma-aminobutyric acid receptor mediated chloride transport in rat brain synaptoneurosomes. *Proc. Natl. Acad. Sci. USA* 1986; 83: 4071–4075.
81. Allan AM, Spuhler KP, and Harris RA. Gamma-aminobutyric acid-activated chloride channels: relationship to genetic differences in ethanol sensitivity. *J. Pharmacol. Exp. Ther.* 1988; 244: 866–870.
82. Yu O, Ito M, Chiu TH, and Rosenberg HC. GABA-gated chloride ion influx and receptor binding studies in C57BL6J and DBA2J mice. *Brain Res.* 1986; 399: 374–378.
83. Harris RA, Allan AM. Neurochemistry of brain chloride channels: Genetic variation in modulation by GABA agonists, alcohol and benzodiazepines. In Biggio, G, Costa E, eds. *Chloride Channels and Their Modulation by Neurotransmitters and Drugs*. NY: Raven 1988; 189–198.
84. Dudek BC, Abbott ME. A biometrical genetic analysis of ethanol response in selectively bred long-sleep and short-sleep mice. *Behav. Genet.* 1984; 14: 1–19.
85. Erwin VG, Heston WDW, and McClearn GE. Effect of hypnotics on mice genetically selected for sensitivity to ethanol. *Pharm. Biochem. Behav.* 1976; 4: 679–683.
86. Baker R, Melchior C, and Deitrich R. The effect of halothane on mice selectively bred for differential sensitivity to alcohol. *Pharm. Biochem. Behav.* 1980; 23(5): 691–695.
87. Koblin DD, Deady JE. Anaesthetic requirement in mice selectively bred for differences in ethanol sensitivity. *Br. J. Anaesth.* 1981; 53: 5–10.
88. Deitrich RA, Draski LJ, and Baker RC. Effect of pentobarbital and gaseous anesthetics on rats selectively bred for ethanol sensitivity. *Pharm. Biochem. Behav.* 1994; 47(3): 721–725.

5 Structure–Activity Relationships of Inhaled Anesthetics

Donald D. Koblin

CONTENTS

0-8493-8555-5/01/$0.00+$.50

5.1 INTRODUCTION — A DEFINITION OF ANESTHESIA

From the title of this chapter, the reader is expected to understand what is meant by the "activity" of an inhaled "anesthetic," yet the terms activity and anesthesia are difficult to define. When an anesthetic is described as being "active," it is implied that varying degrees of activity may exist and that anesthetics can also be "inactive." One situation in which inhaled anesthetics are considered to be inactive is when they are administered at relatively low concentrations (partial pressures) such that they have no obvious behavioral or physiologic effects. An example of an inactive agent might be breathing 0.79 atm nitrogen in room air at atmospheric pressure. However, when air is breathed at hyperbaric pressures greater than 3 atm, the presence of nitrogen in air produces behavioral changes in humans that include euphoria, a slowing of mental activity, and impaired neuromuscular coordination.[1] Should nitrogen be labeled an anesthetic, and if so, at what partial pressure does nitrogen change from being inactive to active?

The dictionary[2] provides a starting point for the definition of general anesthesia: "loss of ability to perceive pain associated with loss of consciousness." However, the loss of ability to perceive pain (analgesia) is considered by some investigators as a desirable supplement to and not a component of general anesthesia,[3,4] and others would include in the definition of clinical anesthesia the capacity to provide immobility in response to surgical incision (in the absence of muscle relaxants) and the production of amnesia.[3,4]

Also lacking from the dictionary definition of general anesthesia is the important component of reversibility. From a clinical viewpoint, it would be a worried group of patients and anesthesiologists if the loss of consciousness produced by inhaled agents was not predictably reversible. From a research viewpoint, the demonstration of reversibility is necessary to distinguish the toxic side effects of experimental agents from their anesthetic properties.

From the above discussion, it is clear that complete agreement does not exist as to what constitutes general anesthesia. Nevertheless, properties associated with the administration of inhaled anesthetics, such as immobility, amnesia, and analgesia can be measured and quantitated as a function of the partial pressure of the anesthetic. The analysis of structure–activity relationships may depend on which behavioral end point is used to measure anesthetic activity. In the present chapter, the discussion of activity will be limited to the potency of inhaled anesthetics in whole organisms.

5.2 MEASUREMENT OF ANESTHETIC ACTIVITY

5.2.1 Determination in Animals

5.2.1.1 MAC (Minimum Alveolar Concentration)

Most measurements of inhaled anesthetic potencies involve the abolishment of movement (either induced or reflexive/spontaneous) as an anesthetic end point. The most often employed measurement of anesthetic potency is MAC, the minimum alveolar concentration, (at 1 atm) of an agent that produces immobility in 50% of animals administered a supramaximal noxious stimulus.[5] In mammals, the noxious stimulus is usually produced by clamping the tail or by passing electrical current through subcutaneous electrodes. The standard is to apply the stimulus for 60 s if no movement occurs (negative response). A "gross purposeful muscular movement," typically of the head or extremities, occuring within the 60-s stimulation period constitutes a positive response.[5] Tail clamp and electrical stimulation give equivalent values for MAC provided care is taken to apply a voltage that is high enough to give a supramaximal electrical stimulus, and yet a voltage low enough to avoid tissue damage and desensitization.[6] For a given inhaled anesthetic, MAC is approximately constant across species, although differences of 30 to 50% in anesthetic potencies within species are not unusual[5,7] (Table 5.1).

An advantage of using the alveolar concentration (estimated by sampling the end-tidal [i.e., end-expiratory] gas) as a measure of anesthetic activity is that this concentration directly represents the partial pressure of anesthetic achieved in the central nervous system and (provided adequate time is allowed for equilibration) is independent of the uptake and distribution of the agent to other tissues.[8] Often, anesthetic potencies are determined (especially in small animals) using inspired rather than alveolar concentrations. However, it must be recognized that this overestimates the "true" MAC and applies best to rapidly equilibrating (poorly blood-soluble) agents. In rats, even after a 3-h equilibration period, inspired concentrations were 9% greater for halothane and 3% greater for isoflurane than alveolar (end-tidal) concentrations.[9]

The standard deviations associated with MAC determinations are typically on the order of 10 to 20% of the mean value (see Table 5.1). Thus, it may be difficult to discern differences in MAC values that are separated by approximately 20% (e.g., MAC differences among strains of the same species[10]). Nearly twofold differences in MAC in the same species can occur over the extremes of age[11] or over a 10°C change in body temperature for a potent volatile agent.[12,13]

From the above discussion, it is clear that several parameters need to be controlled for accurate MAC determinations. The species and age of the animal need to be known. Temperatures should be monitored and maintained at normal body

TABLE 5.1
Anesthetic Requirements (Mean ± SD) for an Alkane (Halothane, $CF_3CBrClH$) and an Ether (Isoflurane, $CF_2HOCClHCF_3$) in Various Adult Species as Meaured by Minimum Alveolar Concentration (MAC)

Species	Halothane MAC (% atm)	Isoflurane MAC (% atm)	Ref.
Human	0.74 ± 0.23	1.23 ± 0.23	87, 91
Monkey	1.15 ± 0.20	1.28 ± 0.18	85
Pig	0.90 ± 0.15	1.49 ± 0.09	86
Horse	0.88 ± 0.09	1.31 ± 0.21	94
Sheep	0.97 ± 0.09	1.58 ± 0.16	93
Goat	1.3 ± 0.1	1.5 ± 0.3	92
Dog	0.87 ± 0.12	1.41 ± 0.18	46, 78
Cat	1.19 ± 0.15	1.61 ± 0.10	79
Rabbit	1.39 ± 0.23	2.05 ± 0.18	79
Ferret	1.01 ± 0.10	1.52 ± 0.10	89
Guinea pig	1.01 ± 0.08	1.15 ± 0.13	90
Rat	1.03 ± 0.04	1.46 ± 0.06	9
Mouse	0.95 ± 0.07	1.34 ± 0.10	22
Crane	ND	1.34 ± 0.14	88

levels. Equilibration with the test anesthetic needs to be assured, and end-tidal or inspired concentrations need to be measured by quantitative techniques (e.g., gas chromatography). The noxious stimulus provided for testing must not produce tissue injury to the extent that the animal is desensitized to the stimulus. Adequate oxygenation needs to be provided and documented, especially with the use of closed-circuit systems. Excess carbon dioxide accumulation in the anesthetic circuit needs to be avoided either by performing the experiments at a high flow rate or by circulating gases in a closed anesthetic system through a carbon dioxide absorbant (e.g., soda lime). Finally, reversibility of anesthesia must be demonstrated.

5.2.1.1.1 Central Nervous System Determinants

Traditionally, most investigators have assumed that surgical anesthesia results from an interaction of anesthetics with certain brain structures. Although this is probably true for the production of amnesia, recent experiments indicate that for the end point measured by MAC (i.e., motor response to noxious peripheral stimulation), the site of anesthetic action resides in the spinal cord, not the brain.

Isoflurane MAC in rats is not altered by decerebration (including removal of thalamus and hippocampus).[14] Furthermore, high thoracic spinal cord transection and functional separation of the rat brain from the spinal cord does not alter isoflurane MAC.[15] These two studies in rats suggest the spinal cord as the site of anesthetic inhibition of motor response.

The goat, which has a cerebral blood supply that allows for the preferential anesthetization of the intact brain, provides an additional model for examining the relative importance of anesthetics in brain versus spinal cord.[16] When isoflurane is isolated to the head and brain, MAC in the goat is more than twice the value found when anesthetic is administered via the native circulation.[16] This finding in a second species confirms the importance of extracranial structures in anesthesia as defined by MAC, although it remains possible that the brain modulates MAC by neurotransmission through descending pathways to the spinal cord.[17]

5.2.1.2 Righting Reflex (Rolling Response)

With small animals (e.g., mice), a commonly used end point of anesthetic requirement is the righting reflex ED_{50}, the anesthetic concentration that abolishes the righting reflex in 50% of animals. For inhaled anesthetics, animals are typically placed in individual cages that can be rotated in a given gaseous environment and are defined to be anesthetized when they meet specific criteria (e.g., mice rolling over twice during five complete turns of a rotator set at 4 rpm[18]). (Strictly speaking, a righting reflex [animal regaining an "all fours" position within a given time once rocked off its feet] may be distinguished from a rolling response tested with a rotating cage. For present purposes, the term "righting reflex" typically refers to assessment of righting ability using a rotating cage.) This method is most applicable to rapidly equilibrating (poorly blood-soluble) agents because only inspired concentrations can be measured as the animals are too small to obtain alveolar samples. Only with adequate equilibration can it be assumed that the partial pressure of the inspired gas is representative of the anesthetic partial pressure at the site of action. The use of small animals and measurement of inspired concentrations facilitates work with agents at extreme environmental conditions (e.g., high pressures). Righting reflex ED_{50} values for inhaled anesthetics can also be quantitated in nonmammalian species in aqueous environments (e.g., newts[19] and tadpoles[20]).

For any given inhaled anesthetic, the righting reflex ED_{50} is less than MAC. For example, righting reflex ED_{50} values in mice for isoflurane and enflurane are approximately 0.7 and 1.2% atm, respectively, values that are about one half the MACs for these anesthetics[21-23] (Table 5.1).

5.2.1.3 Anesthetic Potencies in Very Small Organisms

Investigators of anesthetic mechanisms have employed small, nonmammalian, organisms because of their relative ease in genetic manipulations and well-defined nervous systems. Anesthetic requirements in these organisms have been estimated using a variety of end points, including withdrawal response of snails to gentle mechanical stimulation;[24] swimming activity in the freshwater shrimp initiated by an electrical stimulus;[25] movement of the water flea to a strong light;[26] loss of locomotion in nematodes[27] (Chapter 4), and for the fruit fly *Drosophila melanogaster,* an ability to respond after stimulation with a paint brush[28] or disruption of negative geotaxis (the normal ability of flies to climb upward).[29] The meaningfulness of these various anesthetic end points in small organisms to surgical anesthesia in humans remains to be established. It seems reasonable to assume, however, that for the model to have clinical relevance, the end points in these small organisms should be affected by anesthetic concentrations that are similar to MAC in humans.

5.2.2 Determination in Humans

5.2.2.1 Surgical Incision and Other End Points

Determination of MAC in humans usually involves observation of movement versus no movement in response to surgical incision. These patient studies are more difficult to control than measurements in animals. The patients cannot be taking other drugs and cannot receive preoperative medications or intravenous drugs used for "smoothing" anesthetic induction because such adjuvants may alter the anesthetic requirement (e.g., narcotics can decrease isoflurane MAC by ~50 to 90%[30]). For MAC studies in humans the surgeon must also delay incision until adequate equilibration has been achieved, as documented by the measurement of end-tidal gas samples. Alternatively, MAC can be measured in volunteers using electrical stimulation to the ulnar nerve,[31] but this may provide less than a supramaximal stimulus.[32]

Aside from the six inhaled agents (nitrous oxide, halothane, enflurane, isoflurane, desflurane, and sevoflurane) in current clinical use, MAC in humans has been determined for only six additional anesthetics (Table 5.2). MACs of these agents in humans vary ~500-fold, from a low of 0.16% atm for methoxyflurane to a high of 1.04 atm for nitrous oxide (measured in volunteers in a hyperbaric chamber) (Table 5.2).

Other clinical end points besides response to skin incision have been utilized as measures of anesthetic potency. The partial pressure of isoflurane required to supress

TABLE 5.2
Minimum Alveolar Concentrations (MACs) in Humans

Anesthetic	MAC[a] (% atm)	Chemical Structure	Ref.
Nitrous oxide	104	N_2O	31
Halothane	0.74	$CF_3CBrClH$	87
Enflurane	1.68	CF_2HOCF_2CClFH	5
Isoflurane	1.23	$CF_2HOCClHCF_3$	91
Desflurane	7.25	$CF_2HOCFHCF_3$	32
Sevoflurane	2.05	$CFH_2OCH(CF_3)_2$	95
Methoxyflurane	0.16	$CH_3OCF_2CCl_2H$	5
Cyclopropane	9.2	$c(C_3H_6)$	5
Fluroxene	3.4	$CF_3CH_2OCH{=}CH_2$	5
Ethylene	67	$CH_2{=}CH_2$	5
Diethylether	1.92	$CH_3CH_2OCH_2CH_3$	5
Xenon	71	Xe	97

[a] Typical values for adults, ~30 to 40 years old.

eye opening to verbal command is less than (~40% of) isoflurane MAC, whereas the isoflurane concentration needed to prevent movement in response to tracheal intubation is greater than (~150% of) isoflurane MAC for skin incision.[33]

5.2.2.2 Amnesia

A major clinical goal with the use of inhaled anesthetics is to make the patient oblivious to the surgical procedure. This goal is often not achieved with patients requiring emergency surgery for trauma when only limited amounts of inhaled anesthetics can be administered because of hemodynamic instability, and there is a high incidence (up to 43%) of recall of surgery.[34] The abilities of inhaled agents to produce amnesia can be tested under more controlled conditions in volunteers.[35] Learning tasks used as a measure of memory can be quantitated and are disrupted at inhaled anesthetic concentrations that are lower than those of MAC required to prevent response to surgical incision (e.g., ~0.3 MAC for isoflurane and desflurane and ~0.6 MAC for nitrous oxide).[35,36]

5.3 STRUCTURAL VARIATIONS — IMPLICATIONS FOR LIMITED VERSUS MULTIPLE SITES AND MECHANISMS OF ANESTHETIC ACTION

The chemical structures (see Table 5.2) and physical properties of inhaled agents that produce anesthesia vary considerably. They range from an inert element (xenon), to explosive/flammable compounds (cyclopropane, diethylether), to compounds that undergo extensive biotransformation (e.g., methoxyflurane metabolism results in nephrotoxic levels of fluoride ion).

Given the structural diversity of the inhaled anesthetics, it is difficult to imagine how all of these agents could act on a single site via a single mechanism of action. Exactly how many sites/mechanisms might be involved is an ongoing debate.[37,38] Correlations of anesthetic activities (e.g., MAC) with physicochemical properties of anesthetics have been used to search for common sites and mechanisms of general anesthetic action (Chapter 1). One approach in trying to relate the activity of an anesthetic to its structure and physical properties is to examine anesthetic activities (potencies) in a homologous series of compounds (e.g., n-alkanes). Because of safety and practical limitations, only a limited number of compounds can be administered to humans (Table 5.2), and thus most structure–activity relationships in a homologous series of anesthetics need to be quantitated in animals.

5.4 GASES — INERT AND OTHERWISE

Many beginning investigators are simultaneously excited and confused upon first learning that xenon is an excellent anesthetic, because it is difficult to conceive how such an "inert" element could produce such a profound physiologic effect. Indeed, this phenomenon of "inert gas anesthesia" has even stimulated (and baffled) Nobel Prize–winning chemists.[39,40] Of the noble gases, the most potent anesthetic is xenon, and potency progressively decreases with decreasing atomic weight (Table 5.3). Helium and neon have no detectable anesthetic effect at high pressures (~100 atm), and administration of these high pressures may actually antagonize the effects of conventional inhaled anesthetics and initiate convulsions (Chapter 4). Nevertheless, examinations of differences in anesthetic depth in tadpoles between hydrostatic versus helium and neon pressures are consistent with small inherent anesthetic properties of these noble gases.[41]

Nitrous oxide potency (as determined by MAC) varies more among species than most other anesthetics, and more than a twofold difference in nitrous oxide requirement exists between humans and rodents[21,31,42-44] (see Table 5.3). Although MACs in humans for gases other than nitrous oxide and xenon have not been measured, sulfur hexafluoride and nitrogen impair psychomotor, perceptual, and cognitive abilities in humans at relative potencies that are similar to the anesthetic potencies in animals.[45]

TABLE 5.3
Potencies of Anesthetic Gases (Noncarbon Containing) in Mammals

Gas	Potency (atm)	Anesthetic End Point	Species	Ref.
Xenon (Xe)	0.71	MAC[a]	Human	97
	0.95	MAC	Monkey	98
	1.19	MAC	Dog	46
	0.95	Righting reflex[b]	Mouse	48
Krypton (Kr)	4.5	Righting reflex	Mouse	48
Argon (Ar)	15.2	Righting reflex	Mouse	48
Neon (Ne)	>110	Righting reflex	Mouse	96
Helium (He)	>120	Righting reflex	Mouse	96
Nitrous oxide (N_2O)	1.04	MAC	Human	31
	1.88	MAC	Dog	46
	1.55–2.35	MAC	Rat	42–44
	>2.75	MAC	Mouse	21
	1.5	Righting reflex	Mouse	18, 48
Sulfur hexafluoride (SF_6)	4.9	MAC	Dog	47
	14.6	MAC	Rat	[c]
	6.9	Righting reflex	Mouse	96
Nitrogen (N_2)	>43	MAC	Dog	47
	32.5	Righting reflex	Mouse	96
Hydrogen (H_2)	~200[d]	Righting reflex	Mouse	99

[a] Minimum alveolar concentration.
[b] Righting reflex is equated here to the rolling response, typically measured by placing animals in a rotating cage.
[c] Unpublished; value for SF_6 (14.6 ± 1.3; n = 10, ± SD) based on additivity experiments with desflurane.
[d] Estimated by additivity studies with nitrous oxide.

Because noble gases may constitute the centers of crystallized structures of water molecules (hydrates), it was once proposed that hydrates were important in the production of anesthesia.[39,40] However, such crystal water formations cannot explain the anesthetic properties of all of these gases.[46-48] In general, these gases do obey the Meyer–Overton hypothesis (i.e., the lipid solubilities of these gases correlate with their anesthetic potencies) (Chapter 1).[45-48]

5.5 ALKANES

5.5.1 HYDROCARBONS

5.5.1.1 Normal Alkanes

The normal alkanes constitute a series of compounds for which simple and progressive changes in structure (i.e., increase in carbon chain length) can be related to anesthetic potencies. Early (in the 1920s) qualitative observations defined anesthetic properties of the alkanes methane (CH_4) through octane [$CH_3(CH_2)_6CH_3$], and a tendency was found for increasing chain length to be associated with increased anesthetic potency.[49] In contrast, in 1971 Mullins[50] reported that n-decane [$CH_3(CH_2)_8CH_3$] had no anesthetic effect. This observation was of considerable theoretical importance because it opposes the "lipid theories" of anesthesia (Chapter 1) (as decane is a very lipid-soluble agent), and led to a proposal in which a critical "molecular size" was deemed important in anesthetic action.[50] However, these initial studies with n-decane did not include details concerning the determination of anesthetic requirement nor the difficult measurement of n-decane concentrations.

Recent experiments have reexamined the anesthetic properties of ten consecutive n-alkanes (methane through decane) in rats by measuring the inspired concentrations of n-alkanes required to abolish response to electrical stimulation of the tail.[49,51] Eight compounds (methane through octane) provided anesthesia when administered alone at partial pressures ranging from 9.9 to 0.017 atm.[49] Neither nonane nor decane were anesthetic when administered alone, but they did have an anesthetic effect as demonstrated by an ability to lower isoflurane MAC, and the anesthetizing partial pressures of these compounds were estimated from additivity studies to be 0.0113 and 0.0142 atm, respectively.[49] With the exception of n-decane, anesthetic potencies increased with increasing chain length.[49,51] However, because of the considerable tissue solubilities of the long-chain alkanes there is a large difference in the inspired-to-arterial blood partial pressures, and the inspired partial pressures overestimate the alkane partial pressures at the central nervous system site of action.[52] When end-tidal samples are obtained in tracheotomized rats, the anesthetic requirement (MAC) for n-decane is 0.0024 atm, or about one sixth of the anesthetizing inspired partial pressure.[52]

Potencies of a series of n-alkanes have also been determined in the fruit fly using geotactic behavior as the anesthetic end point.[29] As was found in the studies with rats,[49,51,52] alkane anesthetic potency in the fruit fly increased with increasing chain length but the increase in potency was less than that predicted by the Meyer–Overton rule.[29] Anesthetic effects were also found for undecane [$CH_3(CH_2)_9CH_3$] and dodecane [$CH_3(CH_2)_{10}CH_3$] but were small, variable, and slow to develop.[29]

5.5.1.1.1 "Cut-Off Effect"

The "cut-off effect" describes the phenomenon where there exists in a homologous series of compounds (e.g., the n-alkanes) a progressive increase in anesthetic potency with successive homologues in the series (e.g., sequential addition of methylene groups for the n-alkanes) until a point is reached where there is loss of anesthetic activity. There are two possible reasons for this cut-off phenomenon for inhaled agents.[53] The first possibility relates to physicochemical properties of the inhaled agent that may limit its delivery. This may occur either because of a limitation of volatility (low vapor pressure), a limited aqueous solubility, or conversely because the inhaled compound is so soluble in blood that an unduly long equilibration time would be required for adequate amounts of the inhaled agent to reach its site of anesthetic action (central nervous system).[29,49,52-54] The second possibility relates to efficacy. It may be that the compound is able to reach the anesthetic site of action but is unable to induce the perturbation required to produce anesthesia.[53] For n-decane, the lack of anesthetic potency when administered alone is at least in part related to its limited delivery to the site of action, because of both its low vapor pressure and its high solubility in blood.[29,49,52]

5.5.1.2 Cycloalkanes

Cyclic hydrocarbons are more potent anesthetics than their n-alkane analogs of equal carbon numbers. For example, the MAC of cyclopropane in rats (~0.2 atm)[5] is about one fifth the MAC for n-propane (0.94 atm),[49] and the MAC of cyclopentane (0.053 atm)[55] is less than one half the MAC for n-pentane (0.127 atm).[49] As with the n-alkanes,[49] the anesthetic potencies of the cycloalkanes tend to increase with increasing carbon chain length until a cut-off phenomenon is observed, with cyclooctane having no anesthetic effect in the rat[55] (although cyclooctane did produce anesthesia in the fruit fly as assessed by geotactic behavior[29]).

5.5.1.3 Unsaturated Compounds

Hydrocarbons containing double bonds appear to have a relatively greater anesthetic potency, although only limited information is available. For example, ethylene MAC is 0.67 atm[5] in humans and 1.32 atm[5] in rats, whereas the MAC for ethane in rats is 1.59 atm.[49] Benzene (MAC = 0.0101 atm in rats) is approximately four times as potent as cyclohexane (MAC = 0.042 atm).[55]

5.5.2 Halogenated Alkanes

The unsuitability of inhaled hydrocarbons for clinical anesthesia provided the impetus to search for hydrocarbon alkane derivatives that might be more clinically useful. Although cyclopropane and ethylene were at one time routinely administered to patients and did have some favorable properties (e.g., rapid induction of and emergence from anesthesia), there were distinct disadvantages (e.g., explosiveness/flammability, cardiac arrhythmias). The approach taken to find a safer and more stable inhaled anesthetic was to develop fluorinated compounds, because it was known that the strong chemical bond between fluorine and carbon was nonreactive. In particular, the knowledge that the CF_3 moiety not only was very stable itself but also reduced the reactivity of halogens on an adjacent carbon atom led to the development of halothane.[56]

5.5.2.1 Partially Fluorinated Alkanes

The influence of hydrogenation versus fluorination on anesthetic potency (MAC) was systematically examined for methanes, ethanes, propanes, and butanes in rats.[57] For the methanes, CF_2H_2 is the most potent, with a MAC of 0.72 atm (the MAC of CH_4 [9.9 atm] being approximately tenfold greater). Of the ethanes, CF_2HCF_2H was the most potent (MAC = 0.115 atm). The most potent propanes were $CF_3CFHCFH_2$ (MAC = 0.115 atm) and $CF_2HCF_2CF_2H$ (MAC = 0.146 atm), and $CF_2H(CFH)_2CF_2H$ was the most potent butane.[57] Thus, for the partially fluorinated ethanes, propanes, and butanes, there is a tendency to maximize anesthetic potency when the number of hydrogen atoms equals the number of carbon atoms and when the hydrogen atoms are distributed among the various carbon atoms of the molecules. Most partially fluorinated pentanes, hexanes, and heptanes did not produce anesthesia when administered alone at their vapor pressures.[57]

5.5.2.2 Perfluorinated Alkanes

Anesthetic properties of a series of completely fluorinated (perfluorinated) alkanes from perfluoromethane (CF_4) to perfluorohexane [$CF_3(CF_2)_4CF_3$] were assessed by MAC determinations in rats.[58] Of the perfluoroalkanes, only CF_4 had anesthetic properties when administered alone and its MAC was only slightly above the lethal pressure.[58] By performing additivity studies with a conventional inhaled anesthetic (desflurane), the MAC of CF_4 was estimated at ~66 atm. In contrast, none of the other alkanes [pefluoroethane through perfluorohexane, as well as $CF(CF_3)_3$], when administered at partial pressures near their vapor pressures, produced anesthesia when administered alone nor did any of these agents lower the anesthetic requirements (MACs) for desflurane, isoflurane, or halothane.[58]

The finding that perfluoroethane and perfluoropropane are "nonanesthetic" in the rat[58] differs from previous results in mice where the potencies of these compounds were estimated by righting reflex measurements to be ~18 atm.[48] The reasons for these differences are not entirely clear, although as was noted above, anesthetic requirements as assessed by righting reflex are lower than those determined by noxious stimulation (MAC) measurements. In addition, perfluoroalkanes at high pressures cause respiratory depression, which may be associated with arterial carbon dioxide elevation and relative hypoxia. The sustained coordinated activity of the righting reflex in mice (having a greater metabolic rate than rats) may be more susceptible to carbon dioxide elevation and hypoxia than the brief and minimally coordinated movement required to respond to a noxious stimulus.[58]

5.5.2.2.1 Definition of "Nonanesthetics"

The perfluoroalkanes exhibit a cut-off effect in potency, with perfluoroethane and higher derivatives not producing anesthesia (as defined by MAC) when administered alone at near their vapor pressures.[58] However, the reason for the lack of anesthetic potency of the perfluoroalkanes differs from the reason for the cut-off effect of n-decane[49] described above. Although n-decane does not produce anesthesia when administered alone, n-decane does lower MAC for conventional anesthetics, demonstrating that n-decane does have anesthetic properties.[49] In contrast, the perfluoroalkanes perfluoroethane through perfluorohexane do not lower MAC for conventional anesthetics even when administered at partial pressures near their vapor pressure, partial pressures high enough so that they should have an anesthetic effect as predicted by the Meyer–Overton hypothesis.[58] The lack of anesthetic properties of these perfluoroalkanes is not simply related to pharmacokinetics, because compounds such as perfluoropropane and perfluoropentane do quickly reach the brain, and the arterial and brain partial pressures of these "nonanesthetic" perfluoroalkanes rapidly equilibrate with the inspired partial pressures of these agents.[59]

Thus, for a compound to be called a "nonanesthetic,"[55,58-60] it must (1) not produce anesthesia when administered alone (as defined by MAC); (2) not decrease anesthetic requirement (MAC) of conventional anesthetics; (3) be soluble in blood and tissues and capable of reaching and equilibrating with the central nervous system; and (4) be able to be administered at adequate partial pressures, such that the inhaled partial pressures are sufficient to have an anesthetic effect as predicted by the solubility of the agent in oil (lipid) according to the Meyer–Overton hypothesis. Because MAC is used as the anesthetic end point in this definition, some may argue that it might be more appropriate to call these compounds "nonimmobilizers" rather than "nonanesthetics." It is possible that such nonanesthetics fail to produce immobility in response to a noxious stimulus but do cause amnesia.[61] It is also possible that compounds labeled as "nonanesthetic" have anesthetic properties but that the effects are too small to be measured given the standard deviations associated with the MAC measurements.

Examples of nonanesthetics, their homologues with anesthetic properties, and solubilities in saline and olive oil are given in Table 5.4. Of note is the observation that some of the nonanesthetics have considerable solubility in oil and provide marked exceptions to the Meyer–Overton hypothesis (Chapter 1). Nonanesthetics tend to have relatively low saline/gas partition coefficients, but there is no distinct value of saline/gas partition coefficient that separates nonanesthetics from their homologues with anesthetic properties (Table 5.4).

5.5.2.3 Chlorine and Bromine Substitutions

Initial screening studies performed in mice (using the righting reflex under nonequilibrium conditions) demonstrated that the substitution of a chlorine or bromine into a fluorohydrocarbon resulted in a more potent anesthetic, and that bromine was several times more potent than chlorine in enhancing anesthetic potency.[62,63] For example, the righting reflex ED_{50} values for CF_3CClH_2, CF_3CCl_2H, CF_3CBrH_2, and CF_3CBr_2H, were 8.0, 2.7, 2.8, and 0.4% atm, respectively.[62] However, it was also recognized that increased substitution of hydrogen atoms by either fluorine, chlorine, or bromine atoms was associated with excitory (convulsive) properties and that completely halogenated alkanes were poor anesthetics.[63] Indeed, most of the nonanesthetic alkanes listed in Table 5.4 are completely halogenated, and the

TABLE 5.4
Examples of Inhaled Nonanesthetic Alkanes and Their Homologues with Anesthetic Properties

Nonanesthetics	Saline/Gas Part. Coeff.[a] (37°C)	Oil/Gas Part. Coeff.[a] (37°C)	Compounds with Anesthetic Properties	Saline/Gas Part. Coeff. (37°C)	Oil/Gas Part. Coeff. (37°C)	Rat MAC[a] (atm)	Ref.
CF_3CF_3	0.00135	0.146	CF_2HCF_3	0.055	1.52	1.51	57, 58
			$CClF_2CF_3$	0.0027	1.03	7.8	60
			$CBrF_2CF_3$	0.0077	3.48	2.2	60
$CF_3CF_2CF_3$	0.000674	0.208	CF_3CFHCF_3	0.0215	2.77	0.95	57, 58
$CF_3CClFCF_3$	0.0012	2.08	$CF_3CBrFCF_3$	0.0028	6.71	3.7	60
$CF_3CF_2CF_2CF_3$	0.000136	0.437	$CBrF_2CF_2CF_2CF_3$	0.00102	10.4	2.0	57, 100
$CF_3CClFCClFCF_3$	0.0019	25.0	$CF_2HCF_2CF_2CF_2H$	0.158	30.4	0.058	58, 60
$c(CF_2)_4$	0.0016	1.01	$c(CClFCF_2CH_2CH_2)$	1.58	248	0.014	60
$c(CClFCClFCF_2CF_2)$	0.0119	43.5	$c(CClFCClFCF_2)$	0.0646	49.7	0.222	[b], 60
Cyclooctane	0.054	7010	Cycloheptane	0.075	2780	0.014	55
1,3,5-$(CF_3)_3C_6H_3$	0.013	264	C_6F_6	0.40	251	0.0161	55

Note: Part. Coeff., partition coefficient; $c(CF_2)_4$, perfluorocylclobutane; $c(CClFCClFCF_2CF_2)$, 1,2-dichloroperfluorocyclobutane; $c(CClFCF_2CH_2CH_2)$, 1-chloro-1,2,2-trifluorocyclobutane; $c(CClFCClFCF_2)$, 1,2-dichloroperfluorocyclopropane; 1,3,5-$(CF_3)_3C_6H_3$, 1,3,5-tris(trifluoromethyl)benzene; C_6F_6, hexafluorobenzene.

[a] Most compounds listed here did not produce anesthesia when administered alone (but commonly induce excitatory behavior), and the MAC for these agents is estimated by additivity studies with desflurane.

[b] Unpublished results.

completely halogenated alkanes that do have anesthetic properties (defined by their ability to lower desflurane requirement, Table 5.4) have MACs that are higher than those predicted by the Meyer–Overton hypothesis and commonly produce excitatory behavior when administered alone.

Iodinated alkanes have also been synthesized and tested for their anesthetic potencies, but these iodinated agents tend to be chemically unstable and promote cardiac arrhythmias.[63]

5.5.2.3.1 Influence of Deuteration

Substitution of deuterium for hydrogen tends to decrease the reactivity of compounds, and the anesthetic properties of chloroform versus deuterated chloroform[64] and of halothane versus deuterated halothane[65] have been examined in mice using the righting reflex. Deuteration did not influence anesthetic potency, a finding that needs to be addressed in theories of anesthetic action that speculate on the importance of hydrogen bonding (Chapter 1).

5.5.2.4 Halogenated Cycloalkanes

It was recognized early that completely halogenated cycloalkanes were poor anesthetics and that these compounds were often convulsants.[66] Several completely halogenated cyclobutane derivatives have been reexamined recently for their anesthetic properties (MAC measurements in rats) and have been classified as nonanesthetics[60] (Table 5.4). An example is 1,2-dichloroperfluorocyclobutane [c($CClFCClFCF_2CF_2$)], which produces convulsions at about 5.5% atm and does not lower the requirement for conventional anesthetics.[60] Hydrogen substitutions into halogenated cyclobutane derivatives may result in an anesthetic; for example, 1-chloro-1,2,2-trifluorocyclobutane [c($CClFCF_2CH_2CH_2$)] produces anesthesia when administered alone and is slightly less potent than halothane[60] (Table 5.4). However, many cyclic halogenated compounds are unstable and difficult to study. For instance, 2,2,3-trichloro-3,4,4-trifluorocyclobutane [c($CH_2CCl_2CClFCF_2$)] reacts with soda lime in the anesthesia circuit to produce a toxic compound and 1,2-dibromoperfluorocyclobutane [c($CBrFCBrFCF_2CF_2$)] is lethal at very low concentrations (unpublished observations).

One notable exception among the perhalogenated cyclic compounds is hexafluorobenzene. This compound does not produce convulsions when administered alone and is anesthetic in mice[67] and rats[55] at ~1.6% atm (Table 5.4).

5.6 ETHERS

The introduction of halothane into clinical practice in the 1950s made apparent the advantages of a nonflammable inhaled anesthetic. Nevertheless, halothane was also recognized to be imperfect because of its requirement for additives for stability in storage, its ability to react with soda lime and undergo metabolic breakdown, and the propensity of alkanes to cause cardiac arrhythmias. The search for a superior inhaled agent involved a systematic examination of halogenated ethers.[63,68]

5.6.1 Influence of Carbon Chain Length and Branching

5.6.1.1 Diethyl Ethers

Because diethylether had been in clinical use since the 1840s, it was reasonable to expect that halogenated derivatives of diethylether might provide safer and nonflammable inhaled anesthetics. However, when sufficient halogenation of diethylethers was achieved to limit nonflammability, the halogenated diethylethers were found to be poor anesthetics (as assessed by qualitative screening studies of the righting reflex in mice) and tended to produce convulsive activity.[63] Unsaturated derivatives (vinyl ethers) enhanced anesthetic potency but were also associated with irritation and instability.[63] For example, fluroxene ($CF_3CH_2OCH{=}CH_2$; see Table 5.2) was banned from clinical use because of toxic components produced from metabolic breakdown and its strong emetic properties.

5.6.1.2 Methyl Ethyl Ethers

Examination of a large series of halogenated methyl ethyl ethers in the 1960s and 1970s led to the conclusion that the compounds having the most favorable anesthetic properties contain either (1) one hydrogen with two halogens other than fluorine or (2) two or more hydrogens with at least one bromine or one chlorine.[69,70] This generalization was consistent with the favorable anesthetic properties of isoflurane ($CF_2HOCClHCF_3$) and enflurane (CF_2HOCF_2CClFH), two agents that were developed from these investigations[69,70] and remain in current clinical use. The conclusion also fit methoxyflurane ($CH_3OCF_2CCl_2H$), now banned from clinical practice because of its nephrotoxic effects. However, the generalization from these earlier studies did not predict the favorable anesthetic properties of desflurane ($CF_2HOCFHCF_3$), in clinical use since 1992.

5.6.1.3 Isopropyl Methyl Ethers

Screening studies for the anesthetic properties of a series of halogenated isopropyl methylethylethers demonstrated that compounds containing a chlorine atom on the methyl group were relatively unstable, whereas substitution of chlorine for fluorine or hydrogen on the isopropyl group enhanced anesthetic effect but also resulted in irritating and toxic side effects.[71] Sevoflurane [$CFH_2OCH(CF_3)_2$], containing no chlorine atoms, is the only isopropyl methyl ether in current clinical use.[72]

5.6.1.4 "Cyclic" Ethers

Although certain cyclic ethers might be expected to be potent anesthetics and be reasonably stable, none have been in clinical use and only limited quantitative information is available on anesthetic potencies. Dioxychlorane [4,5-dichloro-2,2-difluoro-1,3-dioxylane, c($OCF_2OCClHCClH$)], which contains two ether linkages in a five-member cyclic ring, is an order of magnitude more potent than isoflurane in dogs and has a MAC of 0.11% atm.[73] Aliflurane [1-chloro-2-methoxy-1,2,3,3-tetrafluorocyclopropane, c($CClFCF_2CF_2$)-O-CH_3)] contains an ether linkage that connects a perhalogenated cyclopropane ring to a methyl group. It has a potency between that of isoflurane and enflurane (MAC = 1.84% atm in dogs).[74]

5.6.2 Influence of Chemical Substitutions

5.6.2.1 Thioethers

Qualitative screening studies in mice showed that thioethers tended to be more potent than their oxygen analogues, but would probably not be clinically useful compounds because of their unpleasant odor, greater toxicity, and limited volatility.[63] The anesthetic requirement (MAC) for thiomethoxyflurane ($CH_3SCF_2CCl_2H$) in dogs was 0.035% atm, about seven times more potent than methoxyflurane ($CH_3OCF_2CCl_2H$).[75] Because thiomethoxyflurane was about seven times more soluble in oil than methoxyflurane, the anesthetic potency of this thioether was consistent with the predicted value from the Meyer–Overton rule.[75]

5.6.2.2 Chlorine and Bromine Substitutions

Initial screening studies in mice revealed that chlorine or bromine substitution into ethers enhances anesthetic potency and that insertion of bromine is more potent than chlorine.[63,69,70] Quantitation of this effect is seen by comparing the MAC values in rats for desflurane ($CF_2HOCFHCF_3$), isoflurane ($CF_2HOCClHCF_3$), and an investigational agent I-537 ($CF_2HOCBrHCF_3$), which only differ by F, Cl, and Br placement at a single molecular position. Replacement of fluorine (desflurane) for chlorine (isoflurane) increases anesthetic potency more than fourfold, and potency is enhanced nearly threefold further by bromine (I-537) replacement of Cl.[76]

5.6.2.3 Deuteration

As with anesthetic alkanes,[64,65] replacement of hydrogen by deuterium does not appear to change the anesthetic requirement of ethers. The MAC for enflurane in dogs (2.3% atm) does not differ from MAC for D-enflurane (2.2% atm).[77]

5.6.3 Isomers

5.6.3.1 Structural

The best-known pair of anesthetic ether structural isomers is isoflurane ($CF_2HOCClHCF_3$) and enflurane (CF_2HOCF_2CClFH), empirical formula $C_3ClF_5H_2O$, because these agents are in routine clinical use. In mammals (e.g., humans,[5] dogs,[78] rabbits,[79] and rats[80]), the MAC for enflurane is consistently greater than (40 to 89% higher) the MAC of isoflurane. Because isoflurane and enflurane have similar solubilities in oil (as well as in octanol and lipids),[80] these two structural isomers represent a minor deviation from the Meyer–Overton rule.

The anesthetic properties of two additional structural isomers of isoflurane and enflurane have been quantitated in dogs. The MAC of chlorofluoromethyl-1,2,2,2-fluoroethyl ethyl ($CClFHOCFHCF_3$) is 2.24% atm and is similar to isoflurane, whereas the MAC of chlorodifluoromethyl-2,2,2-fluoroethyl ethyl ($CClF_2OCH_2CF_3$)

is 12.5% atm and is five to ten times greater than values for the other three isomers.[78] This finding agrees with earlier qualitative screening studies in mice in which it was shown that methylethyl ethers with completely halogenated end-methyl groups tended to be relatively poor anesthetics.[69,70]

5.6.3.2 Optical (Stereoisomers)

Because optical isomers of volatile anesthetics can be isolated only in limited quantities at great expense, most experiments with these agents have involved *in vitro* systems (Chapter 12). However, limited data are available concerning the potencies of isoflurane isomers in whole organisms. In tadpoles, (+) and (–) isomers of isoflurane are equipotent, as evaluated by loss of righting reflex at the anesthetic end point.[81] Mice injected intraperitoneally with liquid agent demonstrated a modest increase in sleep time with the (+) isomer compared to the (–) isomer of isoflurane, but insufficient amounts of the isomers precluded an accurate determination of anesthetic potencies under equilibrium conditions.[82] In the rat, complete MAC determinations have been performed after obtaining adequate quantities of the isoflurane stereoisomers, and the (+) isomer (MAC = 1.06% atm) is 53% more potent than the (–) isomer (MAC = 1.62% atm).[83]

5.6.4 Convulsant Ethers

As noted previously, ethers containing end-methyl groups that are completely halogenated often are poor anesthetics and are commonly associated with convulsive activity.[69,70,78] The prototype of convulsant ethers is flurothyl (hexafluorodiethylether, $CF_3CH_2OCH_2CF_3$), which has been used clinically to produce convulsions and employed as a substitute for electroconvulsive therapy. Also, as noted for the completely halogenated alkanes,[60] certain agents with convulsive activity may also have anesthetic properties as demonstrated by an ability to decrease the anesthetic requirement of a conventional anesthetic. However, for flurothyl (which produces convulsions at ~0.1% atm),[84] end-tidal concentrations of 3 to 4% atm are associated with only a marginal and variable decrease of isoflurane MAC in the dog (a decrease to 81 ± 35% of the background isoflurane MAC).[84] In contrast, Iso-Indoklon [$(CF_3)_2CHOCH_3$], an isopropyl methylether structural isomer of flurothyl, was devoid of convulsant properties and was an excellent anesthetic in dogs.[84]

5.7 CONCLUSION

Knowledge of the structure–activity relationships of inhaled anesthetics provides bases for the theoretical study of anesthetic mechanisms and for the practical development of an ideal clinical anesthetic. The most common method of determining anesthetic potency is by measurement of the MAC, the partial pressure of anesthetic at which 50% of subjects do not respond to a supramaximal noxious stimulus. MAC is an equilibrium measurement that primarily involves the spinal cord as the site of anesthetic inhibition of the motor response. Clinically useful (nonflammable, stable, potent) inhaled alkanes and ethers have two to four carbon atoms and are partially but not completely halogenated. Substitution of chlorine or bromine for fluorine or hydrogen on anesthetic alkanes or ethers enhances anesthetic potency, with bromine being more potent than chlorine. Complete halogenation of an alkane or ether results in a compound that is a poor anesthetic and typically has convulsive properties. "Nonanesthetics" do not produce anesthesia when administered alone and do not decrease the anesthetic requirement of conventional anesthetics, in spite of being soluble in blood and tissues and capable of reaching and equilibrating with the central nervous system. Such nonanesthetics may be useful tools in testing theories and mechanisms of anesthetic action.

REFERENCES

1. Behnke, A.R., Thomson, R.M., and Motley, E.P., The psychologic effects of breathing air at 4 atmospheres pressure, *Am. J. Physiol.*, 112, 554, 1935.
2. *Stedman's Medical Dictionary*, 23rd edition, Williams & Wilkins, Baltimore, 1978.
3. Eger E.I., II, and Kissin, I., What is general anesthetic action?, *Anesth. Analg.*, 77, 408, 1993.
4. Prys-Roberts, C., Anaesthesia: A practical or impractical construct?, *Br. J. Anaesth.*, 59, 1341, 1987.
5. Quasha, A.L., Eger, E.I., II, and Tinker, J.H., Determination and applications of MAC, *Anesthesiology*, 53, 315, 1980.
6. Laster, M.J., Liu, J., Eger, E.I., II, and Taheri, S., Electrical stimulation as a substitute for the tail clamp in the determination of MAC, *Anesth. Analg.*, 76, 1310, 1993.
7. Travis, C.C., and Bowers, J.C., Interspecies scaling of anesthetic potency, *Toxicol. Ind. Health*, 7, 249, 1991.
8. Eger, E.I., II, *Anesthetic Uptake and Action*, Williams & Wilkins, Baltimore, 1974, chaps. 1 and 4.
9. White, P.F., Johnston, R.R., and Eger, E.I., II, Determination of anesthetic requirement in rats, *Anesthesiology*, 40, 52, 1974.
10. Russell, G.B., and Graybeal, J.M., Differences in anesthetic potency between Sprague-Dawley and Long-Evans rats for isoflurane but not nitrous oxide, *Pharmacology*, 50, 162, 1995.
11. Mapleson, W.W., Effect of age on MAC in humans: A meta-analysis, *Br. J. Anaesth.*, 76, 179, 1996.
12. Vitez, T.S., White, P.F., and Eger, E.I., II, Effects of hypothermia on halothane MAC and isoflurane MAC in the rat, *Anesthesiology*, 41, 80, 1974.
13. Franks, N.P., and Lieb, W.R., Temperature dependence of the potency of volatile general anesthetics: Implications for *in vitro* experiments, *Anesthesiology*, 84, 716, 1996.
14. Rampil, I.J., Mason, P., and Singh, H., Anesthetic potency (MAC) is independent of forebrain strucures in the rat, *Anesthesiology*, 78, 707, 1993.
15. Rampil, I.J., Anesthetic potency (MAC) is not altered after hypothermic spinal cord transection in rats, *Anesthesiology*, 80, 606, 1994.
16. Antognini, J.F., and Schwartz, K., Exaggerated anesthetic requirements in the preferentially anesthetized brain, *Anesthesiology*, 79, 1244, 1993.
17. Borges, M., and Antognini, J.F., Does the brain influence somatic responses to noxious stimuli during isoflurane anesthesia?, *Anesthesiology*, 81, 1511, 1994.

18. Koblin, D.D., Dong, D.E., and Eger, E.I., II, Tolerance of mice to nitrous oxide, *J. Pharmacol. Exp. Ther.*, 211, 317, 1979.
19. Miller, K.W., Paton, W.D.M., Smith, R.A., and Smith, E.B., The pressure reversal of general anesthesia and the critical volume hypothesis, *Mol. Pharmacol.*, 9, 131, 1973.
20. Firestone, L.L., Sauter, J.F., Braswell, L.M., and Miller, K.W., Actions of general anesthetics on acetylcholine receptor-rich membranes from *Torpedo californica*, *Anesthesiology*, 64, 694, 1986.
21. Deady, J.E., Koblin, D.D., Eger, E.I., II, Heavner, J.E., and D'Aoust, B., Anesthetic potencies and the unitary theory of narcosis, *Anesth. Analg.*, 60, 380, 1981.
22. Mazze, R.I., Rice, S.A., and Baden, J.M., Halothane, isoflurane, and enflurane MAC in pregnant and nonpregnant female and male mice, *Anesthesiology*, 62, 339, 1985.
23. Ichinose, F., Huang, P.L., and Zapol, W.M., Effects of targeted neuronal nitric oxide synthase gene disruption and nitro^{G}-L-arginine methylester on the threshold for isoflurane anesthesia, *Anesthesiology*, 83, 101, 1995.
24. Girdlestone, D., Cruikshank, G.H., and Winlow, W., The actions of three volatile general anesthetics on withdrawal responses of the pond-snail *Lymnaea stagnalis* (L.), *Comp. Biochem. Physiol.*, 92C, 39, 1989.
25. Smith, E.B., Bowser-Riley, F., Daniels, S., Dunbar, I.T., Harrison, C.B., and Paton, W.D.M., Species variation and the mechanism of pressure-anaesthetic interactions, *Nature*, 311, 56, 1984.
26. McKenzie, J.D., Calow, P., and Nimmo, W.S., A model to test the potency of inhalation anaesthetics, *Br. J. Anaesth.*, 63, 489, 1989.
27. Morgan, P.G., and Sedensky, M.M., Mutations conferring new patterns of sensitivity to volatile anesthetics in *Caenorhabditis elegans*, *Anesthesiology*, 81, 888, 1994.
28. Gamo, S., Ogaki, M., and Nakashima-Tanaka, E., Strain differences in minimum anesthetic concentrations in *Drosophila melanogaster*, *Anesthesiology*, 54, 289, 1981.
29. Allada, R., and Nash, H.A., *Drosophila melanogaster* as a model for study of general anesthesia: The quantitative response to clinical anesthetics and alkanes, *Anesth. Analg.*, 77, 19, 1993.
30. Westmoreland, C.L., Sebel, P.S., and Gropper, A., Fentanyl or alfentanil decreases the minimum alveolar anesthetic concentration of isoflurane in surgical patients, *Anesth. Analg.*, 78, 23, 1994.
31. Hornbein, T.F., Eger, E.I., II, Winter, P.M., Smith, G., Wetstone, D., and Smith, K.H., The minimum alveolar concentration of nitrous oxide in man, *Anesth. Analg.*, 61, 553, 1982.
32. Rampil, I.J., Lockhart, S.H., Zwass, M.S., Peterson, N., Yasuda, N., Eger, E.I., II, Weiskopf, R.B., and Damask, M.C., Clinical characteristics of desflurane in surgical patients: Minimum alveolar concentration, *Anesthesiology*, 74, 429, 1991.
33. Zbinden, A.M., Maggiorini, M., Petersen-Felix, S., Lauber, R., Thomson, D.A., and Minder, C.E., Anesthetic depth defined using multiple noxious stimuli during isoflurane/oxygen anesthesia, *Anesthesiology*, 80, 253, 1994.
34. Bogetz, M.S., and Katz, J.A., Recall of surgery for major trauma, *Anesthesiology*, 61, 6, 1984.
35. Dwyer, R., Bennett, H.L., Eger, E.I., II, and Heilbron, D., Effects of isoflurane and nitrous oxide in subanesthetic concentrations on memory and responsiveness in volunteers, *Anesthesiology*, 77, 888, 1992.
36. Chortkoff, B.S., Gonsowski, C.T., Bennett, H.L., Levinson, B., Crankshaw, D.P., Dutton, R.C., Ionescu, P., Block, R.I., and Eger, E.I., II, Subanesthetic concentrations of desflurane and propofol suppress recall of emotionally charged information, *Anesth. Analg.*, 81, 728, 1995.

37. Eckenhoff, R.G., Eger, E.I., II, Koblin, D.D., and Halsey, M.J., Tests of anesthesia relevance, *Anesth. Analg.*, 81, 431, 1995.
38. Morgan, P.G., Koblin, D.D., Eger, E.I., II, and Halsey, M.J., 1 + (–1) = 0, or not all anesthetic sites are created equal, *Anesth. Analg.*, 82, 218, 1996.
39. Pauling, L., A molecular theory of general anesthesia, *Science*, 134, 15, 1961.
40. Miller, S.L., A theory of gaseous anesthetics, *Proc. Natl. Acad. Sci. USA*, 47, 1515, 1961.
41. Dodson, B.A., Furmaniuk, Z.W., and Miller, K.W., The physiological effects of hydrostatic pressure are not equivalent to those of helium pressure on *Rana pipiens*, *J. Physiol. (Lond.)*, 362, 233, 1985.
42. Russell, G.B., and Graybeal, J.M., Direct measurements of nitrous oxide MAC and neurologic monitoring in rats under hyperbaric conditions, *Anesth. Analg.*, 75, 995, 1992.
43. Antognini, J.F., Lewis, B.K., and Reitan, J.A., Hypothermia minimally decreases nitrous oxide anesthetic requirements, *Anesth. Analg.*, 79, 980, 1994.
44. Gonsowski, C.T., and Eger, E.I., II, Nitrous oxide minimum alveolar anesthetic concentration in rats is greater than previously reported, *Anesth. Analg.*, 79, 710, 1994.
45. Ostlund, A., Linnarsson, D., Lind, F., and Sporrong, A., Relative narcotic potency and mode of action of sulfur hexafluoride and nitrogen in humans, *J. Appl. Physiol.*, 76, 439, 1994.
46. Eger, E.I., II, Brandstater, B., Saidman, L.J., Regan, M.J., Severinghaus, J.W., and Munson, E.S., Equipotent alveolar concentrations of methoxyflurane, halothane, diethylether, fluroxene, cyclopropane, xenon and nitrous oxide in the dog, *Anesthesiology*, 26, 771, 1965.
47. Eger, E.I., II, Lundgren, L., Miller, S.L., and Stevens, W.C., Anesthetic potencies of sulfur hexafluoride, carbon tetrafluoride, chloroform and Ethrane in dogs: Correlation with the hydrate and lipid theories of anesthetic action, *Anesthesiology*, 30, 129, 1969.
48. Miller, K.W., Paton, W.D.M., Smith, E.B., and Smith, R.A., Physicochemical approaches to the mode of action of general anesthetics, *Anesthesiology*, 36, 339, 1972.
49. Liu, J., Laster, M.J., Taheri, S., Eger, E.I., II, Koblin, D.D., and Halsey, M.J., Is there a cut-off in anesthetic potency for the normal alkanes?, *Anesth. Analg.*, 77, 12, 1993.
50. Mullins, L.J., Anesthetics, in *Handbook of Neurochemistry*, vol 6, Lajtha, A., Ed., Plenum Press, New York, 1971, 395.
51. Taheri, S., Laster, M.J., Liu, J., Eger, E.I., II, Halsey, M.J., and Koblin, D.D., Anesthesia by n-alkanes not consistent with the Meyer-Overton hypothesis: Determinations of the solubilities of alkanes in saline and various lipids, *Anesth. Analg.*, 77, 7, 1993.
52. Liu, J., Laster, M.J., Taheri, S., Eger, E.I., II, Chortkoff, B., and Halsey, M.J., Effect of n-alkane kinetics in rats on potency estimations and the Meyer-Overton hypothesis, *Anesth. Analg.*, 79, 1049, 1994.
53. Raines, D.E., and Miller, K.W., On the importance of volatile agents devoid of anesthetic action, *Anesth. Analg.*, 79, 1031, 1994.
54. Franks, N.P., and Lieb, W.R., Mapping of general anaesthetic target sites provides a molecular basis for cutoff effects, *Nature*, 316, 349, 1985.
55. Fang, Z., Sonner, J., Laster, M.J., Ionescu, P., Kandel, L., Koblin, D.D., Eger, E.I., II, and Halsey, M.J., Anesthetic and convulsant properties of aromatic compounds and cycloalkanes: Implicatons for mechanisms of narcosis, *Anesth. Analg.*, 83, 1097, 1996.
56. Suckling, C.W., Some chemical and physical factors in the development of fluothane, *Br. J. Anaesth.*, 29, 466, 1957.

57. Eger, E.I., II, Liu, J., Koblin, D.D., Laster, M.J., Taheri, S., Halsey, M.J., Ionescu, P., Chortkoff, B., and Hudlicky, T., Molecular properties of the ideal inhaled anesthetic: Studies of fluorinated methanes, ethanes, propanes, and butanes, *Anesth. Analg.,* 79, 245, 1994.
58. Liu, J., Laster, M.J., Koblin, D.D., Eger, E.I. II, Halsey, M.J., Taheri, S., and Chortkoff, B., A cut-off in potency exists in the perfluoroalkanes, *Anesth. Analg.,* 79, 238, 1994.
59. Chortkoff, B., Laster, M.J., Koblin, D.D., Taheri, S., Eger, E.I., II, and Halsey, M.J., Pharmacokinetics do not explain the absence of an anesthetic effect of perfluoropropane or perfluoropentane, *Anesth. Analg.,* 79, 234, 1994.
60. Koblin, D.D., Chortkoff, B., Laster, M.J., Eger, E.I., II, Halsey, M.J., and Ionescu, P., Polyhalogenated and perfluorinated compounds that disobey the Meyer-Overton hypothesis, *Anesth. Analg.,* 79, 1043, 1994.
61. Kandel, L., Chortkoff, B., Sonner, J., Laster, M.J., and Eger, E.I., II, Nonanesthetics can suppress learning, *Anesth. Analg.,* 82, 321, 1996.
62. Robbins, B.H., Preliminary studies of the anaesthetic activity of fluorinated hydrocarbons, *J. Pharmacol. Exp. Ther.,* 86, 197, 1946.
63. Rudo, F.G., and Krantz, J.C., Anaesthetic molecules, *Br. J. Anaesth.,* 46, 181, 1974.
64. Wood, S., Wardley-Smith, B., Halsey, M.J., and Green, C.J., Hydrogen bonding in mechanisms of anaesthesia tested with chloroform and deuterated chloroform, *Br. J. Anaesth.,* 54, 387, 1982.
65. Vulliemoz, Y., Triner, L., Verosky, M., Hamm, M.W., and Krishna, G., Deuterated halothane — anesthetic potency, anticonvulsant activity, and effect on cerebellar cyclic guanosine 3′,5′-monophosphate, *Anesth. Analg.,* 63, 495, 1984.
66. Burns, T.H.S., Hall, J.M., Bracken, A., Gouldstone, G., and Newland, D.S., An investigation of new fluorine compounds in anaesthesia, *Anaesthesia,* 16, 3, 1961.
67. Burns, T.H.S., Hall, J.M., Bracken, A., and Gouldstone, G., An investigation of new fluorine compounds in anaesthesia (3). The anaesthetic properties of hexafluorobenzene, *Anaesthesia,* 16, 333, 1961.
68. Terrell, R.C., Physical and chemical properties of anaesthetic agents, *Br. J. Anaesth.,* 56, 3S, 1984.
69. Terrell, R.C., Speers, L., Szur, A.J., Treadwell, J., and Ucciardi, T.R., General anesthetics. 1. Halogenated methyl ethyl ethers as anesthetic agents, *J. Med. Chem.,* 14, 517, 1971.
70. Terrell, R.C., Speers, L., Szur, A.J., Ucciardi, T.R., and Vitcha, J.F., General anesthetics. 3. Fluorinated methyl ethyl ethers as anesthetic agents, *J. Med. Chem.,* 15, 604, 1972.
71. Speers, L., Szur, A.J., Terrell, R.C., Treadwell, J., and Ucciardi, T.U., General anesthetics. 2. Halogenated methyl isopropyl ethers, *J. Med. Chem.,* 14, 593, 1971.
72. Smith, I., Nathanson, M., and White, P.F., Sevoflurane — A long-awaited volatile anaesthetic, *Br. J. Anaesth.,* 76, 435, 1996.
73. Eger, E.I., II, Koblin, D.D., and Collins, P.A., Dioxychlorane: A challange to the correlation of anesthetic potency and lipid solubility, *Anesth. Analg.,* 60, 201, 1981.
74. Munson, E.S., Schick, L.M., Chapin, J.C., Kushins, L.G., and Navarro, A.A., Determination of the minimum alveolar concentration (MAC) of aliflurane in dogs, *Anesthesiology,* 51, 545, 1979.
75. Tanifuji, Y., Eger, E.I., II, and Terrell, R.C., Some characteristics of an exceptionally potent inhaled anesthetic: Thiomethoxyflurane, *Anesth. Analg.,* 56, 387, 1977.
76. Targ, A.G., Yasuda, N., Eger, E.I., II, Huang, G., Vernice, G.G., Terrell, R.C., and Koblin, D.D., Halogenation and anesthetic potency, *Anesth. Analg.,* 68, 599, 1989.

77. Tinker, J.H., Milde, J.H., and Noback, C.R., Deuterated enflurane: No MAC or cardiorespiratory or isotope effect, *Anesthesiology,* 55, A3, 1981.
78. Koblin, D.D., Eger, E.I., II, Johnson, B.H., Collins, P., Harper, M.H., Terrell R.C., and Speers, L., Minimum alveolar concentrations and oil/gas partition coefficients of four anesthetic isomers, *Anesthesiology,* 54, 314, 1981.
79. Drummond, J.C., MAC for halothane, enflurane, and isoflurane in the New Zealand white rabbit: And a test for the validity of MAC determinations, *Anesthesiology,* 62, 339, 1985.
80. Taheri, S., Halsey, M.J., Liu, J., Eger, E.I., II, Koblin, D.D., and Laster, M.J., What solvent best represents the site of action of inhaled anesthetics in humans, rats, and dogs?, *Anesth. Analg.,* 72, 627, 1991.
81. Firestone, S., Ferguson, C., and Firestone, L., Isoflurane's optical isomers are equipotent in *Rana pipiens* tadpoles, *Anesthesiology,* 77, A758, 1992.
82. Harris, B., Moody, B., and Skolnick, P., Isoflurane anesthesia is stereoselective, *Eur. J. Pharmacol.,* 217, 215, 1992.
83. Lysko, G.S., Robinson, J.L., Casto, R., and Ferrone, R.A., The stereospecific effects of isoflurane isomers in vivo, *Eur. J. Pharmacol.,* 263, 25, 1994.
84. Koblin, D.D., Eger, E.I., II, Johnson, B.H., Collins, P., Terrell R.C., and Speers, L., Are convulsant gases also anesthetics?, *Anesth. Analg.,* 60, 464, 1981.
85. Tinker, J.H., Sharbrough, F.W., and Michenfelder, J.D., Anterior shift of the dominant EEG rhythm during anesthesia in the Java monkey: Correlation with anesthetic potency, *Anesthesiology,* 46, 252, 1977.
86. Tranquilli, W.J., Thurmon, J.C., Benson, G.J., and Steffey, E.P., Determination of halothane MAC in swine, *Anesth. Analg.,* 67, 597, 1988.
87. Fahey, M.R., Sessler, D.I., Cannon, J.E., Brady, K., Stoen, R., and Miller, R.D., Atracurium, vecuronium, and pancuronium do not alter the minimum alveolar concentration of halothane in humans, *Anesthesiology,* 71, 53, 1989.
88. Ludders, J.W., Rode, J., and Mitchell, G.S., Isoflurane anesthesia in sandhill cranes (Grus canadenis): Minimal anesthetic concentration and cardiopulmonary dose-response during spontaneous and controlled breathing, *Anesth. Analg.,* 68, 511, 1989.
89. Murat, I., and Housmans, P.R., Minimum alveolar concentrations (MAC) of halothane, enflurane, and isoflurane in ferrets, *Anesthesiology,* 68, 783, 1988.
90. Seifen, A.B., Kennedy, R.H., Bray, J.P., and Seifen, E., Estimation of minimum alveolar concentration (MAC) for halothane, enflurane and isoflurane in spontaneously breathing guinea pigs, *Lab. Anim. Sci.,* 39, 579, 1989.
91. McEwan, A.I., Smith, C., Dyar, O., Goodman, D., Smith, L.R., and Glass, P.S.A., Isoflurane minimum alveolar concentration reduction by fentanyl, *Anesthesiology,* 78, 864, 1993.
92. Antognini, J.F., and Eisele, P.H., Anesthetic potency and cardiopulmonary effects of enflurane, halothane, and isoflurane in goats, *Lab. Anim. Sci.,* 39, 579, 1989.
93. Palahnuik, R.J., Shnider, S.M., and Eger, E.I., II, Pregnancy decreases the requirement for inhaled anesthetic agents, *Anesthesiology,* 41, 82, 1974.
94. Steffey, E.P., Howland, D., Giri, S., and Eger, E.I., II, Enflurane, halothane, and isoflurane potency in horses, *Am. J. Vet. Res.,* 38, 1037, 1977.
95. Scheller, M.S., Saidman, L.J., and Partridge, B.L., MAC of sevoflurane in humans and the New Zealand white rabbit, *Can. J. Anaesth.,* 35, 153, 1988.
96. Miller, K.W., Paton, W.D.M., and Smith, E.B., The anaesthetic pressures of certain fluorine containing gases, *Br. J. Anaesth.,* 39, 910, 1967.

97. Cullen, S.C., Eger, E.I., II, Cullen, B.F., and Gregory, P., Observations on the anesthetic effect of the combination of xenon and halothane, *Anesthesiology,* 31, 305, 1969.
98. Whitehurst, S.L., Nemoto, E.M., Yao, L., and Yonas, H., MAC of xenon and halothane in Rhesus monkeys, *J. Neurosurg. Anesth.,* 6, 275, 1994.
99. Halsey, M.J., Eger, E.I., II, Kent, D.W., and Warne, P.J., High pressure studies of anesthesia, in *Progress in Anesthesiology, Volume 1, Molecular Mechanisms of Anesthesia,* Fink, B.R., Ed., Raven Press, New York, 1975, 353.
100. Koblin, D.D., Chortkoff, B.S., Laster, M.J., Eger, E.I., II, Halsey, M.J., and Ionescu, P., Nonanesthetic polyhalogenated alkanes and deviations from the Meyer-Overton hypothesis, *Prog. Anesth. Mech.,* 3, 451, 1995.

6 Volatile Anesthetic Effects on Calcium Channels

Hugh C. Hemmings, Jr.

CONTENTS

Changes in intracellular Ca^{2+} regulate multiple cellular functions including stimulus–secretion coupling, excitation–contraction coupling, neuronal plasticity, gene expression, and cell death. In most cell types the concentration of intracellular Ca^{2+} is maintained at extremely low levels (e.g., <0.1 μ*M* in neurons). Specific physiologic stimuli can lead to rapid and marked increases in cytosolic Ca^{2+} through multiple routes of entry. Ca^{2+} then acts as an intracellular second messenger ion that mediates cellular responses *via* interactions with Ca^{2+} binding proteins such as calmodulin to regulate enzyme and protein function. Intracellular Ca^{2+} concentration is tightly controlled by several complex systems, including influx through voltage-dependent Ca^{2+} channels and ligand-gated Ca^{2+} channels, release from intracellular stores, extrusion by membrane Ca^{2+}-ATPase and Na^+/K^{2+} exchange, uptake into intracellular organelles, and binding to high-affinity and high-capacity Ca^{2+} binding proteins. Each of these processes is a potential target for volatile anesthetic effects. This chapter reviews the effects of volatile anesthetics on voltage-dependent Ca^{2+} channels and intracellular Ca^{2+} release, which have been investigated in detail. Volatile

0-8493-8555-5/01/$0.00+$.50

anesthetic effects on Ca^{2+}-ATPase and ligand-gated Ca^{2+} channels are discussed in Chapters 8 and 10, respectively.

6.1 PHYSIOLOGIC AND PHARMACOLOGIC CLASSIFICATION OF Ca^{2+} CHANNELS

Voltage-dependent Ca^{2+} channels are multimeric integral membrane proteins that play important roles in excitable cells, including membrane excitability, excitation–contraction coupling, neurotransmitter release, and gene expression. Ca^{2+} channels open in response to membrane depolarization and allow the rapid entry of Ca^{2+} ($\sim10^6$ ions per second per channel), an important intracellular second messenger. Biophysical and pharmacologic properties such as single-channel conductance, voltage- and time-dependent kinetics, and sensitivity to inhibitors have been used to identify the diversity of Ca^{2+} channel types present in vertebrates. This classification has distinguished four classes of Ca^{2+} channels, termed T-, N-, L-, and P-type Ca^{2+} channels.[1,2] Low-threshold or low-voltage-activated (LVA) Ca^{2+} channels (T-type, for transient or tiny) are transiently activated (i.e., decay rapidly) by relatively small depolarizations (i.e., at more negative membrane potentials) from hyperpolarized holding potentials, are inactivated at positive potentials, and have a small ion conductance. These channels mediate a transient Ca^{2+} current that is important in determining action potential generation in neurons and cardiac myocytes. The other three Ca^{2+} channel classes require larger depolarizations for activation (high-threshold or high-voltage-activated [HVA]) and have distinguishing electrophysiologic features. L-type channels (for long lasting or large), which have large ion conductances and decay slowly during maintained depolarization, are the best characterized. These channels are sensitive to 1,4-dihydropyridine (DHP) compounds and are expressed primarily in muscle, but are found in virtually all excitable cells and in many nonexcitable cells. They mediate the Ca^{2+} influx necessary for excitation–contraction coupling in cardiac and smooth muscle cells and the potential change coupled to contraction in skeletal muscle cells. N-type channels (for neuronal or neither) have intermediate ion conductances and incompletely inactivate, and P-type channels (for Purkinje cell) have intermediate conductances and do not inactivate (Table 6.1). N- and P-type Ca^{2+} channels are expressed primarily in neurons, where they regulate neurotransmitter release and gene expression.

Differential sensitivity to a variety of pharmacologic probes has been used to further distinguish HVA Ca^{2+} channels.[3-5] L-type channels are sensitive to DHP agonists and antagonists; N-type channels are irreversibly blocked by ω-conotoxin GVIA (ω-CTx GVIA), a peptide toxin from the cone shell mollusk *Conus geographus*; and P-type channels are blocked by ω-agatoxin IVA (ω-Aga IVA), a peptide toxin from the venom of the funnel web spider *Agelenopsis aperta*. The toxin ω-conotoxin MVIIC from *Conus magus* blocks N- and P-type channels, as well as an additional channel termed the Q-type channel, which is also blocked by high concentrations of ω-Aga IVA. Many neurons possess a Ca^{2+} current that is resistant to all these blockers, known as the R-type channel.

TABLE 6.1
Properties of Calcium Currents and Channels

Name	Type	Voltage-Dependent Inactivation	Steady-State Inactivation V_{50} (mV)	Single-Channel Conductance (pS)	Pharmacology: Selective Blockers	Distribution	α_1 Subunit
L	HVA	None (Ca^{2+} dependent) ($\tau > 500$ ms)	–20	22–27	DHP	Skeletal muscle	α_{1S}
					DHP, PAA, BTZ	Cardiac muscle, brain, aorta, lung fibroblast	α_{1C}
					DHP	Neuroendocrine, kidney, brain	α_{1D}
P	HVA	Very slow	–5	10–18	ω-Aga IVA (<10 n*M*)	Neurons, kidney	$\alpha_{1A?}$
Q	HVA	Intermediate	–45	?	No specific blockers ω-Aga IVA (≥10 n*M*) ω-CTx MVIIC (≥100 n*M*; also blocks N,P)	Neurons, kidney	α_{1A}
N	HVA	Intermediate ($\tau = 50$–80 ms)	–50	13–20	ω-CTx GVIA (≤100 n*M*) ω-CTx MVIIC (>100 n*M*)	Neurons	α_{1B}
R	H/LVA	Fast ($\tau = 20$–30 ms)	–15	?	No specific blockers Ni^{2+} (≤30 μ*M*)	Neurons, heart	$\alpha_{1E?}$
T	LVA	Fast ($\tau = 20$–40 ms)	–70	~8	No specific blockers Ni^{2+}, octanol, amiloride, carbamazepine, phenytoin	Muscle, neurons, neuroendocrine	$\alpha_{1E?}$

Note: τ, time constant of inactivation; DHP, dihydropyridines; PAA, phenylalkylamines; BTZ, benzothiazepine.

Source: Modified from Dolphin, A. C., *Exp Physiol* 80, 1, 1995.

6.2 STRUCTURE

The functional and pharmacologic classification of Ca^{2+} channels has been supplemented with structural information provided by molecular cloning.[4,6] This was made possible by the purification and extensive biochemical characterization of the skeletal muscle L-type Ca^{2+} channel.[7] This channel in skeletal and cardiac muscle consists of five subunits: α_1, α_2, β, γ, and δ. The α_1 subunit contains receptor sites for Ca^{2+} channel-blocking drugs. The α_2 and δ subunits, which are linked by a disulfide bond, are encoded by a single gene and are formed by posttranslational processing. The β subunit has an intracellular location, and the hydrophobic γ subunit appears to reside in the membrane (Figure 6.1). Diversity of L-type channels arises by multiple genes and alternative splicing, which can be developmentally regulated. For other Ca^{2+} channel types, the subunit composition is less clear.

Molecular cloning has revealed at least six Ca^{2+} channel α_1 genes, which form a multigene family (Table 6.1). The α_1 subunit of several cloned Ca^{2+} channels has been shown to function as the Ca^{2+} ion pore and the voltage sensor for channels with properties consistent with L-, N-, P/Q-, and R-type channels. The deduced amino acid sequences of the Ca^{2+} channel α_1 subunits indicate an overall conserved structure and evolutionary similarities to voltage-dependent Na^+ and K^+ channels. The generalized secondary structure consists of four repeated motifs (I through IV), each comprised of six hydrophobic α-helical transmembrane segments (S1 through S6). Transmembrane segment S4, which contains a positively charged amino acid every three to four residues, is the voltage sensor in Na^+ and K^+ channels, and probably in Ca^{2+} channels as well. The connecting loop between S5 and S6 forms a hairpin "P" loop that bends back into the membrane to form the lining of the channel pore, and imparts Ca^{2+} selectivity. A number of factors modify α_1 subunit's function including alternative splicing, auxiliary subunits, G proteins, Ca^{2+} itself, and protein phosphorylation.[8]

Photoaffinity labeling, peptide mapping, and chimeric Ca^{2+} channels have been used to identify the binding sites for the three major classes of L-type Ca^{2+} channel blockers on the pore-forming α_1 subunit.[9,10] The functional properties of L-type Ca^{2+} channels are modulated by several distinct classes of clinically useful Ca^{2+} channel antagonists. The DHPs (e.g., nifedipine, nitrendipine, isradipine), phenylalkylamines (e.g., verapamil), and benzothiazepines (e.g., diltiazem) have separate, but allosterically linked, binding sites in close proximity to the high-affinity Ca^{2+} binding site in the S5 to S6 connecting loop and part of S6 in domain IV (Figure 6.1). Recent studies show that the S5 segments of domains III and IV also contribute to the DHP binding site.[11] The common, but not identical, molecular determinants of these three drug classes in segment S6 of domain IV[12-14] provide a molecular basis for their noncompetitive interactions. Drug binding influences Ca^{2+} channel function by interacting with the pore-lining region of the channel. The Ca^{2+} channel antagonists exert distinct pharmacologic effects. Blockade by the phenylalkylamines is enhanced by repetitive stimulation, which is consistent with increased access to an intracellular receptor site through the open channel pore. The DHPs, which can act as either inhibitors (nifedipine) or activators (Bay K 8644) of L-type Ca^{2+} channels, modulate voltage-dependent gating. Inhibition is enhanced by prolonged depolarization due

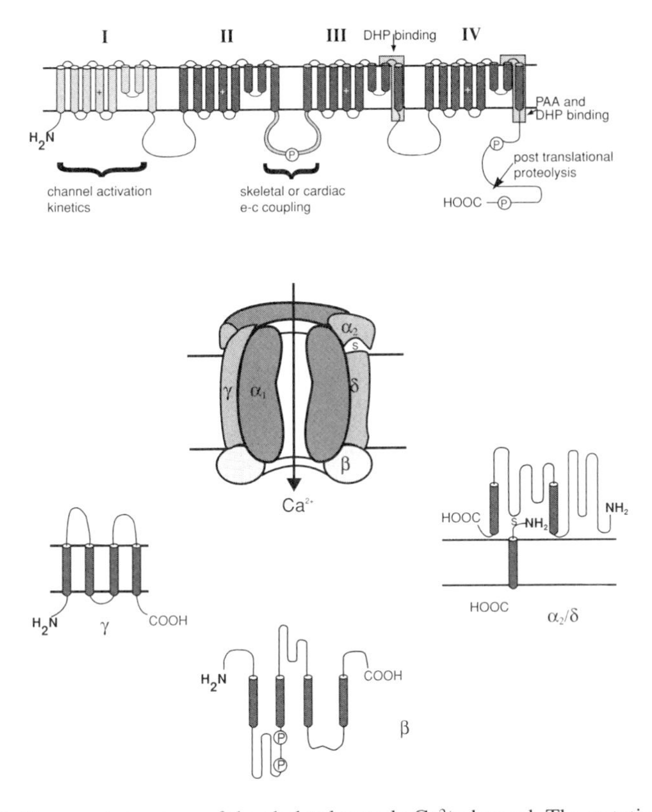

FIGURE 6.1 Proposed structure of the skeletal muscle Ca^{2+} channel. The putative transmembrane configuration of individual subunits is taken from the hydropathicity analysis of the primary sequences. The suggested structure of the α_1 subunit is shown at the top. I, II, III, and IV are proposed repeats of the Ca^{2+} channel α_1 subunit, each composed of six transmembrane segments (S1 to S6, from left to right, shown as the long barrels) and a linker between S5 and S6 (the SS1 to SS2 region or "P" loop, suggested to be a part of the channel pore, shown as the short barrels). +, transmembrane amphipathic α-helix, which is the proposed voltage sensing helix of the channel; P, sites phosphorylated *in vitro* by cAMP-dependent protein kinase; DHP and PAA, dihydropyridine and phenylalkylamine binding sites in segment IVS6 and the SS1 to SS2 region (benzothiazepines also bind in this region); *e-c coupling*, domain involved in excitation–contraction coupling. The brackets indicate parts of the protein that are responsible for the skeletal (α_{1S}) or cardiac (α_{1C}) properties of the channel. The dash at the C terminus indicates the site where the α_1 subunit is processed posttranslationally. The suggested structures of the γ, β, and α_2/δ subunits are also shown. *S*, disulfide bridge between the transmembrane δ and the extracellular α_2 subunit. The quaternary structure of the multimeric channel showing the central ion pore is modeled in the center of the figure. The extracellular space is above the horizontal lines. (From Hofmann, F., Biel, M. and Flockerzi, V., *Annu Rev Neurosci* 17, 399, 1994. With permission.)

to selective (state-dependent) binding to the inactivated state of the channel, which occurs in a hydrophobic region of the extracellular surface. Evidence suggests that the structure of T-type channels is distinct from L-type channels; further elucidation of the molecular properties of T-type channels awaits further purification, molecular cloning, and expression.

6.3 LOCALIZATION AND FUNCTION

L-type Ca^{2+} channels are essential for excitation–contraction coupling in skeletal, cardiac, and smooth muscle. In skeletal muscle, L-type Ca^{2+} channels are located in the T tubule sarcolemma and are closely apposed to oligomeric ryanodine receptor complexes, which are located in the sarcoplasmic reticulum membrane.[15] Channel activation is indirectly coupled to excitation–contraction coupling in skeletal muscle. After channel activation, a voltage potential change rather than Ca^{2+} influx mediates ryanodine receptor opening by a poorly understood mechanism, which releases Ca^{2+} from intracellular stores, resulting in myofibril contraction. In cardiac tissues, L- and T-type channels are present in atrial and ventricular myocytes, sinoatrial nodal cells, and Purkinje cells. Influx of extracellular Ca^{2+} through Ca^{2+} channels contributes to contraction while also activating intracellular Ca^{2+} release in cardiac muscle (see below). L-type channels contribute to the slow inward current in ventricular myocytes and play a major role in allowing the influx of external Ca^{2+} coupled to contraction. The role of T-type channels is less clear, but they appear to be important for pacemaking and impulse conduction. In smooth muscle, influx of extracellular Ca^{2+} through Ca^{2+} channels is directly coupled to contraction.

Neuronal Ca^{2+} channel subtypes exhibit heterogeneous cellular and subcellular localizations within the brain as revealed by immunocytochemistry. Class C and D L-type Ca^{2+} channels are preferentially located in neuronal cell bodies and proximal dendrites, where they are involved in regulating Ca^{2+}-dependent protein phosphorylation, enzyme activity, and gene expression.[16] Class D channels are distributed fairly evenly, whereas class C channels are concentrated in clusters. N-type Ca^{2+} channels have a largely complementary distribution in neurons. They are concentrated in dendrites, where they mediate Ca^{2+}-dependent membrane potential changes and conduction, and in certain nerve terminals, where they contribute to the rapid Ca^{2+} influx coupled to neurotransmitter release.[17] P/Q-type Ca^{2+} channels are highly expressed in the cerebellum along the length of Purkinje cell dendrites, where the P channel is responsible for dendritic currents.[18] Different Ca^{2+} channel types can coexist in the same neuron, as demonstrated by the elegant pharmacologic dissection of Ca^{2+} currents in rat cerebellar granule cells by Randall and Tsien[19] (Figure 6.2).

The expression of different Ca^{2+} channel types varies between different species, tissues, brain regions, neuron types, and neuronal processes. This differential distribution has led to the concept that different Ca^{2+} types are associated with specific functions. For example, the Ca^{2+} channels involved in excitation–contraction coupling in muscle are distinct from those involved in excitation–secretion coupling in nerve terminals.[20] These associations are not universal, however, because different Ca^{2+} channels are coupled to Ca^{2+} influx into presynaptic nerve terminals and neurotransmitter release depending on the neuron type, brain region, species, and

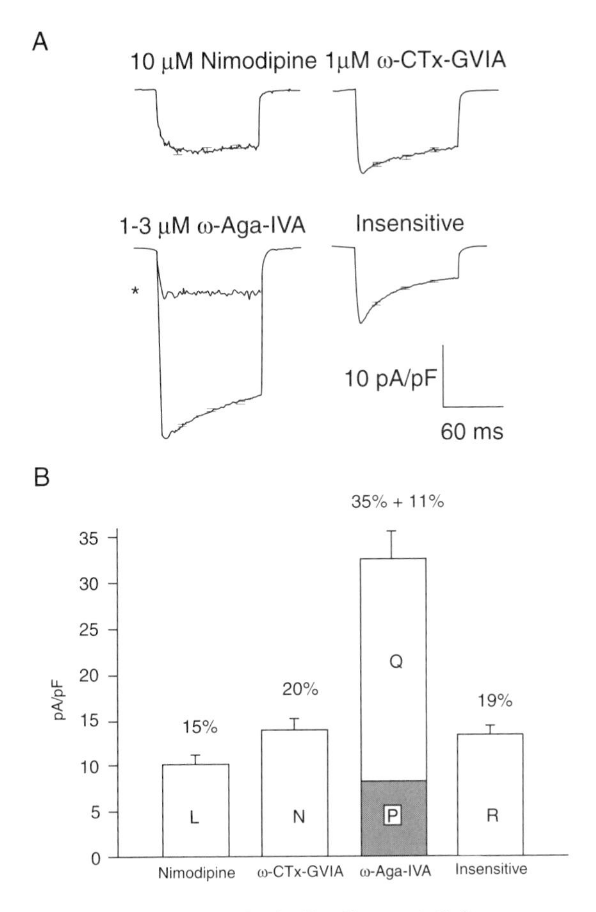

FIGURE 6.2 Comparison of pharmacologically dissected Ca^{2+} current components in rat cerebellar granule cell neurons. (A) Pharmacologically defined Ca^{2+} current components from individual cells were normalized to peak current (measured with Ba^{2+}), averaged, then scaled by their mean peak current density. L-type currents were dissected with 10 μ*M* nimodipine; N-type currents with 1 μ*M* ω-CTx GVIA; and P- and Q-type currents with 1 to 2 μ*M* ω-Aga IVA after blockade of N- and L-type currents. R-type current is estimated as the current that remains unblocked in the combined presence of 10 μ*M* nimodipine, 1 μ*M* ω-CTx GVIA, and 3 μ*M* ω-Aga IVA. (B) Pooled results for mean current density of the four Ca^{2+} current components shown in A. The number above for each bar denotes the percentage contribution the current makes to the global Ca^{2+} current. The bar denoting ω-Aga IVA sensitivity is divided into fractions corresponding to the amount of P-type (shaded) and Q-type (unshaded) current. The asterisk in A marks the time course and estimated magnitude of the corresponding P-type current (waveform determined by application of 1.5 n*M* ω-Aga IVA). (From Randall, A. and Tsien, R. W., *J Neurosci* 15, 2995, 1995. With permission.)

neurotransmitter.[21] L-type Ca^{2+} channels do not appear to be involved in neurotransmitter release in the central nervous system (CNS). However, they are coupled to hormone secretion in a variety of neuroendocrine cells. N-type Ca^{2+} channels are involved in exocytosis from CNS nerve terminals, because the release of a variety of transmitters can be inhibited by ω-CTx GVIA. Only 20 to 30% of the release can be inhibited by this antagonist, which implies that other Ca^{2+} channels are also involved. ω-Aga IVA, which blocks P-type Ca^{2+} channels, also results in incomplete inhibition of neurotransmitter release (50 to 70%), which is additive with ω-CTx GVIA. However, ω-CTx GVIA and ω-Aga IVA together do not block Ca^{2+}-dependent neurotransmitter release completely, suggesting the existence of unidentified toxin "resistant" exocytotic Ca^{2+} channels. A consensus view is that multiple Ca^{2+} channel types coexist in nerve terminals to regulate exocytosis jointly, with both cell-specific and transmitter-specific differences.

6.4 REGULATION

Modulation of Ca^{2+} channel function can have profound effects on physiologic function, as in the ionotropic effect of catecholamines due to β-adrenoceptor–mediated enhancement of Ca^{2+} currents. Cyclic AMP (cAMP) increases cardiac Ca^{2+} currents following β-adrenoceptor activation by increasing channel open probability (P_o) as well as mean open time.[22-24] Both the α_1 and β subunits, which can associate to form a functional Ca^{2+} channel when coexpressed, are phosphorylated. Evidence suggests that Ca^{2+} channels must be phosphorylated to respond to membrane depolarization,[25] and the Ca^{2+} flux of purified channels is enhanced by phosphorylation.[26,27] Cardiac and skeletal muscle and neurons express homologous but nonidentical α_1 subunits that exist in two size forms. The truncated form, which is present at high levels in muscle, is formed by proteolytic cleavage of the C terminal tail. Only the full-length form is phosphorylated by cAMP-dependent protein kinase on the longer C terminal tail[28-30] (Figure 6.1). Thus channels containing the two size forms can undergo differential regulation. In hippocampal neurons, *N*-methyl-D-aspartate (NMDA)-receptor–induced proteolytic processing of class C L-type Ca^{2+} channels (by the Ca^{2+}-dependent protease calpain) converts the long form to the short form,[31] which exhibits about fivefold greater ion conductance.[32] The evidence for the regulation of neuronal Ca^{2+} channels by cAMP is less consistent, possibly due to higher basal adenylyl cyclase activity in neurons.[33]

Ca^{2+} channels are phosphorylated by protein kinases in addition to cAMP-dependent protein kinase.[8] Activators of protein kinase C facilitate Ca^{2+} currents in cardiac myocytes and sympathetic neurons. The α_{1B} subunit can be phosphorylated by protein kinase C, cAMP-dependent protein kinase, Ca^{2+}/calmodulin-dependent protein kinase II, and cGMP-dependent protein kinase, a component of the nitric oxide (NO) signal transduction pathway.[29] Recent evidence suggests that Ca^{2+} channels can also be regulated by phosphorylation of tyrosine residues. The effects of these and possibly other posttranslational modifications on Ca^{2+} channel function remain to be clarified, but it is clear that Ca^{2+} channels are subject to complex regulation by multiple messengers acting *via* phosphorylation.

Regulation of Ca^{2+} channels is also mediated by heterotrimeric G proteins.[8] In addition to the role of G_s in activating adenylyl cyclase, this G protein also directly enhances cardiac L-type Ca^{2+} currents, which may be responsible for the beat-to-beat regulation of Ca^{2+} currents by sympathetic stimulation. Ca^{2+} currents can also be inhibited by a number of neurotransmitters acting through the G protein G_o, including norepinephrine acting at α_2 adrenoceptors, opioids acting at μ, κ, and δ receptors, GABA acting at $GABA_B$ receptors, adenosine acting at A_1 receptors, and acetylcholine acting at M_2 or M_4 muscarinic receptors.[34] This form of regulation has been demonstrated for N-, P-, and Q-type channels in neurons, and for L-type channels in nonneuronal secretory cells. These effects appear to be mediated by a direct interaction of the α subunit of G_o with the Ca^{2+} channel, although a role for $\beta\gamma$ subunits is emerging. There is less consistent evidence for modulation of T-type channel currents by neurotransmitters by these mechanisms.

Ca^{2+} currents can be modulated by membrane potential in a reversible manner[33]. Current is facilitated by depolarizing prepulses, which involves the appearance of an L-type channel current that is normally silent and requires phosphorylation for activation. Voltage-dependent facilitation apparently involves the phosphorylation of a site that is exposed during the prepulse depolarization. This site is slowly dephosphorylated at the holding potential, such that the interval between the prepulse and test pulse is insufficient for complete dephosphorylation.[35]

6.5 INTRACELLULAR CA^{2+} CHANNELS

Ca^{2+} homeostasis is regulated not only by voltage-dependent Ca^{2+} channels in the plasma membrane, but also by release from internal Ca^{2+} stores.[36,37] Intracellular Ca^{2+} is stored within specialized zones of the endoplasmic reticulum (and sarcoplasmic reticulum in muscle) and is rapidly released and taken up in response to appropriate stimuli. Intracellular Ca^{2+} concentration is maintained at extremely low levels by the action of membrane-associated Ca^{2+}-adenosine triphosphatases (Ca^{2+} pumps), which pump Ca^{2+} against a large concentration gradient into the two major Ca^{2+} sinks, the extracellular space and the endoplasmic reticulum. There are apparently at least two pools of Ca^{2+} stores in the endoplasmic reticulum; release from each one is mediated by distinct mechanisms involving two distinct classes of Ca^{2+} channels (Figure 6.3). Inositol 1,4,5-trisphosphate ($InsP_3$) is an intracellular second messenger generated along with diacylglycerol by G protein–coupled receptor activation of phospholipase C. $InsP_3$ binds to a receptor on a component of the endoplasmic reticulum or a more specialized structure ("calciosome") to mobilize Ca^{2+} from intracellular stores, whereas diacylglycerol activates protein kinase C. $InsP_3$ receptors are $InsP_3$-gated Ca^{2+} channels that are widely distributed in both excitable and nonexcitable tissues, reflecting the ubiquitous nature of the $InsP_3$/diacylglycerol signaling pathway.[38] Intracellular Ca^{2+} can also be released by a second mechanism mediated by ryanodine receptors (named after a specific plant alkaloid agonist), which have been extensively studied in muscle, where they play a critical role in excitation–contraction coupling.[39] Ryanodine receptors are highly expressed in electrically excitable tissues and have recently been identified in nonexcitable cells.

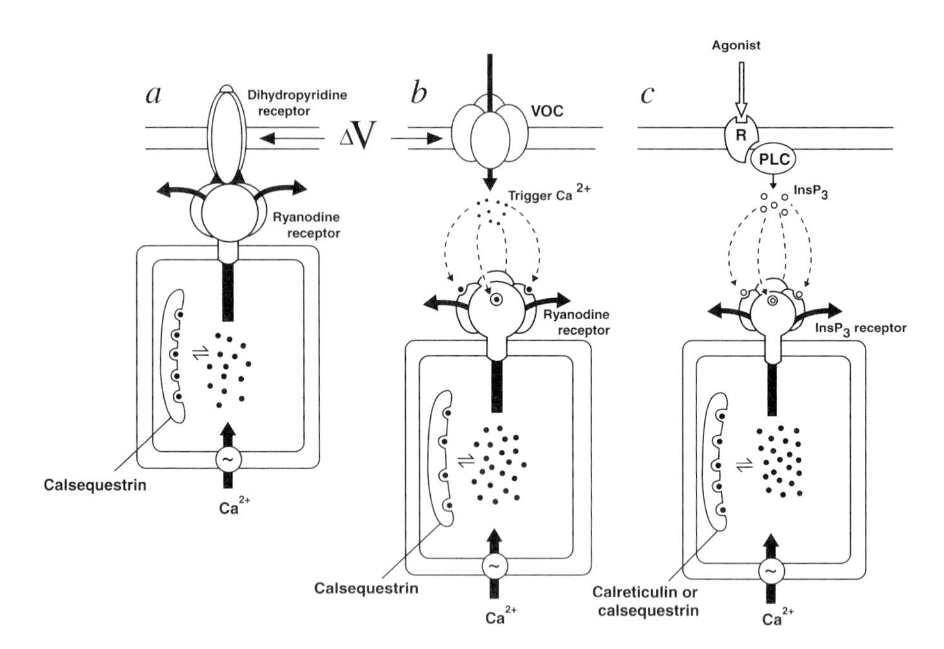

FIGURE 6.3 Control of Ca^{2+} release by intracellular Ca^{2+} channels. (a) Ca^{2+} release from the sarcoplasmic reticulum. Ryanodine receptors located in the sarcoplasmic reticulum of skeletal muscle contribute to the T tubule foot structure responsible for excitation–contraction coupling. The L-type channel (dihydropyridine receptor) in the plasma membrane senses a change in voltage (ΔV) and undergoes a conformational change, which is transmitted through the bulbous head of ryanodine receptors to open the Ca^{2+} channel in the sarcoplasmic reticulum. Calsequestrin, a Ca^{2+} binding protein, serves as a Ca^{2+} sink. (b) Ca^{2+}-induced calcium release in cardiac muscle and neurons. A voltage-operated channel (VOC) responds to ΔV by gating a small amount of trigger Ca^{2+}, which then activates the ryanodine receptor to release stored Ca^{2+}. (c) Agonist-induced Ca^{2+} release. Signal transduction at the cell surface generates inositol trisphosphate ($InsP_3$), which diffuses into the cell to release Ca^{2+} by binding to inositol trisphosphate receptors. (From Berridge, M. J., *Nature* 361, 315, 1993. With permission.)

The $InsP_3$ and ryanodine receptors share considerable structural and functional similarities. Ryanodine receptors are sarcolemmal Ca^{2+} release channels involved in regulating excitation–contraction coupling in both skeletal and cardiac muscle. The ryanodine receptor comprises the microscopic foot proteins observed between the sarcoplasmic reticulum and the T tubules at the triadic junction.[40] The skeletal and cardiac ryanodine receptors are distinct but homologous proteins with specialized modes of regulation. In skeletal muscle, the ryanodine receptor undergoes voltage-sensitive activation involving the L-type Ca^{2+} channel as a voltage sensor for sarcolemmal depolarization; Ca^{2+} influx through the L-channel is unnecessary. In cardiac muscle, trans-sarcolemmal Ca^{2+} influx through L-channels, which is itself insufficient to activate myofilament contraction, activates further Ca^{2+} release from the sarcoplasmic reticulum (known as Ca^{2+}-induced Ca^{2+} release). Although the $InsP_3$ receptors are the major intracellular release channels in neurons, evidence also indicates the presence of both the skeletal and cardiac muscle ryanodine receptor isoforms. A third ryanodine receptor isoform has been identified that is present in the endoplasmic reticulum of many tissues, including neurons. The cardiac isoform appears to be the most abundant in brain, where it may regulate Ca^{2+}-induced Ca^{2+} release. Recent evidence suggests that cyclic ADP-ribose acts as an intracellular messenger to release Ca^{2+} from $InsP_3$-insensitive and ryanodine-sensitive intracellular Ca^{2+} stores in a number of tissues, probably by an action on ryanodine receptors.[41]

6.6 VOLATILE ANESTHETIC EFFECTS

6.6.1 Cardiac Ca^{2+} Channels

The effects of volatile anesthetics on cardiac Ca^{2+} channels have been analyzed extensively. The critical role of Ca^{2+} channels in cardiac muscle excitation–contraction coupling and impulse conduction suggested that they may be an important target for the negative inotropic and negative chronotropic effects of volatile anesthetics observed clinically.[42] For example, anesthetics inhibit sinoatrial node automaticity,[43] prolong conduction time through the AV node and the His-Purkinje pathway,[44] and depress myocardial contractility.[45,46] Early reports demonstrated reductions by halothane and enflurane of action potential overshoot, plateau phase peak and duration, and the slow inward Ca^{2+} current without a change in resting potential in guinea pig ventricular muscle.[47-50] Direct evidence for an inhibitory effect of volatile anesthetics on L-type Ca^{2+} channels was obtained by whole-cell voltage-clamp recording. Halothane, isoflurane, and enflurane were found to reduce Ca^{2+} currents in single isolated canine,[51] rat,[52] and guinea pig[53-58] ventricular myocytes. This was evident in reduced amplitude and increased inactivation of inward Ca^{2+} currents produced by isoflurane[54] or enflurane, but not isoflurane or halothane.[51] Using isolated bullfrog atrial myocytes, which lack sarcoplasmic reticulum (SR), isoflurane and sevoflurane reduced both peak whole-cell Ca^{2+} current amplitude and the inactivation time constant (τ_f). At lower anesthetic concentrations, the reduction in τ_f was sufficient to explain the reduction in amplitude.[59] Similar results were obtained for isoflurane in guinea pig myocytes.[58] Reductions in cAMP at higher anesthetic concentrations[60] were hypothesized as a mechanism for additional reductions in Ca^{2+} current. Single-channel measurements made using the cell-attached patch-clamp technique in guinea pig ventricular myocytes indicated that high concentrations of enflurane, propofol,[61] halothane, and isoflurane[58] inhibited L-type channels by stabilizing the closed inactive state of the channel and decreasing open probability without reducing the single channel conductance. This was associated with an increase in the apparent rate of the slow component of inactivation.[58,62]

The electrophysiologic effects of volatile anesthetics on L-type Ca^{2+} channel function are paralleled by their dose-dependent and reversible inhibition of DHP and phenylalkylamine binding to cardiac sarcolemmal membranes.[63-67] Halothane inhibited nitrendipine[63,64] and gallopamil[66] (D600; a phenylalkylamine) binding to purified bovine sarcolemmal membranes by reducing the number of ligand binding sites (B_{max}) without affecting their affinity (K_d). Of interest, the effects of volatile anesthetics on myocardial function are reproduced and potentiated by the L-type Ca^{2+} channel blocker nifedipine.[68] Inhibition of isradipine binding has also been demonstrated for halothane, enflurane, and isoflurane in both cardiac and skeletal muscle membranes.[69,70] This phenomenon has been confirmed for halothane in the intact Langendorff-perfused rat heart preparation, in which halothane protected isradipine binding to L-type Ca^{2+} channels *in situ*.[71] The observation that the "protected" channels had a reduced affinity for isradipine suggested that halothane led to an unidentified channel modification, possibly phosphorylation.

Halothane, isoflurane, and enflurane depressed Ca^{2+} current amplitude similarly at equianesthetic concentrations in canine ventricular myocytes[51] and cardiac Purkinje cells[72] by whole-cell voltage-clamp analysis (~30% inhibition at 1 MAC). These three volatile anesthetics also inhibited verapamil-sensitive $^{54}Mn^{2+}$ uptake during electrical stimulation (an index of Ca^{2+} channel activity) into rat cardiomyocytes with similar potencies relative to their MAC values.[73] In contrast, a recent study by Pancrazio[58] demonstrated greater potency for halothane than isoflurane in the inhibition of whole-cell Ca^{2+} currents (apparently due to an agent-specific kinetic effect of halothane), which supports earlier single microelectrode studies.[48,53,54] However, there are important quantitative differences in the depressant effects of specific volatile anesthetics on myocardial contractility,[46] because halothane and enflurane depress cardiac function more than isoflurane at equianesthetic concentrations.[74,75] These differences are most likely due to differences in their actions at other cellular targets (e.g., sarcoplasmic reticulum Ca^{2+} release and sequestration or Ca^{2+} pumps).

At relevant clinical concentrations, volatile anesthetics have modest effects on myofibrillar actomyosin function and Ca^{2+} sensitivity.[76-82] The weaker negative inotropic effect of isoflurane compared to halothane was associated with less depression of peak intracellular free Ca^{2+} measured by aequorin luminescence in guinea pig ventricular papillary myocytes[80] (Figure 6.4) and in canine cardiac Purkinje fibers,[83] probably due to a more potent effect of halothane and enflurane in depressing sarcoplasmic Ca^{2+} uptake and release.[84,85] Depression of myofibrillar Ca^{2+} sensitivity did not appear to play a major role in the negative inotropic effects of halothane or enflurane. A small inhibitory effect of isoflurane on myofibrillar Ca^{2+} sensitivity, which was apparently compensated for by its reduced effect on intracellular Ca^{2+},[80] was not observed in a separate study from the same laboratory.[83]

The quantitative and qualitative effects of sevoflurane on myocardial contractility are similar to those of isoflurane.[86,87] Sevoflurane inhibited sarcolemmal Ca^{2+} channels in isolated guinea pig ventricular myocytes (25% inhibition of peak Ca^{2+} current at 3.4 vol%) and increased the apparent rate of inactivation.[87] A similar study in canine ventricular myocytes[60] reported greater inhibition (27% inhibition of peak Ca^{2+} current at 2 vol%), possibly due to differences in holding potentials used, which would lead to greater channel inactivation. Sevoflurane did not have significant effects on Na^+ currents or intracellular cAMP levels,[60] although a delayed outward K^+ current was markedly depressed.[87]

In contrast to the 1,4-dihydropyridines, volatile anesthetics are not selective blockers of L-type channels. Halothane (Figure 6.5), isoflurane, and enflurane produced similar depression of both L- and T-type Ca^{2+} channel currents at equianesthetic concentrations by whole-cell voltage-clamp analysis in canine cardiac Purkinje cells,[72] with similar potency to their effects on ventricular myocyte L-type Ca^{2+} channels.[51] Similarly, sevoflurane was equally effective in blocking L- and T-type

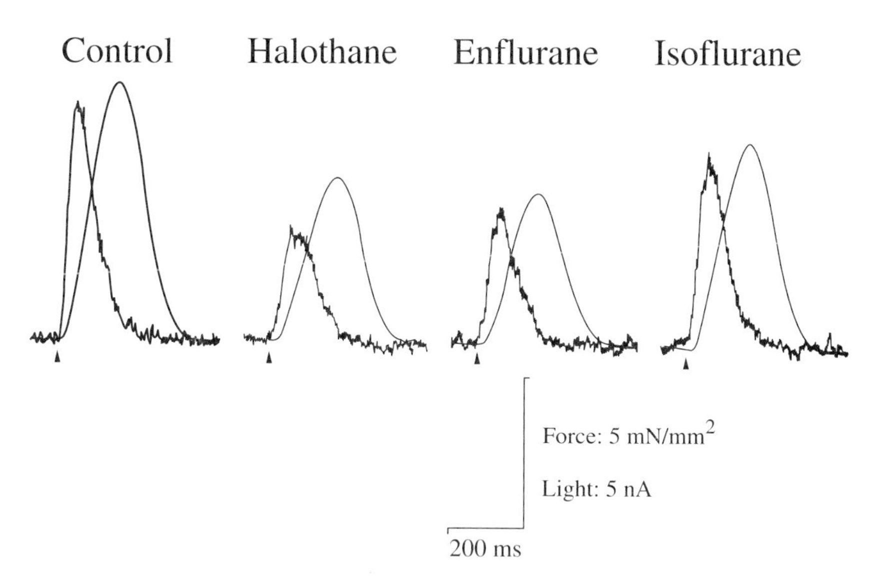

FIGURE 6.4 Volatile anesthetic effects on cardiac cell Ca^{2+} and contractility. Effects of halothane (1.1 vol%), enflurane (2.2 vol%), and isoflurane (1.6 vol%) on aequorin signal (Ca^{2+} signal) and isometric contractions were determined on a single isolated guinea pig papillary muscle fiber. Anesthetic depression of contractile force was accompanied by depression of the intracellular Ca^{2+} signal. Depression of the Ca^{2+} transients in the presence of the isoflurane was less than that produced in the presence of halothane or enflurane ($p < 0.05$). Pacing rate, 1 Hz (at the arrowhead), 30°C. (From Bosnjak et al.[80] With permission.)

Ca^{2+} channel currents in guinea pig ventricular myocytes.[87] In contrast, isoflurane and enflurane, but not halothane, inhibited KCl depolarization-induced increases in intracellular [Ca^{2+}] ([Ca^{2+}]$_i$) in rat cardiac cell suspensions, which may indicate different effects on peak Ca^{2+} current (usually measured by voltage-clamp analysis) versus sustained Ca^{2+} current (measured by prolonged KCl depolarization) mediated by different Ca^{2+} channel types.[88]

The differential effects of volatile anesthetics on myocardial contractility are not explained completely by their effects on Ca^{2+} channels, which appear to be comparable at equianesthetic concentrations.[51,58] Rather, these differences are likely due to different actions on Ca^{2+} release from the sarcoplasmic reticulum (SR). Considerable evidence supports volatile anesthetic alterations of myocardial contractility by effects on SR Ca^{2+} uptake and release.[84,85,89-91] Volatile anesthetics also increase the rate of Ca^{2+} release from skeletal muscle SR.[92-94] Halothane and enflurane reduced the Ca^{2+} loading capacity of SR in chemically skinned rabbit myocardial fiber bundles more than did isoflurane.[77,78] The observation that the depression by halothane of caffeine-induced tension generation in skinned myocardial fibers was blocked by ruthenium red, an antagonist of SR Ca^{2+} release channels, provided indirect evidence that halothane stimulates SR Ca^{2+} release.[95] This would lead to depletion of SR Ca^{2+} and consequently reduced Ca^{2+}-induced Ca^{2+} release and a smaller [Ca^{2+}]$_i$ transient upon excitation. This mechanism of myocardial depression is supported by an analysis of anesthetic effects on sarcoplasmic [Ca^{2+}] using Ca^{2+}-sensitive fluorescent dyes in

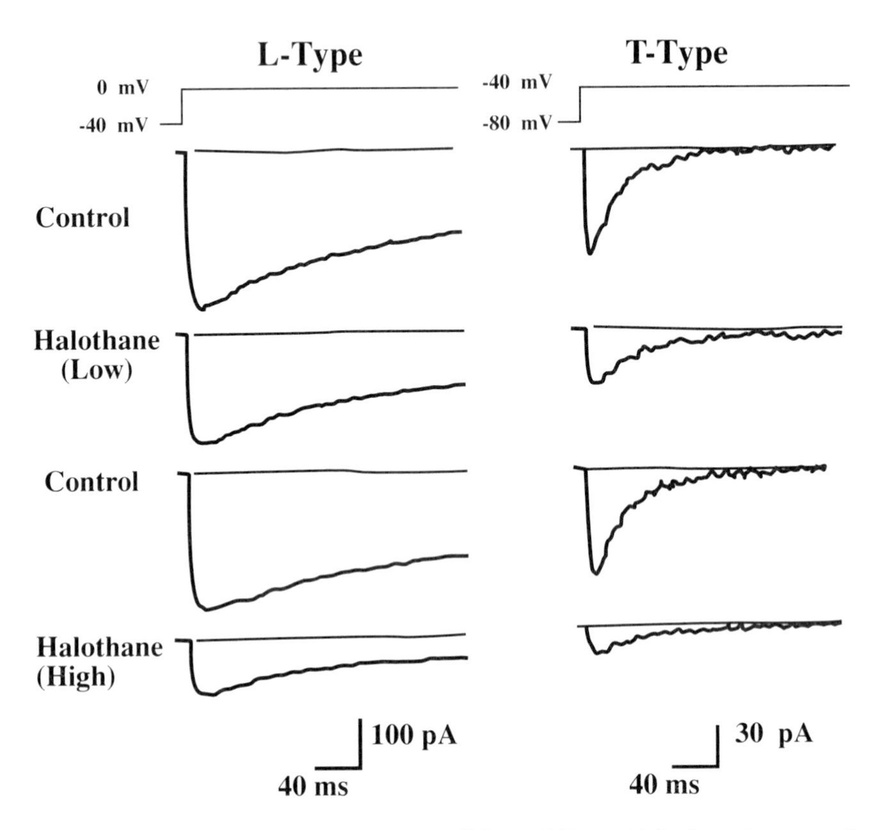

FIGURE 6.5 Effects of halothane on whole-cell L- and T-type Ca^{2+} channel currents in a single canine cardiac Purkinje cell. L-type current was elicited by depolarizing the cell from –40 to 0 mV (*left*). T-type currents were elicited by depolarizing the cell from –80 to –40 mV (*right*). Exposure of the cell to 0.7 vol% (low) and 1.5 vol% (high) halothane similarly depressed both L- and T-type currents in a concentration-dependent and reversible manner. (From Eskinder et al.[72] With permission.)

which halothane and enflurane caused greater depression of caffeine-induced [Ca^{2+}] transients than did isoflurane.[96]

A direct interaction between volatile anesthetics and the Ca^{2+} release channel was first suggested by the ability of halothane and enflurane, but not isoflurane, to increase the binding of ryanodine, which binds to the open state of the channel, to porcine[97] and canine[98] SR vesicles. These results are consistent with an effect of halothane and enflurane to open the Ca^{2+} release channel. Anesthetic concentrations of halothane and enflurane, but not isoflurane, were subsequently found to reduce SR Ca^{2+} content[88,96] and to activate porcine cardiac Ca^{2+} release channels without affecting channel conductance using single-channel recordings in artificial lipid bilayers.[99] The greater negative inotropic effects of halothane and enflurane thus appear to be due both to activation of SR Ca^{2+} channels with depletion of SR Ca^{2+} stores available for excitation–contraction coupling and to sarcolemmal Ca^{2+} channel inhibition. This depletion would be enhanced and maintained by the depression of

Ca^{2+} influx. The reduced contractile depression of isoflurane results from inhibition of Ca^{2+} influx through sarcolemmal Ca^{2+} channels only.

Some of the vascular effects of volatile anesthetics are also mediated by actions on Ca^{2+} channels. The contractile force of arterial smooth muscle, which determines peripheral vascular resistance, is regulated by cytoplasmic [Ca^{2+}]. Halothane and isoflurane have been shown to inhibit macroscopic Ca^{2+} channel currents in voltage-clamped canine coronary artery,[101] canine middle cerebral artery,[102] and guinea pig[103] and rabbit[104] portal vein smooth muscle cells. These effects were demonstrated at relatively high anesthetic concentrations (1.5% halothane and 2.6 to 3.0% isoflurane) and may underlie the direct vasodilatory effects of volatile anesthetics on these vessels. Some of the vascular effects of volatile anesthetics may also be mediated by actions on Ca^{2+} release from intracellular stores (see below). For example, enflurane has been shown to constrict canine mesenteric artery[105] and rat[106] or rabbit[107] thoracic aorta, possibly due to transient increases in intracellular Ca^{2+} from intracellular stores.

Inhibition of Ca^{2+} channels by volatile anesthetics may also mediate their direct relaxant effect on airway smooth muscle. Halothane, isoflurane, and sevoflurane inhibited the macroscopic voltage-activated Ca^{2+} current in isolated porcine tracheal smooth muscle cells analyzed by whole-cell patch-clamp recording.[104] This current was sensitive to nifedipine, which suggests that it is mediated by L-type channels.

6.6.2 Neuronal Ca^{2+} Channels

The important role of neuronal Ca^{2+} channels in the regulation of neuronal excitability, neurotransmitter release, and intracellular signaling suggests these channels are potential target sites for general anesthetic effects. Inhibition of neuronal Ca^{2+} channels could explain the inhibition of synaptic transmission produced by general anesthetics in a number of systems.[108] The effects of volatile anesthetics on Ca^{2+} channels have been analyzed in various neuronal preparations, including cell lines, isolated neurons, brain subcellular fractions, and brain slices. Each preparation has particular advantages and limitations that must be considered in the interpretation of the resulting data. Despite their many similar properties, the evidence for volatile anesthetic effects on Ca^{2+} channels in neuronal systems is less consistent than the convincing evidence that inhibition of Ca^{2+} channels is involved in the myocardial depressant properties of volatile anesthetics.

Krnjevic and Puil[109] demonstrated that halothane (1 to 3 vol%) reversibly inhibited both HVA and LVA Ca^{2+} currents in hippocampal brain slice pyramidal neurons in a dose-dependent manner. Peak current was reduced with no change in the threshold voltage for activation, and inactivation was accelerated. The specific Ca^{2+} channel types involved could not be determined at that time. Subsequent studies confirmed that neuronal Ca^{2+} currents are sensitive to volatile anesthetics at clinically relevant concentrations. Ca^{2+} currents in unclamped presynaptic axons from rat olfactory cortex were inhibited by halothane ($IC_{50} \cong 1$ mM).[110] Halothane reversibly inhibited LVA (T-type) Ca^{2+} currents analyzed by whole-cell patch-clamp recording in neonatal rat dorsal root ganglion neurons ($IC_{50} \cong 0.1$ mM) and HVA Ca^{2+} currents ($IC_{50} \cong 1.5$ mM).[111] A more detailed analysis of anesthetic effects in adult rat sensory neurons revealed that isoflurane (IC_{50} = 0.30 m*M*) and halothane (IC_{50} = 0.66 m*M*) reversibly and completely inhibited T-type Ca^{2+} channels.[112] The different sensitivity of T-type Ca^{2+} channels in neonatal compared to adult rat dorsal root ganglion neurons may be methodologic or due to developmental differences. The difference in the anesthetic sensitivities of LVA and HVA Ca^{2+} currents in neurons contrasts with their similar sensitivities to isoflurane, enflurane, and halothane in cardiac Purkinje cells[72] (Figure 6.5). Halothane, isoflurane, and enflurane partially inhibited the transient increase in $[Ca^{2+}]_i$ in cultured rat hippocampal neurons measured fluorimetrically with fura 2 evoked by 50 m*M* KCl, which activates Ca^{2+} channels by depolarization. This is consistent with inhibition of Ca^{2+} channels.[113]

The observation that L-type Ca^{2+} channel blockers (i.e., DHPs, phenylalkylamines, and benzothiazepines) lack general anesthetic properties suggests that blockade of this Ca^{2+} channel type does not contribute to the anesthetic properties of volatile anesthetics. This is perhaps not surprising, because L-type channels, which have major roles in cardiovascular physiology, are not principally involved in neurotransmitter release.[21] A small anesthetic-sparing effect for L-type Ca^{2+} channel blockers has been reported, however.[114,115] Ca^{2+} currents in the CNS are carried out by multiple Ca^{2+} channel types, which appear to have specialized functions. In contrast, nonneuronal excitable cells have one or two types of Ca^{2+} channel: L-type channels, which mediate slow inward currents, and T-type channels, which mediate small transient currents (see above; Table 6.1).

Analysis of the sensitivities of specific neuronal Ca^{2+} channel types to volatile anesthetics was first reported by Study[116] in isolated rat hippocampal pyramidal neurons. Using the whole-cell patch-clamp technique, Study identified macroscopic Ca^{2+} currents mediated by T-, L-, N-type, and other (probably P-type) channels (Figure 6.6). Isoflurane reversibly inhibited all LVA (T-type) and sustained HVA (L-, N-, and other type) Ca^{2+} currents with similar potencies at clinically relevant concentrations; the IC_{50} values were ~2 vol% (0.78 m*M*) for peak current and 1 vol% (0.39 m*M*) for sustained current at 22°C. Decay of both the transient and sustained components of the HVA current was accelerated. T-type channels were readily distinguished by their biophysical properties. N- and L-type channels were identified pharmacologically using ω-CTx GVIA and nitrendipine, respectively. The small current resistant to these two blockers was also inhibited by isoflurane; the contribution of P-type channels to this current could not be confirmed because ω-Aga IVA was not available. An accompanying study by Hall et al.[118] employed ω-Aga IVA to identify P-type channels (and possibly Q-type as well) in dissociated rat cerebellar Purkinje cells, which contain significantly more P-type channels than hippocampal neurons (91% inhibition of Ca^{2+} current by ω-Aga IVA). In these neurons, halothane inhibited peak P-type channel currents with modest potency (IC_{50} = 1.2 m*M*); isoflurane was similar in potency. These data suggest that P-type channels are relatively insensitive to volatile anesthetics, although the P-type channels present in cerebellar neurons may not be identical to those present in other neurons. The residual isoflurane-sensitive current described by Study[116] may reflect a different P-type channel variant or another Ca^{2+} channel type, such as Q- or R-type, that is resistant to both ω-CTx VIA and ω-Aga IVA. The effects of volatile anesthetics on single neuronal Ca^{2+} channels and on Ca^{2+} currents carried by Q- and R-type channels

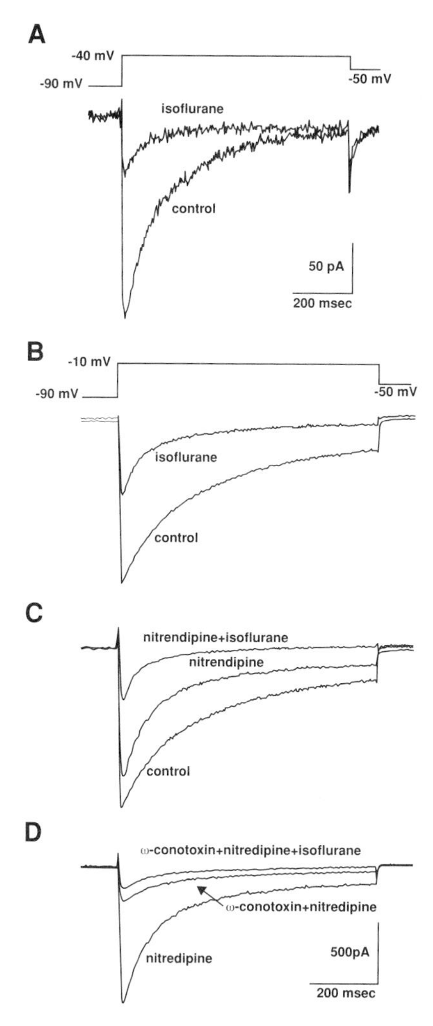

FIGURE 6.6 Inhibition of neuronal Ca^{2+} channels by isoflurane. (A) Isoflurane (2.5 vol%) inhibits the LVA transient Ca^{2+} current (T-type) during a voltage step of –90 to –40 mV, where the HVA currents are not significantly activated. (B–D) Isoflurane (2.5 vol%) inhibition of Ca^{2+} currents in the presence and absence of nitrendipine and ω-conotoxin GVIA. Isoflurane (2.5 vol%) was applied by puffer pipette. (B) Isoflurane inhibits the HVA current resulting from a depolarization from –90 to –10 mV. (C) The same cell before and after a supramaximal concentration of 10 μ*M* nitrendipine was added to block L-type channels, leaving N- and P-type (as well as other undefined) Ca^{2+} channels. (D) The same cell with 0.5 μ*M* ω-conotoxin GVIA added to nitrendipine, to eliminate N- and L-type channels. This cell had no T current. (Modified from Study.[116] With permission.)

remain to be studied. Electrophysiologic analysis of anesthetic effects on single cloned Ca^{2+} channel types should help clarify this issue. Single channel analysis of L-type Ca^{2+} channels in human SH-SY5Y neuroblastoma cells showed that halothane inhibited peak (IC_{50} = 0.80 m*M*) and sustained (IC_{50} = 0.69 m*M*) Ba^{2+} current by reducing open probability and enhancing inactivation.[119] Halothane also inhibited Ca^{2+} channel (N- and/or L-type) dependent [^{3}H]norepinephrine release from SH-5Y5Y cells evoked by elevated KCl (IC_{50} = 0.35 m*M*).[117]

A number of studies have used neuroendocrine cell lines or isolated adrenal chromaffin cells as model systems to study anesthetic effects on Ca^{2+} channels. The effects of halothane on whole-cell Ca^{2+} currents[120] and peptide secretion[121] in a transformed pituitary neuroendocrine cell line (GH_3 cells) have been reported. Halothane (0.5 to 5 m*M*) inhibited both transient LVA Ca^{2+} current ($IC_{50} \cong$ 1.3 m*M* for peak current) and HVA Ca^{2+} current ($IC_{50} \cong$ 0.8 m*M* for peak current); there was little effect on the voltage dependence of activation or inactivation of either current. Ca^{2+}-mediated inactivation of HVA Ca^{2+} channels did not appear to be a factor because Ca^{2+} chelators did not prevent halothane inhibition. Inhibition by halothane of Ca^{2+} currents correlated with inhibition of the sustained phase (extracellular Ca^{2+} dependent) of thyrotropin-releasing hormone (TRH)-induced prolactin secretion ($IC_{50} \cong$ 0.4 m*M*) and $[Ca^{2+}]_i$ increase ($IC_{50} \cong$ 0.4 m*M*). The early $InsP_3$-mediated phase of secretion and $[Ca^{2+}]_i$ increase was less sensitive. Halothane (0.5 m*M*) also inhibited KCl-induced (10 m*M*) prolactin secretion and the corresponding rise in $[Ca^{2+}]_i$. L-type Ca^{2+} channels were involved in the effect of halothane, because the TRH-induced $[Ca^{2+}]_i$ rise was blocked by nimodipine. These findings may not be comparable to anesthetic effects on fast synaptic transmission in the CNS, because the Ca^{2+} sensitivity and specific Ca^{2+} channel types coupled to peptide release may differ.[21]

A similar study carried out in rat pheochromocytoma (PC12) cells showed that methoxyflurane, halothane, isoflurane, and enflurane dose dependently inhibited dopamine and norepinephrine secretion evoked by nicotinic cholinergic receptor stimulation or by high KCl (56 m*M*), with IC_{50} values in the clinical range (<1 m*M*), but not secretion evoked by muscarinic cholinergic receptor stimulation.[122] Inhibition of KCl-evoked secretion involved L-type channels, because nifedipine completely blocked both secretion and the rise in $[Ca^{2+}]_i$. Inhibition of nicotinic cholinergic receptor-evoked release resulted from blockade of nifedipine-insensitive receptor-mediated Ca^{2+} influx, probably through the receptor itself. Whole-cell patch-clamp recording in rat medullary thyroid C cells, a neuroendocrine cell line with both LVA and HVA Ca^{2+} channels, showed that isoflurane and enflurane reversibly inhibited LVA currents (T-type channels; ~20% inhibition at 1 MAC [minimum alveolar concentration]) with greater potency than HVA currents (L- and N-type channels, as indicated by their DHP and ω-CTx GVIA sensitivities), whereas halothane was equipotent against both LVA and HVA currents.[123] Furthermore, L- and N-type Ca^{2+} channels were equally sensitive to isoflurane, and channel activation and inactivation kinetics were not affected. The different sensitivities of LVA currents to anesthetics in these studies may reflect differences between neuroendocrine and neuronal Ca^{2+} channels, a consideration that hampers the generalization of such results to neuronal Ca^{2+} channels.

Isolated bovine adrenal chromaffin cells have been used by several groups as a model system for correlating anesthetic effects on Ca^{2+} signaling and catecholamine secretion.[124] Chromaffin cells are neural crest derivatives homologous to sympathetic postganglionic neurons that express both L- and N-type Ca^{2+} channels.[125] Halothane inhibited catecholamine release evoked by nicotinic cholinergic stimulation in isolated perfused dog adrenals (IC_{50} = 0.8 vol%); inhibition of Ca^{2+} channels was postulated as a mechanism.[126] Clinically relevant concentrations of halothane, methoxyflurane, isoflurane, and enflurane all inhibited catecholamine secretion evoked by nicotinic cholinergic stimulation or by depolarization with high KCl (77 m*M*).[127] Although catecholamine release evoked by cholinergic stimulation was more sensitive to methoxyflurane (≥0.2 m*M*) than that evoked by high KCl (≥0.4 m*M*), inhibition of release in both cases correlated with inhibition of $^{45}Ca^{2+}$ influx.[127] This suggested that, in addition to a less potent direct inhibition of Ca^{2+} channels, volatile anesthetics inhibited the action of cholinergic agonists at their receptor. An effect on Ca^{2+} exocytosis coupling was ruled out by the absence of anesthetic effects on secretion from electropermeabilized cells. Although the interpretation of these data is not straightforward given the multiple mechanisms of Ca^{2+} entry involved, receptor-mediated secretion is apparently more sensitive to anesthetics than secretion evoked by direct Ca^{2+} channel activation. A similar study in bovine adrenal chromaffin cells demonstrated that KCl-evoked (56 m*M*) catecholamine secretion and $^{45}Ca^{2+}$ influx were insensitive to 2 vol% halothane, but that nicotinic cholinergic agonist-evoked catecholamine release and $^{45}Ca^{2+}$ influx were inhibited by halothane (0.5 to 2 vol%).[128]

Further analysis of the effects of methoxyflurane (0.2 to 1.6 m*M*) and halothane (0.5 to 2 m*M*) on Ca^{2+} currents in bovine adrenal chromaffin cells by whole-cell patch-clamp analysis revealed concentration-dependent and reversible inhibition of peak HVA currents ($IC_{50} \cong$ 1.3 m*M* and 2.5 m*M*, respectively) and an increase in the rate of decay (inactivation), with no change in the voltage–current relationship.[129] Both L- and N-type channels were present because Ca^{2+} current was blocked by nicardipine (67%) and ω-CTx GVIA (44%). The specific Ca^{2+} channel type(s) inhibited by the anesthetics could not be determined, however, because >50% inhibition of current was not demonstrated for either anesthetic. In a similar study, halothane (0.9 m*M*) or isoflurane (0.78 m*M*) had minimal effects on chromaffin cell Ca^{2+} currents,[130] whereas enflurane (1.7 m*M*) inhibited peak inward Ca^{2+} current by 60%.[125] These results are in contrast to the greater sensitivity of Ca^{2+} channels to anesthetics reported in GH_3 pituitary cells[121] and PC12 cells,[122] in which equipotent anesthetic effects on inhibition of KCl- or receptor-evoked secretion were reported also. The reasons for these discrepancies are unknown, but they highlight the complexity of the role of Ca^{2+} channels in neurosecretion. In summary, studies in neurosecretory cells have revealed relatively low sensitivity of Ca^{2+} channels to volatile anesthetics in bovine adrenal chromaffin cells, and somewhat greater anesthetic sensitivity of LVA and HVA Ca^{2+} channels in PC12 or GH_3 cells. These studies provided early evidence that Ca^{2+} channels coupled to transmitter release may be an important target for volatile anesthetic effects on synaptic transmission.

Additional evidence for interactions between volatile anesthetics and neuronal L-type channels has been obtained by ligand binding studies. Halothane inhibited the binding of isradipine to crude rat brain cerebrocortical membranes by reducing B_{max} without affecting K_d;[132] isoflurane and enflurane had inconsistent effects, as reported previously for cardiac membranes.[63] Halothane (1.9 vol%) was associated with a maximal reduction in B_{max} of 48%. Significant but nonstereoselective effects of (+) and (–) isoflurane on isradipine binding to crude rat brain cerebrocortical membranes (IC_{50} = 0.4 vol%) were demonstrated subsequently.[133] This effect was due to a reduction in B_{max} and an increase in K_d. These studies suggest an interesting noncompetitive interaction of volatile anesthetics and DHP binding to neuronal L-type Ca^{2+} channels.

6.7 NEUROTRANSMITTER RELEASE

A critical role of Ca^{2+} channels in the CNS is their regulation of neurotransmitter release.[134] Thus, the effects of anesthetics on neurotransmitter release may indirectly reflect their effects on Ca^{2+} channels. Nerve terminal depolarization is coupled to Ca^{2+} entry and neurotransmitter release by a highly organized supramolecular complex (synaptic core complex) consisting of synaptic vesicles and multiple vesicular and presynaptic proteins clustered at specialized zones of the presynaptic membrane.[135] Both synaptotagmin, the putative Ca^{2+} sensor, and syntaxin, a plasma membrane docking protein for fusion-competent vesicles, interact with N-type Ca^{2+} channels, a Ca^{2+} channel type that is coupled to neurotransmitter release in a number of neuronal systems.[21]

Considerable evidence supports a presynaptic locus for the inhibitory effects of general anesthetics on synaptic transmission in the CNS. Quantal analysis of electrophysiologic data revealed inhibition of excitatory neurotransmitter release in the spinal cord by ether[136] and halothane.[137,138] Halothane inhibited glutamate-mediated postsynaptic currents ($IC_{50} \cong 0.6$ m*M*), probably by inhibition of glutamate release presynaptically, determined by whole-cell recordings from CA1 pyramidal cells in the mouse hippocampal slice.[139] Indirect support for inhibition of glutamate release by a presynaptic mechanism by halothane ($IC_{50} \cong$ 1.3 vol%) or isoflurane ($IC_{50} \cong$ 1.0 vol%) was also obtained using extracellular recordings in the rat hippocampal slice.[140] Inhibition of Ca^{2+} channels was proposed as a possible mechanism for the anesthetic effects in both studies.

The effects of volatile anesthetics on Ca^{2+}-dependent neurotransmitter release have also been examined directly. Isoflurane inhibited the KCl-evoked release of glutamate, the predominant excitatory neurotransmitter in the CNS, from rat cerebral cortex slices ($IC_{50} \cong$ 3 vol% at 32°C).[141] The use of brain slices for this analysis is complicated by the possibility of glutamate release from nonneuronal cells and anesthetic effects on intact neuronal circuits and glutamate reuptake, as well as very high background rates of release. Subsequent studies employed the synaptosome preparation, an isolated subcellular nerve terminal fraction, which is less encumbered by these limitations.[142] Halothane, isoflurane, and enflurane inhibited glutamate release from rat cerebrocortical synaptosomes stimulated with 4-aminopyridine or veratridine, but not with ionomycin or elevated KCl.[144] These data suggested that volatile anesthetics were acting via a presynaptic mechanism involving Na^+ channels, because 4-aminopyridine- and veratridine-evoked release, which is tetrodotoxin sensitive, was inhibited, whereas KCl- and ionomycin-evoked release, which is tetrodotoxin insensitive, was not. Halothane did inhibit the KCl-evoked increase in intrasynaptosomal [Ca^{2+}], which indicated inhibition of Ca^{2+} channels, but apparently not of those closely coupled to glutamate release. A similar study employing guinea pig cerebrocortical synaptosomes reported that halothane, isoflurane, and enflurane inhibited both KCl-evoked glutamate release and $[Ca^{2+}]_i$ increase, consistent with blockade of Ca^{2+} channels; other secretogogues were not analyzed.[145] It is unclear why halothane did not inhibit KCl-evoked glutamate release in the prior study. Possibilities include a species difference or a difference in the analysis techniques that may have preferentially detected different pools of glutamate released.[146]

6.8 INTRACELLULAR CA^{2+} RELEASE

The recognition of the important role of intracellular Ca^{2+} regulation in normal and pathologic cell function has led to a number of studies of the effects of anesthetics on intracellular Ca^{2+} channels and $[Ca^{2+}]_i$. The role of Ca^{2+} release from myocardial SR in determining the negative inotropic potency of various volatile anesthetics was discussed above. Anesthetic effects on intracellular Ca^{2+} release have also been demonstrated in noncardiac cells. Volatile anesthetics also augment Ca^{2+} release from the SR in vascular smooth muscle.[106,143] There is good evidence that the pathogenesis of malignant hyperthermia involves anesthetic effects on skeletal muscle Ca^{2+}-induced Ca^{2+} release mediated by the ryanodine receptor.[147] A single mutation in the skeletal muscle ryanodine receptor has been identified as the cause of malignant hyperthermia in pigs[148] and in some human lineages, although there is genetic heterogeneity in the human disease. Halothane increases the open probability and conductance of ryanodine-sensitive Ca^{2+} release channels in tissue from humans susceptible to malignant hyperthermia but not of those from control subjects.[149] Halothane and caffeine potentiate, whereas dantrolene may inhibit, SR Ca^{2+} release, probably through direct effects on the ryanodine receptor. In contrast, there appears to be no difference in the sensitivities of malignant hyperthermia-susceptible and normal pig skeletal muscle L-type Ca^{2+} channels to volatile anesthetic inhibition of DHP binding, which correlates with channel activity.[70] Halothane, enflurane, and isoflurane inhibited Ca^{2+} mobilization in stimulated neutrophils in the absence or presence of extracellular Ca^{2+}, which suggested an inhibitory effect on intracellular Ca^{2+} release.[33] However, a high concentration of halothane (5.7 m*M*) produced a rapid increase in $[Ca^{2+}]_i$ in peripheral blood mononuclear cells from normal or malignant hyperthermia-susceptible humans or pigs.[151] In bovine aortic endothelial cells, halothane and enflurane, but not isoflurane, inhibited agonist-induced $[Ca^{2+}]_i$ transients, including the initial peak that is due to intracellular Ca^{2+} mobilization,[152-154] although another study[155] found that halothane did not affect the agonist-induced increase in cytoplasmic Ca^{2+}. In permeabilized rat hepatocytes, however, halothane, enflurane, and to a lesser extent isoflurane (at concentrations equivalent to or less than 1 MAC) stimulated $^{45}Ca^{2+}$ release from intracellular stores;[153] a similar effect of sevoflurane was observed in permeabilized hepatocytes, but no increase in $[Ca^{2+}]_i$ was observed in aequorin-loaded intact hepatocytes.[155] Halothane, enflurane, and isoflurane produced immediate transient increases in $[Ca^{2+}]_i$, measured using fura-2 fluorescence or aequorin (a Ca^{2+}-sensitive photoprotein) luminescence in cultured rat hepatocytes.[157] Using the L_6 skeletal muscle cell line, a high concentration of halothane (5.7 m*M*) increased $[Ca^{2+}]_i$, measured by indo-1 spectrofluorimetry.[158] This effect was observed both in the presence and absence of external Ca^{2+}, indicating a partial dependence on intracellular Ca^{2+} release. The effect of halothane on spontaneous and $InsP_3$-induced $^{45}Ca^{2+}$ release and $[Ca^{2+}]_i$ measured by aequorin luminescence was enhanced in permeabilized hepatocytes prepared from malignant hyperthermia-susceptible versus normal animals.[159] Although these studies are consistent with a direct effect on intracellular Ca^{2+} channels, the precise molecular target for these effects remains to be identified. Evidence for a direct interaction is an important consideration because volatile anesthetics have been suggested to have both stimulatory[160] and inhibitory[161,162] effects on agonist-stimulated $InsP_3$ formation in nonneuronal cells. In contrast, halothane (<4 m*M*) had no significant effects on Ca^{2+} signaling in pancreatic acinar cells, although it inhibited cholecystokinin-induced $InsP_3$ production and $[Ca^{2+}]_i$ oscillations, but not $InsP_3$-induced Ca^{2+} release.[163] Nor do clinical concentrations of halothane appear to affect the phosphatidylinositol phosphate signaling pathway or IP_3-induced Ca^{2+} mobilization in bovine aortic endothelial cells[156] or in neuronal cells.[121,122] However, the effects of volatile anesthetics on cellular IP_3 levels have not been measured directly.

Anesthetic effects on intracellular Ca^{2+} mobilization have also been investigated in neuroendocrine and neuronal cells. Halothane and isoflurane, at clinically relevant concentrations, accelerated Ca^{2+} release from intracellular stores in the GH_3 pituitary cell line, which produced an initial increase in $[Ca^{2+}]_i$, followed by a time-dependent inhibition of agonist-induced Ca^{2+} release due to depletion of intracellular Ca^{2+} stores.[164] Although the $InsP_3$ receptor was not directly identified as the anesthetic target, this possibility was favored in view of the significant sequence similarity between the $InsP_3$ receptor and ryanodine receptors, which are stimulated by volatile anesthetics.[99] Halothane (1 to 4 vol%), methoxyflurane (0.5 to 2 vol%), isoflurane (1.5 vol%), and enflurane (2.2 vol%) potentiated phorbol ester-evoked [^{3}H]norepinephrine release from PC12 cells (by ~80% at 1 MAC).[165] This was apparently due to an increase in basal $[Ca^{2+}]_i$ by an undefined mechanism, although a direct effect on protein kinase C cannot be ruled out.[166] Halothane does not appear to affect $InsP_3$ formation in GH_3 cells[121] or rat brain slices.[167] An early report described an increase in resting $[Ca^{2+}]_i$ in mouse whole brain synaptosomes induced by high concentrations of halothane (3 to 6 m*M*) and diethylether,[168] but subsequent studies using rat cortical synaptosomes did not reproduce this finding using clinically relevant anesthetic concentrations.[144,145] Halothane, isoflurane, and enflurane also failed to affect basal $[Ca^{2+}]_i$ in cultured rat hippocampal neurons, even above clinically relevant concentrations.[113] In contrast, Mody et al.[169] presented indirect evidence that halothane tonically increases $[Ca^{2+}]_i$ using whole-cell patch-clamp recordings of rat hippocampal neurons in brain slices. This interpretation was based on the ability of the Ca^{2+} chelator BAPTA or the Ca^{2+} release inhibitor dantrolene to inhibit the halothane-induced (1.2 vol%) increase in the decay time constant (τ_D) of $GABA_A$-receptor–mediated IPSCs. However, a similar study using cultured rat hippocampal neurons found that intracellular BAPTA did not affect volatile anesthetic prolongation of $GABA_A$-receptor–mediated IPSCs.[170] Likewise, the effect of GABA on $GABA_A$

receptors expressed in *Xenopus* oocytes was potentiated by various anesthetics even in the presence of the Ca^{2+} chelator EGTA.[171] Thus, anesthetic effects on $[Ca^{2+}]_i$ are apparently not involved in their actions on $GABA_A$ receptors. Further studies will be required to clarify the actions of volatile anesthetics on neuronal $InsP_3$ and ryanodine receptors and their influence on resting $[Ca^{2+}]_i$.

6.9 CONCLUSIONS

Considerable evidence indicates significant effects of volatile anesthetics on both voltage-dependent Ca^{2+} channels and intracellular Ca^{2+} channels in muscle. Studies of anesthetic effects on Ca^{2+} channels in other tissues, and neurons in particular, are just beginning to unravel the complex interactions of anesthetics with the multiple Ca^{2+} channel types and their possible sites of interaction with the channels. The evidence for volatile anesthetic inhibition of neuronal T-type, L-type, and N-type Ca^{2+} channels is convincing, whereas the evidence for effects on other Ca^{2+} types or on intracellular Ca^{2+} channels is either inconsistent or unavailable. Further work is clearly indicated to determine the relative sensitivities of different Ca^{2+} channel types to volatile anesthetics, and the role of these effects in anesthetic-sensitive cell functions such as synaptic transmission.

ACKNOWLEDGMENTS

I am grateful to Dr. Thomas J. J. Blanck for his critical reading of this manuscript, and to Lisa Cabrera and Christina Kim for expert secretarial assistance.

REFERENCES

1. Bean, B. P., Classes of calcium channels in vertebrate cells, *Annu Rev Physiol* 51, 367, 1989.
2. Hess, P., Calcium channels in vertebrate cells, *Annu Rev Neurosci* 13, 337, 1990.
3. Spedding, M. and Paoletti, R., Classification of calcium channels and the sites of action of drugs modifying channel function, *Pharmacol Rev* 44, 363, 1992.
4. Zhang, J. F., Randall, A. D., Ellinor, P. T., Horne, W. A., Sather, W. A., Tanabe, T., Schwarz, T. L. and Tsien, R. W., Distinctive pharmacology and kinetics of cloned neuronal Ca^{2+} channels and their possible counterparts in mammalian CNS neurons, *Neuropharmacology* 32, 1075, 1993.
5. Birnbaumer, L., Campbell, K. P., Catterall, W. A., Harpold, M. M., Hofmann, F., Horne, W. A., Mori, Y., Schwartz, A., Snutch, T. P., Tanabe, T. and Tsien, R. W., The naming of voltage-gated calcium channels, *Neuron* 13, 505, 1994.
6. Hofmann, F., Biel, M. and Flockerzi, V., Molecular basis for Ca^{2+} channel diversity, *Annu Rev Neurosci* 17, 399, 1994.
7. Catterall, W. A., Structure and function of voltage-sensitive ion channels, *Science* 242, 50, 1988.
8. Dolphin, A. C., Voltage-dependent calcium channels and their modulation by neurotransmitters and G proteins, *Exp Physiol* 80, 1, 1995.
9. Catterall, W. A. and Striessnig, J., Receptor sites for Ca^{2+} channel antagonists, *Trends Pharmacol Sci* 13, 256, 1992.

10. Varadi, G., Mori, Y., Mikala, G. and Schwartz, A., Molecular determinants of Ca^{2+} channel function and drug action, *Trends Pharmacol Sci* 16, 43, 1995.
11. Grabner, M., Wang, Z., Hering, S., Striessnig, J. and Glossmann, H., Transfer of 1,4-dihydropyridine sensitivity from L-type to Class A (BI) calcium channels, *Neuron* 16, 207, 1996.
12. Hockerman, G. H., Johnson, B. D., Scheuer, T. and Catterall. W. A., Molecular determinants of high affinity phenylalkylamine block of L-type calcium channels, *J Biol Chem* 270, 22119, 1995.
13. Peterson, B. Z., Tanada, T. N. and Catterall, W. A., Molecular determinants of high affinity dihydropyridine binding in L-type calcium channels, *J Biol Chem* 271, 5293, 1996.
14. Hering, S., Aczél, S., Grabner, M., Döring, F., Berjukow, S., Mitterdorfer, J., Sinnegger, M. J., Striessnig, J., Degtiar, V. E., Wang, Z. and Glossmann, H., Transfer of high sensitivity for benzothiazepines from L-type to class A (BI) calcium channels, *J Biol Chem* 271, 24471, 1996.
15. Striessing, J., Berger, W. and Glossmann, H., Molecular properties of voltage-dependent Ca^{2+} channels in excitable tissues, *Cell Physiol Biochem* 3, 295, 1993.
16. Hell, J. W., Westernbroek, R. E., Warner, C., Ahlijanian, M. K., Prystay, W., Gilbert, M. M., Snutch, T. P. and Catterall, W. A., Identification and differential subcellular localization of the neuronal class C and class D L-type calcium channel $\alpha 1$ subunits, *J Cell Biol* 123, 949, 1993.
17. Westenbroek, R. E., Hell, J. W., Warner, C., Dubel, S. J., Snutch, T. P. and Catterall, W. A., Biochemical properties and subcellular distribution of an N-type calcium channel $\alpha 1$ subunit, *Neuron* 9, 1099, 1992.
18. Westenbroek, R. E., Sakurai, T., Elliott, E. M., Hell, J. W., Starr, T. V. B., Snutch, T. P. and Catterall, W. A., Immunochemical identification and subcellular distribution of the α_{1A} subunits of brain calcium channels, *J Neurosci* 15, 6403, 1995.
19. Randall, A. and Tsien, R. W., Pharmacological dissection of multiple types of Ca^{2+} channel currents in rat cerebellar granule neurons, *J Neurosci* 15, 2995, 1995.
20. Tsien, R. W., Ellinor, P. T. and Horne, W. A., Molecular diversity of voltage-dependent Ca^{2+} channels, *Trends Pharmacol. Sci* 12(9), 349, 1991.
21. Dunlap, K., Luebke, J. I. and Turner, T. J., Exocytotic Ca^{2+} channels in mammalian central neurons, *Trends Neurosci* 18, 89, 1995.
22. Cachelin, A. B., De Peyer, J. E., Kokubun, S. and Reuter, H., Ca^{2+} channel modulation by 8-bromocyclic AMP in cultured heart cells, *Nature* 304, 462, 1983.
23. Bean, B. P., Nowycky, M. C. and Tsien R. W., β-Adrenergic modulation of calcium channels in frog ventricular heart cells, *Nature* 307, 371, 1984.
24. Yue, D. T., Herzig, S. and Marban, E., β-Adrenergic stimulation of calcium channels occurs by potentiation of high-activity gating modes, *Proc Natl Acad Sci USA* 87, 753, 1990.
25. Armstrong, D. L. and Eckert, R., Voltage-activated calcium channels that must be phosphorylated to respond to membrane depolarization, *Proc Natl Acad Sci USA* 84, 2518, 1987.
26. Hymel, L., Striessing, J., Glossmann, H. and Schindler, H., Purified skeletal muscle 1,4-dihydropyridine receptor forms phosphorylation-dependent oligomeric calcium channels in planar bilayers, *Proc Natl Acad Sci USA* 85, 4290, 1988.
27. Nunoki, K., Florio, V. and Catterall, W. A., Activation of purified calcium channels by stoichiometric protein phosphorylation, *Proc Natl Acad Sci USA* 86, 6816, 1989.
28. Hell, J. W., Yokoyama, C. T., Wong, S. T., Warner, C., Snutch, T. P. and Catterall, W. A., Differential phosphorylation of two size forms of the neuronal class C L-type calcium channel $\alpha 1$ subunit, *J Biol Chem* 268, 19451, 1993.

29. Hell, J. W., Appleyard, S. M., Yokoyama, C. T., Warner, C. and Catterall, W. A., Differential phosphorylation of two size forms of the N-type calcium channel α1 subunit which have different COOH termini, *J Biol Chem* 269, 7390, 1994.
30. Rotman, E. L., De Jongh, K. S., Florio, V., Lai, Y. and Catterall, W. A., Specific phosphorylation of a COOH-terminal site on the full-length form of the alpha 1 subunit of the skeletal muscle calcium channel by cAMP-dependent protein kinase, *J Biol Chem* 267, 16100, 1992.
31. Hell, J. W., Westenbroek, R. E., Breeze, L. J., Wang, K. K. W., Chavkin, C. and Catterall, W. A., *N*-Methyl-D-aspartate receptor-induced proteolytic conversion of postsynaptic class C L-type calcium channels in hippocampal neurons, *Proc Natl Acad Sci USA* 93, 3362, 1996.
32. Wei, X., Neely, A., Lacerda, A. E., Olcese, R., Stefani, E., Perez-Reyes, E. and Birnbaumer, L., Modification of Ca^{2+} channel activity by deletions at the carboxyl terminus of the cardiac alpha 1 subunit, *J Biol Chem* 269, 1635, 1994.
33. Dolphin, A. C., Facilitation of Ca^{2+} current in excitable cells, *TINS* 19, 35, 1996.
34. Schneider, T., Ingelmund, P., Hescheler, J., G protein interaction with K^+ and Ca^{2+} channels, *Trends Pharmacol Sci* (England), 18(1), 8, 1997.
35. Artalejo, C. R., Rossie, S., Perlman, R. L. and Fox, A. P., Voltage-dependent phosphorylation may recruit Ca^{2+} current facilitation in chromaffin cells, *Nature* 358, 63, 1992.
36. Clapham, D. E., Calcium signaling, *Cell* 80, 259, 1995.
37. Ghosh, A. and Greenberg, M. E., Calcium signaling in neurons: Molecular mechanisms and cellular consequences, *Science* 268, 239, 1995.
38. Berridge, M. J., Inositol trisphosphate and calcium signalling, *Nature* 361, 315, 1993.
39. McPherson, P. S. and Campbell, K. P., The ryanodine receptor/Ca^{2+} release channel, *J Biol Chem* 268, 13765, 1993.
40. Lai, F. A., Erickson, L. P., Rousseau, E., Liu, Q.-Y. and Meissner, G., Purification and reconstitution of the calcium release channel from skeletal muscle, *Nature* 331, 315, 1988.
41. Sitsapesan, R., McGarry, S. J. and Williams, A. J., Cyclic ADP-ribose, the ryanodine receptor and Ca^{2+} release, *Trends Pharmacol Sci* 16, 386, 1995.
42. Delaney, T. J., Kistner, J. R., Lake, C. L. and Miller, E. D., Jr, Myocardial function during halothane and enflurane anesthesia in patients with coronary artery disease, *Anesth Analg* 59, 240, 1980.
43. Bosnjak, Z. L. and Kampine, J. P., Effects of halothane, enflurane and isoflurane on the SA node, *Anesthesiology* 58, 314, 1983.
44. Atlee, J. L. and Alexander, S. C., Halothane effects on conductivity of the AV node and His-Purkinje system, *Anesth Analg* 56, 378, 1977.
45. Brown, R. B., Jr., and Crout J. R., A comparative study of five general anesthetics on myocardial contractility, *Anesthesiology* 34, 236, 1971.
46. Rusy, B. F. and Komai, H., Anesthetic depression of myocardial contractility: A review of possible mechanisms, *Anesthesiology* 67, 745, 1987.
47. Hauswirth, O., Effects of halothane on single atrial, ventricular, and Purkinje fibers, *Circ Res* 24, 745, 1969.
48. Lynch, C., Vogel, S. and Sperelakis, N., Halothane depression of myocardial slow action potentials, *Anesthesiology* 55, 360, 1981.
49. Lynch, C., Vogel, S., Pratila, M. and Sperelakis, N., Enflurane depression of myocardial slow action potentials, *J Pharmacol Exp Ther* 222, 405, 1982.
50. Hirota, K., Ito, Y., Masuda, A. and Momose, Y., Effects of halothane on membrane ionic currents in guinea pig atrial and ventricular myocytes, *Acta Anaesthesiol Scand* 33, 239, 1989.

51. Bosnjak, Z. J., Supan, F. D. and Rusch, N. J., The effects of halothane and isoflurane on calcium current in isolated canine ventricular cells, *Anesthesiology* 74, 340, 1991.
52. Ikemoto, Y., Yatani, A., Arimura, H. and Yoshitake, J., Reduction of the slow inward current of isolated rat ventricular cells by thiamylal and halothane, *Acta Anaesthesiol Scand* 29, 583, 1985.
53. Terrar, D. A. and Victory, J. G. G., Influence of halothane on membrane currents associated with contraction in single myocytes isolated from guinea-pig ventricle, *Br J Pharmacol* 94, 500, 1988.
54. Terrar, D. A. and Victory, J. G. G., Isoflurane depresses membrane currents associated with contraction in myocytes isolated from guinea-pig ventricle, *Anesthesiology* 69, 742, 1988.
55. Niggli, E., Rüdisüli, A., Maurer, P. and Weingart, R., Effects of general anesthetics on current flow across membranes in guinea pig myocytes, *Am J Physiol* 256, C273, 1989.
56. Puttick, R. M. and Terrar, D. A., Effects of propofol and enflurane on action potentials, membrane currents and contraction of guinea-pig isolated ventricular myocytes, *Br J Pharmacol* 107, 559, 1992.
57. Baum, V. C., Wetzel, G. T. and Klitzner, T. S., Effects of halothane and ketamine on activation and inactivation of myocardial calcium current, *J Cardiovasc Pharmacol* 23, 799, 1994.
58. Pancrazio, J. J., Halothane and isoflurane preferentially depress a slowly inactivating component of Ca^{2+} channel current in guinea-pig myocytes, *J Physiol* 494, 91, 1996.
59. Hirota, K., Fujimura, J., Wakasugi, M. and Ito, Y., Isoflurane and sevoflurane modulate inactivation kinetics of Ca^{2+} currents in single bullfrog atrial myocytes, *Anesthesiology* 84, 377, 1996.
60. Hatakeyama, N., Momose, Y. and Ito, Y., Effects of sevoflurane on contractile responses and electrophysiologic properties in canine single cardiac myocytes, *Anesthesiology* 82, 559, 1995.
61. Takahashi, H., Puttick, R. M. and Terrar, D. A., The effects of propofol and enflurane on single calcium channel currents of guinea-pig isolated ventricular myocytes, *Br J Pharmacol* 111, 1147, 1994.
62. Hirota, K. and Lambert, D. G., Voltage-sensitive Ca^{2+} channels and anaesthesia, *Br J Anaesth* 76, 344, 1996.
63. Blanck, T. J. J., Runge, S. and Steveson, R.L., Halothane decreases calcium channel antagonist binding to cardiac membranes, *Anesth Analg* 67, 1032, 1988.
64. Drenger, B., Quigg, M. and Blanck, T. J. J., Volatile anesthetics depress calcium channel blocker binding to bovine cardiac sarcolemma, *Anesthesiology* 74, 155, 1991.
65. Nakao, S., Hirata, H. and Kagawa, Y., Effects of volatile anesthetics on cardiac calcium channels, *Acta Anaesthesiol Scand* 33, 326, 1989.
66. Hoehner, P. J., Quigg, M. C. and Blanck, T. J. J., Halothane depresses D600 binding to bovine heart sarcolemma, *Anesthesiology* 75, 1019, 1991.
67. Schmidt, U., Schwinger, R. H. G., Böhm, S., Überfuhr, P., Kreuzer, E., Reichart, B., Meyer, L. V., Erdmann, E. and Böhm, M., Evidence for an interaction of halothane with the L-type Ca^{2+} channel in human myocardium, *Anesthesiology* 79, 332, 1993.
68. Chung, O. Y., Blanck, T. J. J. and Berman, M. R., Depression of myocardial force and stiffness without change in crossbridge kinetics: Effects of volatile anesthetics reproduced by nifedipine, *Anesthesiology* 71, 444, 1989.
69. Blanck, T. J. J., Lee, D. L., Yasukochi, S., Hollmann, C. and Zhang, J., The role of L-type voltage-depenent calcium channels in anesthetic depression of contractility, *Adv Pharmacol* 31, 207, 1994.

70. Louis, C. F., Roghair, T. and Mickelson, J. R., Volatile anesthetics inhibit dihydropyridine binding to malignant hyperthermia-susceptible and normal pig skeletal muscle membranes, *Anesthesiology* 80, 618, 1994.
71. Lee, D. L., Zhang, J. and Blanck, T. J. J., The effects of halothane on voltage-dependent calcium channels in isolated Langendorff-perfused rat heart, *Anesthesiology* 81, 1212, 1994.
72. Eskinder, H., Rusch, N. J., Supan, F. D., Kampine, J. P. and Bosnjak, Z. J., The effects of volatile anesthetics on L- and T-type calcium channel currents in canine cardiac Purkinje cells, *Anesthesiology* 74, 919, 1991.
73. Haworth, R. A. and Goknur, A. B., Inhibition of sodium/calcium exchange and calcium channels of heart cells by volatile anesthetics, *Anesthesiology* 82, 1255, 1995.
74. Housmans, P. R. and Murat, I., Comparative effects of halothane, enflurane, and isoflurane at equipotent anesthetic concentrations on isolated ventricular myocardium of the ferret. I. Contractility, *Anesthesiology* 69, 451, 1988.
75. Stowe, D. F., Monroe, S. M., Marijic, J., Bosnjak, Z. J. and Kampine, J. P., Comparison of halothane, enflurane, and isoflurane with nitrous oxide on contractility and oxygen supply and demand in isolated hearts, *Anesthesiology* 75, 1062, 1991.
76. Su, J. Y. and Kerrick, W. G. L., Effects of halothane on caffeine-induced tension transients in functionally skinned myocardial fibers, *Pflugers Arch* 380, 29, 1979.
77. Su, J. Y. and Kerrick, W. G. L., Effects of enflurane on functionally skinned myocardial fibers from rabbits, *Anesthesiology* 52, 385, 1980.
78. Su, J. Y. and Bell, J., Intracellular mechanism of action of isoflurane and halothane on striated muscle of rabbit, *Anesth Analg* 65, 457, 1986.
79. Blanck, T. J. J., Chiancone, E., Salviati, G., Heitmiller, E. S., Verzilli, D. and Luciani, G., Halothane does not alter Ca^{2+} affinity of troponin C, *Anesthesiology* 76, 100, 1992.
80. Bosnjak, Z. J., Aggarwal, A., Turner, L. A., Kampine, J. M. and Kampine, J. P., Differential effects of halothane, enflurane, and isoflurane on Ca^{2+} transients and papillary muscle tension in guinea pigs, *Anesthesiology* 76, 123, 1992.
81. Murat, I., Veksler, V. I. and Ventura-Clapier, R., Effects of halothane on contractile properties of skinned fibers from cardiomyopathic animals, *J Mol Cell Cardiol* 21, 1293, 1989.
82. Herland, J. S., Julian, F. J. and Stephenson, D. G., Effects of halothane, enflurane, and isoflurane on skinned rat myocardium activated by Ca^{2+}, *Am J Physiol* 264, H224, 1993.
83. Stowe, D. F., Sprung, J., Turner, L. A., Kampine, J. P. and Bosnjak, Z. J., Differential effects of halothane and isoflurane on contractile force and calcium transients in cardiac Purkinje fibers, *Anesthesiology* 80, 1360, 1994.
84. De Traglia, M. C., Komai, H. and Rusy, B. F., Differential effects of inhalation anesthetics on myocardial potentiated-state contractions *in vitro*, *Anesthesiology* 68, 534, 1988.
85. Lynch, C., III, Differential depression of myocardial contractility by volatile anesthetics *in vitro*: Comparison with uncouplers of excitation-contraction coupling, *J Cardiovasc Pharmacol* 15, 655, 1990.
86. Bernard, J. M., Wouters, P. F., Doursout, M. F., Florence, B., Chelly, J. E. and Merin, R. G., Effects of sevoflurane and isoflurane on cardiac and coronary dynamics in chronically instrumented dogs, *Anesthesiology* 72, 659, 1990.
87. Park, W. K., Pancrazio, J. J., Suh, C. K. and Lynch C., III, Myocardial depressant effects of sevoflurane, *Anesthesiology* 84, 1166, 1996.
88. Wheeler, D. M., Katz, A., Rice, T. and Hansford, R. G., Volatile anesthetic effects on sarcoplasmic reticulum Ca content and sarcolemmal Ca flux in isolated rat cardiac cell suspensions, *Anesthesiology* 80, 372, 1994.

89. Blanck, T. J. J. and Thompson, M., Calcium transport by cardiac sarcoplasmic reticulum: Modulation of halothane action by substrate concentration and pH, *Anesth Analg* 60, 390, 1981.
90. Blanck, T. J. J. and Thompson, M., Enflurane and isoflurane stimulate calcium transport by cardiac sarcoplasmic reticulum, *Anesth Analg* 61, 142, 1982.
91. Frazer, M. J. and Lynch III, C., Halothane and isoflurane effects on Ca^{2+} fluxes of isolated myocardial sarcoplasmic reticulum, *Anesthesiology* 77, 316, 1992.
92. Blanck, T. J. J., Peterson, C. V., Baroody, B., Tegazzin, V. and Lou, J., Halothane, enflurane, and isoflurane stimulate calcium leakage from rabbit sarcoplasmic reticulum, *Anesthesiology* 76, 813, 1992.
93. Nelson, T. E. and Sweo, T., Ca^{2+} uptake and Ca^{2+} release by skeletal muscle sarcoplasmic reticulum: Differing sensitivity to inhalational anesthetics, *Anesthesiology* 69, 571, 1988.
94. Andoh, T., Blanck, T. J. J., Nikinorov, I. and Recio-Pinto, E., Volatile anaesthetics effects on calcium conductance of planar lipid bilayers formed with synthetic lipids or extracted lipids from sarcoplasmic reticulum, *J Anaesth* 78, 66, 1997.
95. Herland, J. S., Julian, F. J. and Stephenson, D. G., Halothane increases Ca^{2+} efflux via Ca^{2+} channels of sarcoplasmic reticulum in chemically skinned rat myocardium, *J Physiol* 426, 1, 1990.
96. Wilde, D. W., Davidson, B. A., Smith, M. D. and Knight, P. R., Effects of isoflurane and enflurane on intracellular Ca^{2+} mobilization in isolated cardiac myocytes, *Anesthesiology* 79, 73, 1993.
97. Connelly, T. J., El-Hayek, R., Rusy, B. F. and Coronado, R., Volatile anesthetics selectively alter [^{3}H]ryanodine binding to skeletal and cardiac ryanodine receptors, *Biochem Biophys Res Commun* 186, 595, 1992.
98. Lynch, C., III and Frazer, M. J., Anesthetic alteration of ryanodine binding by cardiac calcium release channels, *Biochim Biophys Acta* 1194, 109, 1994.
99. Connelly, T. J. and Coronado, R., Activation of the Ca^{2+} release channel of cardiac sarcoplasmic reticulum by volatile anesthetics, *Anesthesiology* 81, 459, 1994.
100. Yamazaki, M., Kamitani, Y., Ito, Y. and Momose, Y., Effects of halothane and diltiazem on L-type calcium currents in single smooth muscle cells from rabbit portal veins, *Br J Anaesth* 73, 209, 1994.
101. Buljubasic, N., Rusch, N. J., Marijic, J., Kampine, J. P. and Bosnjak, Z. J., Effects of halothane and isoflurane on calcium and potassium channel currents in canine coronary arterial cells, *Anesthesiology* 76, 990, 1992b.
102. Buljubasic, N., Flynn, N. M., Marijic, J., Rusch, N. J., Kampine, J. P. and Bosnjak, Z. J., Effects of isoflurane on K^+ and Ca^{2+} conductance in isolated smooth muscle cells of canine cerebral arteries, *Anesth Analg* 75, 590, 1992a.
103. Wilde, D. W., Isoflurane reduces Ca^{++} channel current and accelerates current decay in guinea pig portal vein smooth muscle cells, *J Pharmacol Exp Ther* 271, 1159, 1994.
104. Yamakage, M., Hirshman, C. A. and Croxton, T. L., Volatile anesthetics inhibit voltage-dependent Ca^{2+} channels in porcine tracheal smooth muscle cells, *Am J Physiol* 268, L187, 1995.
105. Kakuyama, M., Hatano, Y., Nakamura, K., Toda, H., Terasako, K., Nishiwada, M. and Mori, K., Halothane and enflurane constrict canine mesenteric arteries by releasing Ca^{2+} from intracellular Ca^{2+} stores, *Anesthesiology* 80, 1120, 1994.
106. Namba, H. and Tsuchida, H., Effect of volatile anesthetics with and without verapamil on intracellular activity in vascular smooth muscle, *Anesthesiology* 84, 1465, 1996.
107. Su, J. Y., Chang, Y. I. and Tang, L. J., Mechanisms of action of enflurane on vascular smooth muscle, *Anesthesiology* 81, 700, 1994.

108. Richards, C. D., Actions of general anesthetics on synaptic transmission in the CNS, *Br J Anaesth* 55, 201, 1983.
109. Krnjevic, K. and Puil, E., Halothane suppresses slow inward currents in hippocampal slices, *Can J Physiol Pharmacol* 66, 1570, 1988.
110. McGivern, J. and Scholfield, C. N., Action of general anaesthetics on unclamped Ca^{2+}-mediated currents in unmyelinated axons of rat olfactory cortex, *Eur J Pharmacol* 203, 59, 1991.
111. Takenoshita, M. and Steinbach, J. H., Halothane blocks low-voltage-activated calcium current in rat sensory neurons, *J Neurosci* 11, 1404, 1991.
112. Todorovic, S. M. and Lingle, C., Pharmacological properties of T-type Ca^{2+} current in adult rat sensory neurons: Effects of anticonvulsant and anesthetic agents, *J Neurophysiol* 79, 240, 1997.
113. Bleakman, D., Jones, M. V. and Harrison, N. L., The effects of four general anesthetics on intracellular $[Ca^{2+}]$ in cultured rat hippocampal neurons, *Neuropharmacology* 34, 541, 1995.
114. Maze, M., Mason, D. M. and Kates, R. E., Verapamil decreases MAC for halothane in dogs, *Anesthesiology* 59, 327, 1983.
115. Dolin, S. J. and Little, H. J., Augmentation by calcium channel antagonists of general anaesthetic potency in mice, *Br J Pharmacol* 88, 909, 1986.
116. Study, R. E., Isoflurane inhibits multiple voltage-gated calcium currents in hippocampal pyramidal neurons, *Anesthesiology* 81, 104, 1994.
117. Atcheson, R., Bjornstrom, K., Hirst, R. A., Rowbotham, D. J. and Lambert, D. G., Effect of halothane on K^+ and carbachol stimulated $[^3H]$noradrenaline release and increased $[Ca^{2+}]_i$ in SH-SY5Y human neuroblastoma cells, *Br J Anaesth* 79, 78, 1997.
118. Hall, A. C., Lieb, W. R. and Franks, N. P., Insensitivity of P-type calcium channels to inhalational and intravenous general anesthetics, *Anesthesiology* 81, 117, 1994.
119. Nikonorov, I. M., Blanck, T. J. J. and Recio-Pinto, E., The effects of halothane on single human neuronal L-type calcium channels, *Anesth Analg* 86, 885, 1998.
120. Herrington, J., Stern, R. C., Evers, A. S. and Lingle, C. J., Halothane inhibits two components of calcium current in clonal (GH_3) pituitary cells, *J Neurosci* 11, 2226, 1991.
121. Stern, R. C., Herrington, J., Lingle, C. J. and Evers, A. S., The action of halothane on stimulus-secretion coupling in clonal (GH_3) pituitary cells, *J Neurosci* 11, 2217, 1991.
122. Kress, H. G., Müller, J., Eisert, A., Gilge, U., Tas, P. W. and Koschel, K., Effects of volatile anesthetics on cytoplasmic Ca^{2+} signaling and transmitter release in a neural cell line, *Anesthesiology* 74, 309, 1991.
123. McDowell, T. S., Pancrazio, J. J. and Lynch, C., III, Volatile anesthetics reduce low-voltage-activated calcium currents in a thyroid C-cell line, *Anesthesiology* 85, 1167, 1996.
124. Richards, C. D., The synaptic basis of general anaesthesia, *Eur J Anaesthesiol* 12, 5, 1995.
125. Bossu, J. L., De Waard, M. and Feltz, A., Two types of calcium channels are expressed in adult bovine chromaffin cells, *J Physiol* 437, 621, 1991.
126. Sumikawa, K., Matsumoto, T., Ishizaka, N., Nagai, H., Amenomori, Y. and Amakata, Y., Mechanism of the differential effects of halothane on nicotinic- and muscarinic-receptor-mediated responses of the dog adrenal medulla, *Anesthesiology* 57, 444, 1982.
127. Pocock, G. and Richards, C. D., The action of volatile anaesthetics on stimulus-secretion coupling in bovine adrenal chromaffin cells, *Br J Pharmacol* 95, 209, 1988.

128. Yashima, N., Wada, A. and Izumi, F., Halothane inhibits the cholinergic-receptor-mediated influx of calcium in primary culture of bovine adrenal medulla cells, *Anesthesiology* 64, 466, 1986.
129. Charlesworth, P., Pocock, G. and Richards, C. D., Calcium channel currents in bovine adrenal chromaffin cells and their modulation by anaesthetic agents, *J Physiol* 481, 543, 1994.
130. Pancrazio, J. J., Park, W. K. and Lynch, C., III, Inhalational anesthetic actions on voltage-gated ion currents of bovine adrenal chromaffin cells, *Mol Pharmacol* 43, 783, 1993.
131. Pancrazio, J. J., Park, W. K. and Lynch, C., III, Effects of enflurane on the voltage-gated membrane currents of bovine adrenal chromaffin cells, *Neurosci Lett* 146, 147, 1992.
132. Drenger, B., Heitmiller, E. S., Quigg, M. and Blanck, T. J. J., Depression of calcium channel blocker binding to rat brain membranes by halothane, *Anesth Analg* 74, 758, 1992.
133. Moody, E. J., Harris, B., Hoehner, P. and Skolnick, P., Inhibition of [^{3}H]isradipine binding to L-type calcium channels by the optical isomers of isoflurane, *Anesthesiology* 81, 124, 1994.
134. Smith, S. J. and Augustine, G. J., Calcium ions, active zones and synaptic transmitter release, *Trends Neurosci* 11, 458, 1988.
135. Bajjalieh, S. M. and Scheller, R. H., The biochemistry of neurotransmitter secretion, *J Biol Chem* 270, 1971, 1995.
136. Zorychta, E. and Capek, R., Depression of spinal monosynaptic transmission by diethyl ether: Quantal analysis of unitary synaptic potentials *J Pharmacol Exp Ther* 207, 825, 1978.
137. Takenoshita, M. and Takahashi, T., Mechanisms of halothane action on synaptic transmission in motoneurons of the newborn rat spinal cord *in vitro*, *Brain Res* 402, 303, 1987.
138. Kullmann, D. M., Martin, R. L. and Redman, S. J., Reduction by general anaesthetics of group Ia excitatory postsynaptic potentials and currents in the cat spinal cord, *J Physiol* 412, 277, 1989.
139. Perouansky, M., Baranov, D., Salman, M. and Yaari, Y., Effects of halothane on glutamate receptor-mediated excitatory postsynaptic currents, *Anesthesiology* 83, 109, 1995.
140. MacIver, M. B., Mikulec, A. A., Amagasu, S. M. and Monroe, F. A., Volatile anesthetics depress glutamate transmission *via* presynaptic actions, *Anesthesiology* 85, 823, 1996.
141. Larsen, M., Grondahl, T. O., Haugstad, T. S. and Langmoen, I. A., The effect of the volatile anesthetic isoflurane on Ca^{2+}-dependent glutamate release from rat cerebral cortex, *Brain Res* 663, 335, 1994.
142. Nicholls, D. G., The glutamatergic nerve terminal, *Eur J Biochem* 212, 613, 1993.
143. Tsuchida, H., Namba, H., Seki, S., Fujita, S., Tanaka, S. and Namiki, A., Role of intracellular Ca^{2+} pools in the effects of halothane and isoflurane on vascular smooth muscle contraction, *Anesth Analg* 78, 1067, 1994.
144. Schlame, M. and Hemmings, H. C., Jr., Inhibition by volatile anesthetics of endogenous glutamate release from synaptosomes by a presynaptic mechanism *Anesthesiology* 82, 1406, 1995.
145. Miao, N., Frazer, M. J. and Lynch, C., III, Volatile anesthetics depress Ca^{2+} transients and glutamate release in isolated cerebral synaptosomes, *Anesthesiology* 83, 593, 1995.

146. Herrero, I., Castro, E., Miras-Portugal, M. T. and Sánchez-Prieto, J., Two components of glutamate exocytosis differentially affected by presynaptic modulation, *J Neurochem* 67, 2346, 1996.
147. Louis, C. F., Zualkernan, K., Roghair, T. and Mickelson, J. R., The effects of volatile anesthetics on calcium regulation by malignant hyperthermia-susceptible sarcoplasmic reticulum, *Anesthesiology* 77, 114, 1992.
148. Fujii, J., Otsu, K., Zorzato, F., De Leon, S., Khanna, V. K., Weiler, J. E., O'Brien, P. J. and MacLennan, D. H., Identification of a mutation in porcine ryanodine receptor associated with malignant hyperthermia, *Science* 253, 448, 1991.
149. Nelson, T. E., Halothane effects on human malignant hyperthermia skeletal muscle single calcium-release channels in planar lipid bilayers, *Anesthesiology* 76, 588, 1992.
150. Nakagawara, M., Takeshige, K., Takamatsu, J., Takahashi, S., Yoshitake, J. and Minakami, S., Inhibition of superoxide production and Ca^{2+} mobilization in human neutrophils by halothane, enflurane, and isoflurane, *Anesthesiology* 64, 4, 1986.
151. Klip, A., Mills, G. B., Britt, B. A. and Elliott, M. E., Halothane-dependent release of intracellular Ca^{2+} in blood cells in malignant hyperthermia, *Am J Physiol* 258, C495, 1990a.
152. Loeb, A. L., Longnecker, D. L. and Williamson, J. R., Alteration of calcium mobilization in endothelial cells by volatile anesthetics, *Biochem Pharmacol* 45, 1137, 1993.
153. Loeb, A. L., O'Brien, D. K. and Longnecker, D. L., Halothane inhibits bradykinin-stimulated prostacyclin production in endothelial cells, *Anesthesiology* 81, 931, 1994.
154. Pajewski, T. N., Miao, N., Lynch, C., III and Johns, R. A., Volatile anesthetics affect calcium mobilization in bovine endothelial cells, *Anesthesiology* 85, 1147, 1996.
155. Iaizzo, P. A., Seewald, M. J., Powis, G. and Van Dyke, R. A., The effects of sevoflurane on intracellular Ca^{2+} regulation in rat hepatocytes, *Toxicol Lett* 66, 81, 1993.
156. Simoneau, C., Thuringer, D., Cai, S., Garneau, L., Blaise, G. and Sauvé, R., Effects of halothane and isoflurane on bradykinin-evoked Ca^{2+} influx in bovine aortic endothelial cells, *Anesthesiology*, 85, 366, 1996.
157. Iaizzo, P. A., Olsen, R. A., Seewald, M. J., Powis, G., Stier, A. and Van Dyke, R. A., Transient increases of intracellular Ca^{2+} induced by volatile anesthetics in rat hepatocytes, *Cell Calcium* 11, 515, 1990.
158. Klip, A., Hill, M. and Ramlal, T., Halothane increases cytosolic Ca^{2+} and inhibits Na^+/H^+ exchange in L6 muscle cells, *J Pharmacol Exp Ther* 254, 552, 1990b.
159. Iaizzo, P. A., Seewald, M. J., Olsen, R., Wedel, D. J., Chapman, D. E., Berggren, M., Eichinger, H. M. and Powis, G., Enhanced mobilization of intracellular Ca^{2+} induced by halothane in hepatocytes isolated from swine susceptible to malignant hyperthermia, *Anesthesiology* 74, 531, 1991.
160. Smart, D., Smith, G. and Lambert D. G., Halothane and isoflurane enhance basal and carbachol-stimulated inositol (1,4,5)triphosphate formation in SH-SY5Y human neuroblastoma cells, *Biochem Pharmacol* 47, 939, 1994.
161. Sill, J. C., Uhl, C., Eskuri, S., Dyke, R. V. and Tarara, J., Halothane inhibits agonist-induced inositol phosphate and Ca^{2+} signaling in A7r5 cultured vascular smooth muscle cells, *Mol Pharmacol* 40, 1006, 1991.
162. Kohro, S. and Yamakage, M., Direct inhibitory mechanisms of halothane on human platelet aggregation, *Anesthesiology* 85, 96, 1996.
163. Deutsch, D. E., Williams, J. A. and Yule, D. I., Halothane and octanol block Ca^{2+} oscillations in pancreatic acini by multiple mechanisms, *Am J Physiol* 269, G779, 1995.
164. Hossain, M. D. and Evers, A. S., Volatile anesthetic-induced efflux of calcium from IP_3-gated stores in clonal (GH_3) pituitary cells, *Anesthesiology* 80, 1379, 1994.

165. Tas, P. W. L. and Koschel, K., Volatile anesthetics stimulate the phorbol ester evoked neurotransmitter release from PC12 cells through an increase of the cytoplasmic Ca^{2+} ion concentration, *Biochim Biophys Acta* 1091, 401, 1991.
166. Hemmings, H. C., Jr. and Adamo, A. I. B., Activation of endogenous protein kinase C by halothane in synaptosomes, *Anesthesiology* 84, 652, 1996.
167. Bazil, C. W. and Minneman, K. P., Effects of clinically effective concentrations of halothane on adrenergic and cholinergic synapses in rat brain *in vitro, J Pharmacol Exp Ther* 248, 143, 1989.
168. Daniell, L. C. and Harris, R. A., Neuronal intracellular calcium concentrations are altered by anesthetics: Relationship to membrane fluidization, *J Pharmacol Exp Ther* 245, 1, 1987.
169. Mody, I., Tanelian, D. L. and MacIver, M. B., Halothane enhances tonic neuronal inhibition by elevating intracellular calcium, *Brain Res* 538, 319, 1991.
170. Jones, M. V. and Harrison, N. L., Effects of volatile anesthetics on the kinetics of inhibitory postsynaptic currents in cultured rat hippocampal neurons, *J Neurophysiol* 70, 1339, 1993.
171. Lin, L.-H., Chen, L. L., Zirrolli, J. A. and Harris, R. A., General anesthetics potentiate γ-aminobutyric acid actions on γ-aminobutyric $acid_A$ receptors expressed by *Xenopus* oocytes: Lack of involvement of intracellular calcium, *J Pharmacol Exp Ther* 263, 569, 1992.

7 Inhalation Anesthetic Effects on Neuronal Plasma Membrane Ca^{2+}-ATPase

Piotr K. Janicki

CONTENTS

0-8493-8555-5/01/$0.00+$.50

7.1 ROLE OF Ca^{2+} IN NEUROTRANSMISSION

It is a fundamental tenet of modern biology that small diffusible second messengers act within the cell to provide a link between stimulus and output. Second messenger functions for Ca^{2+} are well established, and the role of intracellular Ca^{2+} in coupling neuronal excitation to the release of neurotransmitters into the synaptic space makes anesthetic modulation of calcium homeostasis an attractive, although disputed, potential mechanism for anesthetic action.[1-4]

Regulation of cytosolic Ca^{2+} concentration ($[Ca^{2+}]_i$) is essential for maintaining a balance between rapid Ca^{2+} signaling and Ca^{2+}-dependent toxicity and must be done against a 10,000:1 gradient of extracellular to intracellular Ca^{2+}. Ca^{2+} channels in the plasma membrane and intracellular release channels in the endoplasmic reticulum are capable of increasing $[Ca^{2+}]_i$ rapidly with appropriate stimulation. Three main systems serve to reduce or maintain $[Ca^{2+}]_i$ at low levels: a plasma membrane Na/Ca exchanger, a plasma membrane Ca^{2+}-ATPase (PMCA) pump, and a smooth endoplasmic reticulum Ca^{2+}-ATPase (SERCA) pump. Mitochondrial uptake and release functions are thought to operate principally at high $[Ca^{2+}]_i$. The role of intracellular Ca^{2+} buffering proteins is unclear at the present time.[5,6] Figure 7.1 (a schematic diagram) and Figure 7.2 (a chart) show various mechanisms involved in the regulation of intracellular Ca^{2+} concentration.

7.2 PMCA AS THE PRINCIPAL Ca^{2+} REMOVAL MECHANISM IN NEURONS

The development in the last decade of powerful methods for recording changes in $[Ca^{2+}]_i$ in single cells, based on electrophysiologic and Ca^{2+}-specific microfluorimetric techniques, has led to a substantial advance in knowledge of the complexities of neuronal $[Ca^{2+}]_i$ regulation and signaling. Ca^{2+} influx via channels and release from intracellular stores have been studied extensively. The modulation of processes that remove Ca^{2+} from neuronal cytosol and their relative importance in neuronal function have also been studied, although to a lesser extent. The consensus has been that the high capacity and low affinity of the Na/Ca exchanger relegates its role to removal

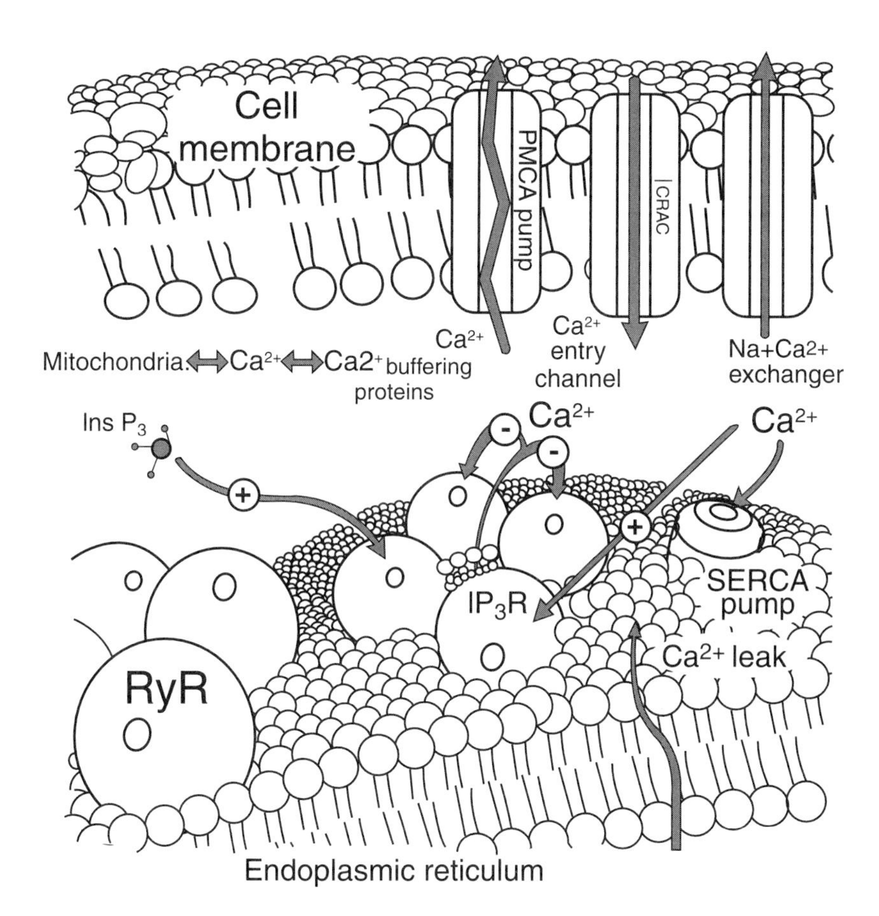

FIGURE 7.1 Cellular localization of structures involved in the regulation of intracellular calcium. PMCA, plasma membrane Ca^{2+}-ATPase; CRAC, Ca^{2+} release activated channels; Ins P_3, inositol 1,4,5-trisphosphate; RyR, ranodin receptor; IP_3R, inositol triphosphate receptor; SERCA, smooth endoplasmic reticulum Ca^{2+}-ATPase.

$[Ca^{2+}]_i \downarrow$	$[Ca^{2+}]_i \uparrow$
Plasma membrane Ca^{2+}-ATPas (PMCA)	**Voltage-operated Ca^{2+} channels (VOCC)**
Smooth endoplasmic retiuclum Ca^{2+}-ATPase (SERCA)	**Receptor-operated Ca^{2+} channels**
Na^+/Ca^{2+} exchanger	**Intracellular Ca^{2+} release channels (ICRC: IP_3R and RyR associated)**
Mitochondrial H^+/Ca^{2+}-ATPase	
Ca^{2+} buffering proteins	

FIGURE 7.2 Intracellular mechanisms responsible for changes in the intracellular calcium concentration in neurons. ICRC, intracellular Ca^{2+} release channels; IP_3R; RyR.

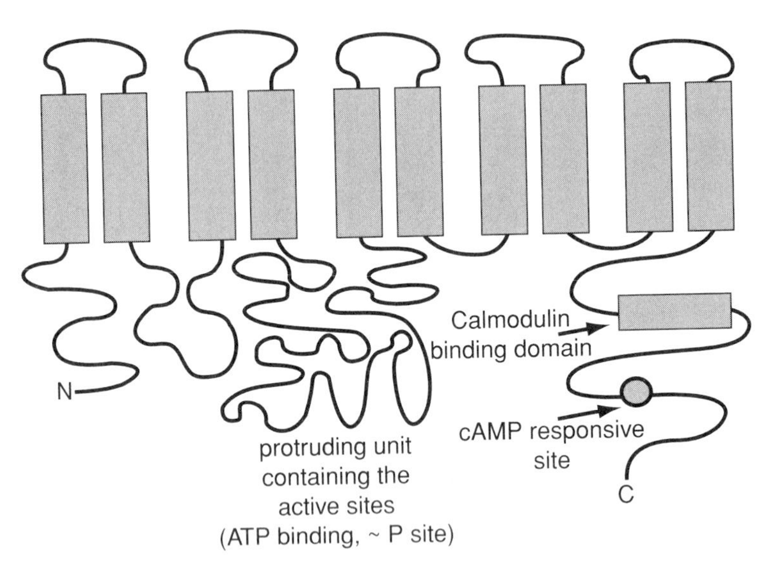

FIGURE 7.3 Schematic representation of the plasma membrane Ca^{2+}-ATPase (PMCA) molecule.

of Ca^{2+} when concentrations are 1 μM or higher, whereas high-affinity PMCA serves as a precise regulator that maintains $[Ca^{2+}]_i$ at low resting levels.[7-9]

PMCA belongs to the family of P-type ATPases, and the general mechanism of the pump is that of all other P-type pumps. Ca^{2+} promotes the transfer of the terminal phosphate of adenosine triphosphate (ATP) to an aspartic acid residue in the pump. By means of gamma-labeled ATP, the approximate molecular mass of the pump protein in erythrocytes has been estimated at about 140,000 Da because Ca^{2+}-dependent incorporation of radioactivity was observed only in a protein corresponding to this mass in electrophoretic gels. The complete primary structure of the pump was deduced in 1989 by Shull and Greeb[10] from complementary DNA isolated from rat brain and by Verma et al.[11] from human teratoma cell libraries. As with all other P-type pumps, PMCA is inhibited by La^{3+} and by vanadate. A comparison of the primary sequences of PMCA with other P-type ATPases confirmed the presumption of a considerable degree of homology. The PMCA pump appears to contain ten transmembrane helices, which have been assigned by the usual hydropathy plots. They are grouped at the N- and C-terminal portions of the pump, leaving a large mid-zone of the enzyme free of hydrophobic interactions. The schematic structure of the PMCA molecule is shown in Figure 7.3. The pump protrudes essentially into the intracellular domain, where about 80% of its mass is located. The domain protruding between putative transmembrane helices 4 and 5 contains the ATP binding site and the site of phosphoenzyme formation (lysine 601 and aspartic acid 475, respectively, in the human teratoma isoform). The moiety protruding out of the last putative transmembrane helix contains several domains (i.e., the calmodulin binding domain [residues 1100-1127 in the human teratoma isoform], an acidic stretch immediately to the N-terminal side of it, and a domain having the typical charac-

FIGURE 7.4 List of factors involved in the modulation of activity of plasma membrane Ca^{2+}-ATPase (PMCA).

teristics of the substrate site for the cAMP-dependent protein kinase). A comparison of the structure of the PMCA pump with its intracellular counterpart (i.e., SERCA) points to two distinct domains of functionally important differences: a large, highly charged PMCA insert between membrane-spanning helices two and three (putative acidic phospholipid binding domain) and an extended PMCA C-terminus comprising the calmodulin binding domain and a site for kinase phosphorylation.[12-14]

The activity of PMCA is modified by a large number of agents that interact either with the enzyme directly (calmodulin, acidic phospholipids) or through integrated regulatory pathways (various kinases, proteolysis, self-association, hormones). A list of known factors affecting the activity of PMCA is shown in Figure 7.4. The major effect of all these regulators is to increase or decrease the V_{max} and the affinity of the pump for Ca^{2+}.[15] Long-term control is achieved through regulation of its gene expression.[16]

Several interesting studies indicate how important PMCA is in removing Ca^{2+} loads in neurons. Benham et al.,[7] using patch-clamp and microfluorimetric techniques in cultured rat dorsal root ganglion (DRG) cells, showed that whereas Na/Ca exchange played only a small part in removing physiologic Ca^{2+} loads (500 n*M*) and no apparent part in maintaining resting $[Ca^{2+}]_i$, inactivation of PMCA with vanadate or high pH resulted in a dramatic slowing of Ca^{2+} removal. They concluded that PMCA is critical for the removal of Ca^{2+} loads of similar amplitude to those generated by the firing of action potentials. Using similar techniques, Bleakman et al.[8] showed that mitochondria, endoplasmic reticulum stores, and the Na/Ca exchanger have little effect on short-term clearance of modest Ca^{2+} loads in rat cultured septal neurons. More recently, Werth et al.,[9] in an extensive study, evaluated Ca^{2+} efflux systems in rat DRG cells and confirmed that PMCA-mediated Ca^{2+} extrusion is the primary process responsible for recovery to basal $[Ca^{2+}]_i$ after stimulation. Their experiments also showed that sequestration of Ca^{2+} in intracellular stores or extrusion of Ca^{2+} from the cell via the Na/Ca exchanger contributed minimally to recovery of $[Ca^{2+}]_i$ to baseline.

Modern studies of inhalation anesthetics (IA) have revealed a wide range of action on many neuronal types. Invertebrate axons, hippocampal pyramidal cells, and DRG cells, among others, have been studied electrophysiologically in the IA context. Involvement of pre- and postsynaptic effects was summarized by Krnjevic,[17] and the roles of K^+, Na^+ , Cl^-, and Ca^{2+} have been discussed by many authors.[18-21] We believe that systematic electrophysiologic studies of IA sensitivity of different groups of neural cells at the critical temperature of 37°C are lacking. Such studies are essential for a reasonable extrapolation of our findings of anesthetic effects on the calcium pump to the production of clinical anesthesia.

In a review of anesthetic effects on neuronal Ca^{2+}, Kress and Tas comment on the paucity of data on systems that eject Ca^{2+} from the cell, whereas anesthetic targeting of Ca^{2+} channels, particularly ligand-gated channels, has received considerable experimental attention.[22,23] However, the reports noted above, underscoring the unique role of PMCA in restoration of $[Ca^{2+}]_i$ to basal levels after physiologic loading, suggest that PMCA may also be worth investigation. In recent years, we have reported studies showing that a wide spectrum of IA, in clinically relevant concentrations, specifically inhibit PMCA pumping in brain synaptic membranes and in cultured cells of neural origin. Parallel decreases or increases in requirements for IA have been demonstrated in rat models with incidentally decreased or increased PMCA pumping activity. In addition, electrophysiologic and microfluorimetric studies utilizing cultured embryonic mouse cortical and spinal cord neurons and adult mouse DRG cells showed delayed repolarization and delayed restoration of $[Ca^{2+}]_i$ to basal levels with exposure to low concentrations of halothane at 37°C.[24,25]

Recent *in situ* hybridization studies indicate that four isoforms of PMCA (PMCA1, 2, 3, and 4), differing in their regulatory domains, are heterogeneously located in the rat brain.[26-28] The hybridization pattern for mRNA for various splice variants of PMCA mRNA isoforms, which also have intrinsic differences in their regulatory domains, has been demonstrated to be heterogeneous in the brain as well.[29] Thus, a variety of molecular types of PMCA are concentrated in different parts of the brain. These differences are in keeping with the possibility that anesthetic action may take place in localized brain areas. Much attention has been focused recently on anatomic sites of IA-induced suppression of movement in response to pain. Studies supporting spinal cord sites or interacting spinal cord and brain sites have been offered.[30,31] Electrophysiologic studies suggest a complex action of anesthetics on pathways conducting somatosensory signals from the spinal cord to specific layers of the cerebral cortex via the ventroposterolateral nucleus of the thalamus.[32] None of these possibilities is incompatible with mediation of anesthetic effect by partial inhibition of synaptic PMCA, given the heterogeneous distribution and varying structure of the calcium pump in the central nervous system. Several arguments may be advanced supporting a pharmacodynamic role for anesthetic inhibition of central neuronal PMCA. Although present in very low amounts in most plasma membranes (less than 0.1% of total membrane protein), PMCA is highly conserved across species, indicating the essential role this pump plays in eukaryocytic function. Comparison of PMCA2 human and rat protein sequences, for example, shows greater than 98% homology. In addition, transcription of isoforms and product expression of isoforms and splice variants is strikingly regional. Isoform 2, in particular, is confined to the central nervous system where it is strategically located in nerve terminals in the brain stem, cerebellum, and cerebrum. These findings suggest specific functions for isoforms and their splice variants in neurons, and may even augur isoform/splice variant differences in susceptibility to anesthetic depression, although location may still be the most important factor.

It is noteworthy that IA in concentrations sufficient to produce anesthesia cause only modest inhibition, usually 20 to 30%, of PMCA pumping activity in brain synaptic membranes, and animal models with small changes in PMCA activity manifest rather large changes in anesthetic requirements. These large effects with small changes are not surprising, considering how critical the pump is to cell function, and they suggest that complete inhibition might be incompatible with life. On the other hand, modest elevations of $[Ca^{2+}]_i$ prevented neuronal responses to depolarizing agents, indicating a possible functional connection between partial PMCA inhibition and diminished synaptic transmission.

7.3 INTERACTIONS BETWEEN PMCA AND PHOSPHOLIPID *N*-METHYLATION

PMCA is very sensitive to the phospholipid milieu in which it is situated. Activation of PMCA by acidic phospholipids has been found to increase both the V_{max} and the affinity for Ca^{2+}.[33] Our continued interest in phospholipid methylation (PLM)–PMCA interactions derives from its enhancement by volatile and gaseous anesthetics, which may produce changes in the membrane lipid microenvironment of PMCA, with associated changes in PMCA pumping activity. PLM, a ubiquitous process occurring in cell membranes, utilizes two enzymes, phospholipid-*N*-methyltransferases I and II (PLMT I and II), which successively methylate phosphatidylethanolamine (PE) to phosphatidyl-*N*-methylethanolamine (PME), phosphatidyl-*N*-dimethylethanolamine (PDE), and phosphatidylcholine (PC). PLMT I, located on the cytoplasmic surface, drives the first, rate-limiting, methylation step, producing PME. Associated with successive methylation is rapid translocation of the methylated product from the inner to the outer surface of the plasma membrane. Franks et al.[34] demonstrated that transmethylation of synaptosomal membrane phospholipids in rat brain was increased by either *in vitro* or prior *in vivo* exposure to halothane and isoflurane. In addition, it was also reported by Horn et al.[35] that clinically relevant concentrations of two less potent anesthetic gases, N_2O and xenon, produced significant increases in PLM in synaptic plasma membrane preparations from rats. We have also observed that IA of varying structure (i.e., halothane, isoflurane, xenon, and N_2O) inhibit PMCA activity. The relationship between enhancement of PLMT I and inhibition of PMCA activity by anesthetics is purely speculative at this juncture. It is possible that anesthetic-induced diversion of PE to synthesis of PME (by stimulating the rate-limiting enzyme, PLMT I), and ultimately PC, diverts PE from conversion to phosphatidylserine (PS), an acidic phospholipid known to enhance PMCA activity. Other evidence for an interaction between PLM and PMCA in the anesthetic response is provided from our studies in diabetic, aged, and hypertensive rats, which were characterized by reduced PMCA activity in a variety of tissues, including brain, and required lower concentrations of IA to prevent movement in response to noxious stimulation. PLMT I activity, as measured by the PLM rates, was elevated in these animals. Panagia et al.[36] have also linked PLM to PMCA changes in streptozocin (STZ)-induced diabetes in rats, proposing possible effects on the lipid environment of the pump. In addition, PS was noted to be decreased in the sciatic nerves of patients with diabetes mellitus. These observations were recently confirmed in our laboratory by the demonstration of reverse changes (i.e., related increase in brain PMCA activity and decrease in PLMT I activity) in rats with *increased* anesthetic requirements (see below). A schematic diagram presenting the possible relationship between PLM and PMCA is presented in Figure 7.5.

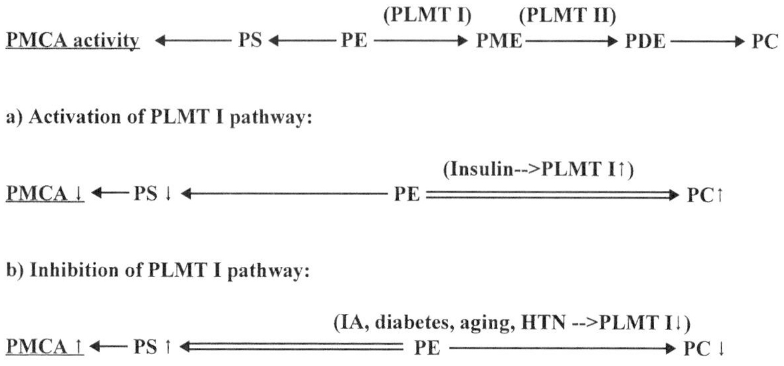

(PLMT I) (PLMT II)
PMCA activity ← PS ← PE → PME → PDE → PC

a) Activation of PLMT I pathway:

(Insulin-->PLMT I↑)
PMCA ↓ ← PS ↓ ← PE ⇒ PC↑

b) Inhibition of PLMT I pathway:

(IA, diabetes, aging, HTN -->PLMT I↓)
PMCA ↑ ← PS ↑ ⇐ PE → PC ↓

FIGURE 7.5 Hypothetical interactions between phospholipid methylation (PLM) and plasma membrane Ca^{2+}-ATPase (PMCA) activity. PMCA, plasma membrane Ca^{2+}-ATPase; PS, phosphatidylserine; PE, phosphotidylethanolamine; PME, phosphatidyl-*N*-methyletyhanolamine; PDE, phosphatidyl-*N*-dimethylethanolamine; PC, phosphatidylcholine; PLMT I/II, phospholipid-*N*-methyltransferases I/II; IA, inhalation anesthetics; HTN, hypertension.

7.4 LOCALIZATION AND FUNCTION OF PMCA IN THE CENTRAL NERVOUS SYSTEM

The techniques of *in situ* hybridization, immunocytochemistry, and histochemistry were combined in an effort to provide a coherent picture of PMCA distribution relevant for further pharmacologic, and particularly, anesthetic studies. We initially focused our localization effort on the PMCA1 isoform, as it represents more than 50% of the total PMCA-related activity in the brain.

7.4.1 *In Situ* Hybridization

Localization of the PMCA1 isoform mRNAs was evaluated by *in situ* hybridization. Uridine 5′[α-(^{35}S)thio]triphosphate-labeled antisense or sense RNA probes were prepared for a SacI/PstI fragment corresponding to nucleotides 384 to 493 of PMCA1 clone RB11-1 (kindly provided by Gary Shull, University of Cincinnati Medical Center). Autoradiographic images were laser scanned and image files were stored on CD-ROM. The highest levels of PMCA1 probe hybridization were found in CA1 pyramidal cells, with lower levels in hippocampal CA3 pyramidal cells and in the granule cells of the dentate gyrus (Figure 7.6b). High levels were also observed in the granular layer and Purkinje cell layer of the cerebellum. Intermediate levels of hybridization were found in the cerebral cortex, thalamic nuclei, olfactory bulb, caudate-putamen, and amygdaloid nuclei. Low levels were noted in the chorioid plexus, hypothalamus, and molecular layer of the cerebellum.

7.4.2 PMCA Immunolocalization

Localization of PMCA was done by direct immunofluorescence on parallel brain sections with 5F10 mouse, anti-human PMCA monoclonal antibody, raised against human erythrocyte PMCA, which is comprised predominantly of isoforms 1 and 4.

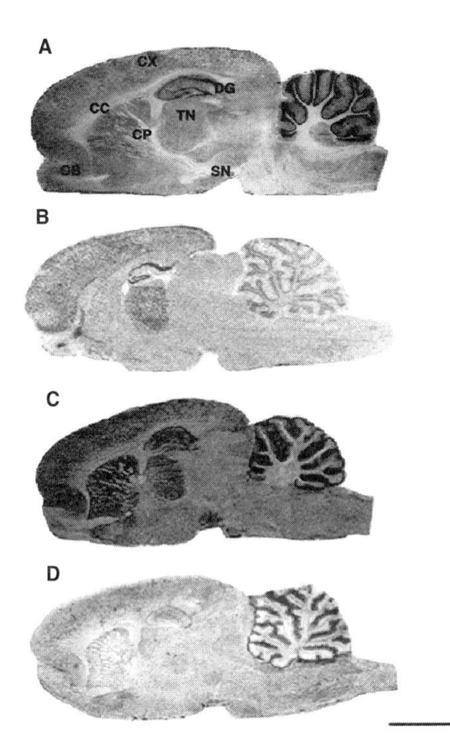

FIGURE 7.6 Localization of PMCA1 isoform mRNA by *in situ* hybridization and lead phosphate product of Ca^{2+}-ATPase activity in sagittal rat brain sections. (A) Identification of selected brain areas in parallel brain sections with cresyl violet staining. OB, olfactory bulb; CX, cerebral cortex; CC, corpus callosum; CP, caudate-putamen; DG, dentate gyrus; TN, thalamic nuclei; SN, substantia nigra. (B) Autoradiographic localization of PMCA1 mRNA isoform by *in situ* hybridization. Dark areas represent highest PMCA1 mRNA expression. (C) Localization of nonspecific Ca^{2+}-ATPase activity as measured by lead phosphate (P_i) staining product (visualized as dark areas). Sections were prepared with unmodified Ando et al.[37] histochemical technique. (D) More specific localization of Ca^{2+}-ATPase reaction product (P_i) in sections preincubated with thio analogues of guanidine phosphate, which inhibit ecto-ATPase, followed by lead phosphate staining. (Original magnification ×6. Bar = 500 μm.)

It has been shown recently that PMCA1 predominates in all tissues tested. Sections were analyzed with a microscope equipped with a fluorescence vertical illuminator. Selected sections were photographed and images were digitalized as noted above. Microscopy of rat brain sections revealed a pattern of immunofluorescence that was heterogeneous and of greatest magnitude in areas rich in synaptic layers (Figure 7.7). Very low fluorescence was observed in layers comprised primarily of cell bodies. The highest level of fluorescence was found in the molecular layer of cerebellum, with considerably less in Purkinje cells and granular layers. In the hippocampus, the highest levels of fluorescence were observed in the molecular layer and the polymorph layer of the dentate gyrus. Significantly less immunofluorescence was seen in the granular layer of dentate gyrus and the pyramidal layers of CA1 to CA3. Intermediate levels were noted in the thalamic nuclei, cerebral

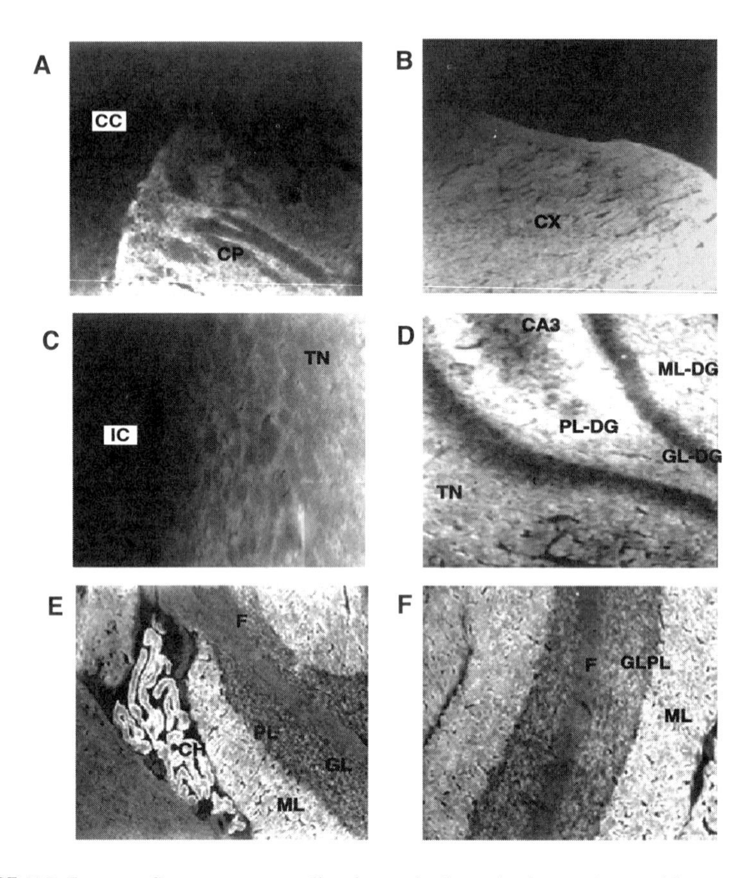

FIGURE 7.7 Immunofluorescence studies in sagittal rat brain sections with mouse, anti-PMCA monoclonal antibodies (light areas indicate higher immunofluorescence and, therefore, higher density of PMCA). (A) Section through caudate-putamen (CP) adjacent to corpus callosum (CC). (B) Fragment of the superficial layers of cortex (CX). (C) Section through thalamic nuclei (TN) adjacent to internal capsula (IC). (D) Section of dentate gyrus (DG) and CA3 of hippocampus. ML-DG, molecular layer of dentate gyrus; GL-DG, granular layer of dentate gyrus; PL-DG, polymorph layer of dentate gyrus. (E) Section through the fourth ventricle, containing chorioid plexus (CH) and the fragments of cerebellum. ML, molecular layer of cerebellar cortex; PL, Purkinje cell layer; GL, granular layer; F, fiber tracts. (F) Fragment of the cerebellar cortex. See part E for abbreviations. (Original magnification ×40. Bar = 100 μm.)

cortex, hypothalamus, and caudate-putamen. In addition to the neuronal tissue, a very high level of specific immunofluorescence was observed in the chorioid plexus of the third and fourth ventricles.

7.4.3 Histochemical Assay of PMCA Activity in Brain Sections

Localized, Ca^{2+}-dependent nucleotide hydrolysis was estimated by staining of the precipitated product, P_i, according to the enzymatic assay of Ando et al.,[37] but with several major modifications. In various experimental protocols examining assay specificity, the incubation medium was modified in one or more of the following respects: (1) omission of $CaCl_2$, $MgCl_2$, and ATP; (2) replacement of ATP by other nucleoside phosphates; (3) replacement of ATP by *p*-nitrophenyl phosphate (NPP) or glucose-6-phosphate (G6P); (4) inclusion of sodium orthovanadate, ouabain, sodium azide, or various nonspecific sulfydryl enzyme inhibitors of ATPase; and (5) preincubation with thapsigargin, a specific inhibitor of the SERCA pump. These studies confirmed that the assay is a measure of Ca^{2+}-dependent nucleotide hydrolysis, exclusive of SERCA. Image analysis of sagittal brain sections showed that nucleotide hydrolytic product was widely distributed in all brain areas (Figure 7.6C), with the highest density of lead phosphate in the caudate-putamen, substantia nigra, thalamic nuclei, cerebral and cerebellar cortex, and CA1 to CA3 hippocampus. The specificity of this staining for PMCA activity is questionable, however. Reports by others and experiments in our laboratory indicate that the phosphate staining pattern of brain sect ions incubated in medium supporting the Ca^{2+}-ATPase reaction, as shown in Figure 7.6, is the sum of several distinct reactions, including that of the ecto-ATPases. Furthermore, P_i release by ecto-ATPase quantitatively exceeds that of PMCA. One potentially useful difference between ecto-ATPases and PMCA is that the latter is ATP specific whereas ecto-ATPases can utilize other nucleotides. To investigate the possible use of nonhydrolyzable nucleotides for differentiating types of Ca^{2+}-ATPase activity, we preincubated brain sections with either guanosine-β-S-diphosphate (GDP-β-S) or guanosine-γ-S-triphosphate (GTP-γ-S), aiming to block subsequent ecto-ATPase binding of ATP, but leaving ATP-specific PMCA activity intact. Results showed that thio analogues greatly reduce the total deposition of P_i and that they alter the distribution of P_i in the brain (Figure 7.6D). Residual staining after pretreatment with guanidine thio analogues was more uniformly distributed, showing a higher density of precipitate in the molecular layer of the cerebellum, the neocortex, and a lower density in the hippocampus, thalamic nuclei, and caudate-putamen. This more specific identification of Ca^{2+}-ATPase hydrolytic activity correlates closely with the distribution of PMCA- specific immunofluorescence, in contrast to the unmodified Ando method wherein ecto-Ca^{2+}-ATPase is not inhibited. Thus evidence from several lines of inquiry indicates that our histochemical procedure primarily monitors PMCA activity.

In summary, this study combined, in sequential brain sections, three different techniques for localizing sites of PMCA synthesis, protein, and function. Most important, the histochemical, enzymatic assay system could be manipulated to reduce ecto-Ca^{2+}-ATPase hydrolysis of ATP, thereby highlighting PMCA functional activity and confirming its location in sites rich in PMCA protein. Because anesthetic depression of PMCA pumping persists in animals sacrificed while anesthetized, this development offers great promise in identifying putative anatomic sites of selective anesthetic action.

7.4.4 Immunolocalization of PMCA in Neurons

In a separate but related study, immunolocalization experiments were carried out, as described above, with cultured embryonic neurons from mouse cortex (CX) and

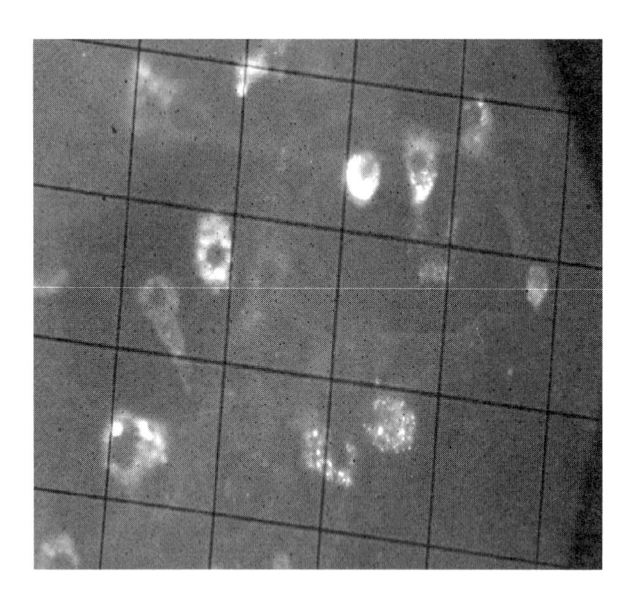

FIGURE 7.8 Immunofluorescence of PMCA in mouse brain cortex neurons in cell culture.

spinal cord (SC) as well as in dissociated neurons from adult mouse DRG. In all cells, an intense immunofluorescence was noted in neuronal plasmalemma and cell processes (Figure 7.8), indicative of sites of PMCA concentration.

7.5 INHIBITION OF PMCA BY INHALATION ANESTHETICS *IN VITRO*

7.5.1. Inhalation Anesthetic Effects on Isolated Brain Synaptic PMCA Activity

Experiments described below were performed in purified fractions of synaptic plasma membranes (SPM) prepared from various parts of dissected rat brain. SPM purification was confirmed by electron microscopy, enzymatic analysis, and specificity of Ca^{2+} transport. PMCA pumping activity was assessed by measuring $^{45}Ca^{2+}$ uptake into everted SPM vesicles. For studies of anesthetic effect, SPM were exposed to halothane, isoflurane, xenon, and nitrous oxide at partial pressures ranging from 0 to 1.6 MAC equivalents. Dose-dependent inhibition of Ca^{2+}-ATPase activity was observed with each member of this structurally disparate group of anesthetic agents. One MED (minimum effective dose, defined as the lowest partial pressure of IA required to prevent movement in response to pain in the intact rat) reduced Ca^{2+} transport by about 30% compared with control.[38]

7.5.2 Stable Inhibition of Brain Synaptic PMCA in Rats Anesthetized with Halothane

PMCA pumping was examined in SPM obtained from rats sacrificed during and after halothane anesthesia. Three treatment groups were studied: (1) control rats decapitated without anesthetic exposure, (2) anesthetized rats exposed to 1 MED for 20 min and decapitated, and (3) "recovered" rats exposed to 1 MED for 20 min and then decapitated after recovery from anesthesia (defined as beginning to groom). PMCA transport of Ca^{2+} into SPM vesicles from anesthetized rats was reduced by 30%, compared with control, whereas transport in the recovered group returned to control levels. Thus PMCA inhibition persisted after elution of the precipitating anesthetic agent during membrane isolation.[39,40]

7.5.3 Inhalation Anesthetic Effects on PMCA Pumping in Tissue Culture Cells

Several lines of neural cells (i.e., C6 rat glioma cells, B104 rat neuroblastoma cells, and PC12 rat pheochromocytoma cells) demonstrated PMCA pumping that was inhibited in a dose-related fashion by halothane or xenon.

In summary, we have found that not only halothane, but also isoflurane, xenon, and nitrous oxide, in pharmacologic concentrations, inhibited PMCA pumping activity. We also showed that halothane and xenon inhibited PMCA in cultured C6 rat glioma cells, B104 rat neuroblastoma cells, and PC12 rat pheochromocytoma cells.[41]

7.6 Changes in Anesthetic Requirements in Animal Models with Altered PMCA Activity

Because of these consistent findings of inhibition of PMCA and enhancement of PLMT I by a variety of IA, we have examined several rat models reported previously to have altered PMCA activity in a variety of nonneural tissues.

7.6.1 Diminished PMCA Pumping and Increased PLM in Brain SPM Vesicles from Rats with Streptozocin-Induced Diabetes: Association with Reduced Anesthetic Requirements

Because chronic hyperglycemia inhibits PMCA in cells from several tissues, diabetic rats could manifest chronic inhibition of brain synaptic PMCA and thus provide a model for testing the hypothesis that synaptic PMCA plays a role in IA pharmacodynamics. We found that PMCA pumping was about 10% less frequent in cerebral SPM from rats with STZ-induced diabetes (duration 13 days) than in SPM from normoglycemic rats. (As noted below, a further decrease in PMCA activity occurred with prolongation of the diabetic state.) The halothane requirement for STZ-hyperglycemic rats was dramatically reduced to about 65% of control. Halothane and xenon MEDs correlated strongly with percent glycated hemoglobin (GHb) in all treatment groups. When PLM was evaluated in SPM taken from diabetic rats 2 weeks after STZ injection, PLMT I activity was enhanced nearly 30% compared with normoglycemic controls. Thus PLMT I activity was also shifted in the expected direction in this rat model with reduced anesthetic requirements.[40]

7.6.2 Effect of Insulin Treatment on Altered Brain Synaptic PMCA/PLM and Anesthetic Requirements in Streptozocin-Induced Diabetes

MEDs for isoflurane, enflurane, and desflurane were substantially reduced in diabetic rats, compared with placebo-injected controls. Insulin treatment, begun 2 weeks after induction of diabetes and continued for 3 more weeks, resulted in weight gain and partial correction of hyperglycemia and hemoglobin glycation. Isoflurane MED was also partially corrected to 67% of control for the insulin-treated group compared with 52% of control in the untreated group. Similarly, PMCA pumping was 87% of control in the treated and 75% of control in the untreated group, whereas PLMT I activity was 111% of control in the treated and 131% of control in the untreated group. These data suggest that partial insulin treatment of diabetic rats, measured by decreased GHb compared with untreated diabetic rats, was associated with the partial normalization of anesthetic requirements for isoflurane, and was accompanied by partial normalization of both PMCA and PLMT I activity.[42]

7.6.3 Diminished Brain Synaptic PMCA Activity in Spontaneously Hypertensive Rats Associated with Reduced Anesthetic Requirements

Reduced erythrocytic PMCA activity has been reported in spontaneously hypertensive rats (SHR). We measured Ca^{2+} transport in SPM prepared from cerebrum and diencephalon–mesencephalon of SHR and from control Wystar-Kyoto rats (WKY). PMCA pumping in SHR was reduced to 85 to 90% of the control rats. MEDs for halothane, isoflurane, and desflurane in SHR were 90% of control.[35,43]

7.6.4 The Zucker Rat — A Paradigm of Increased Anesthetic Requirements

The Zucker homozygous (fa/fa) rat is a well-established model of obesity that exhibits peripheral insulin resistance manifested by hyperinsulinemia and normal to moderately elevated levels of glucose. Insulin stimulates PMCA activity in a variety of tissues, but insulin effects on brain PMCA activity had not been previously reported. Because of the possibility that high insulin levels might be associated with resistance to IA, we compared the anesthetic requirements in obese Zucker homozygous rats, with lean, normoinsulinemic Zucker heterozygotes (fa/+) serving as controls, and examined synaptic PLMT I activity and the response of synaptic PMCA to insulin added to brain tissue from both genotypes.[44] In addition we measured endogenous insulin concentrations in extracts of several brain areas and endogenous brain insulin in intact rats by microdialysis. The Zucker obese rat is unique as an animal model; not bred for anesthetic resistance, it requires substantially higher partial pressures of IA for prevention of movement in response to pain. This difference was observed with all IA examined, but was greatest with halothane. The mean MED for halothane in obese rats was 41% higher than in lean rats. Similarly, mean MEDs for isoflurane, enflurane, and desflurane were 27, 14, and 11%, respectively, higher in obese rats than in lean controls. The reason for the variability in the obese Zucker rats' responses to different anesthetic agents, greatest with halothane and least with desflurane, is not clear. Such differences were not observed in our other animal models. One possible explanation is variability of the insulin secretory response among different agents, with varying degrees of change in brain insulin. It does appear that these differences between lean and obese rats or among different agents cannot be explained by differences in IA pharmacokinetics. MED differences persisted when obese and lean rats were anesthetized for as long as 4 h, ruling out altered IA uptake or distribution as a factor. In addition there is an association between decreased anesthetic susceptibility and increased activity of PLMT I. Although PLMT I activity was decreased as predicted in obese rats, synaptic PMCA activity did not differ from lean controls or from normal Sprague-Dawley rats. It is noteworthy, however, that stimulation of PMCA by insulin has been reported to occur in several tissues and that obese Zucker rats have significant hyperinsulinemia. Because it seemed likely that endogenous brain insulin, if present, would be eluted from brain membranes during the process of SPM isolation, we added insulin to the preparative and assay media, and we were able to demonstrate dose-dependent stimulation of PMCA pumping. Addition of insulin to the assay medium alone had no effect, presumably because Ca^{2+} uptake is measured in everted synaptic membranes whereas insulin receptors are located on the inner vesicular surface. Hyperinsulinemia, as found in obese Zucker rats, can lead to elevated brain levels only if insulin crosses the blood–brain barrier. That such passage occurs is suggested by the previous findings demonstrating that peripheral insulin infusion in Zucker rats produced a marked increase in brain insulin (measured in hypothalamic microdialysates), with significantly greater increases in obese rats than in lean controls. We observed the presence of endogenous insulin in tissue extracts from several brain areas and found concentrations in obese rats that were significantly higher than in lean controls. In addition, there is evidence that general anesthesia itself may further increase insulin secretion and distribution. It was reported previously that anesthesia produced by inhaled agents was associated with significant increases in plasma insulin concentrations when compared with the preoperative state and with epidural anesthesia. We studied the effect of halothane anesthesia on brain insulin levels in several subcortical areas, estimating tissue concentrations by radioimmunoassay of microdialysates. Microdialysis of three subcortical regions (ventromedial hypothalamic nuclei, ventroposterolateral hypothalamic nucleus, and deep mesencephalic nuclei) showed marked increases in insulin levels with halothane (1 MED) exposure in obese rats, compared with lean controls.

In conclusion, we found significantly increased anesthetic requirements in the Zucker homozygous obese rat. Although measured brain synaptic PMCA activity did not differ from that found in lean heterozygotes and in normal Sprague-Dawley rats, it appears that this finding is due to elution of endogenous insulin from brain tissue of Zucker rats during membrane isolation. The preponderance of evidence suggests that synaptic PMCA activity is high in intact, obese rats because of increased concentrations of insulin in the brain: insulin stimulates PMCA in brain membranes *in vitro*; brain insulin concentrations are higher in the obese than in the lean genotype; and anesthetic-induced increases in brain insulin in obese rats exceed those of lean controls. The role of PLMT I in the anesthetic response is uncertain, although it is clear that activity is enhanced by IA. Furthermore, PLMT I activity is increased in animal models susceptible to IA and decreased in the IA-resistant Zucker rat. We have previously raised the possibility that PLMT I modulates the membrane lipid microenvironment of PMCA, thus affecting PMCA pumping activity. Taken together, these observations of enzyme activity in an anesthetic-resistant rat model lend further support to the hypothesis that the calcium pump plays a functional role in the production of the anesthetic state.

7.6.5 Diminished Brain Synaptic PMCA Activity in Aged Rats: Association with Reduced Anesthetic Requirements

Well-established studies in humans show that anesthetic MAC diminishes almost linearly with increasing age. The relationship between age and MAC is less obvious in rats. MAC may not decrease so regularly in this species, but instead may fall off rapidly after a finite period of stability. For example, it was reported that halothane MAC was 17% lower in 24-month-old than in 5-month-old Fischer-344 rats. In addition, although synaptic PMCA pumping activity has not been measured in old rats, diminished Ca^{2+}-dependent ATP hydrolysis in brain membranes has been reported in an Alzheimer mouse model. However, PMCA comprises a relatively small fraction of Ca^{2+}-dependent ATPases in the brain, and ATP hydrolysis is a poor measure of activity unless particular precautions are taken to reduce nonpumping ecto-ATPase nucleotide hydrolysis.

We measured brain synaptic PMCA activity in 2- and 25-month-old Fischer-344 rats. We found that Ca^{2+} pumping was significantly decreased in the aged rats, averaging 91% of the younger controls in cortical SPM and 82% in SPM from diencephalon-mesencephalon preparations.[45] For aged Fischer-344 rats, MEDs for halothane, desflurane, isoflurane, and xenon were reduced to 81, 82, 67, and 86%, respectively, of young controls. Halothane MED was also measured in 2- and 30-month old F344/BNF rats, a strain that undergoes aging with less debilitation. For F344/BNF rats, MED for halothane was reduced to 87% and PMCA activity was diminished to 90% in cortical SPM and 72% in SPM from diencephalon-mesencephalon preparations. In both brain regions and in both strains of rats the activity of PLMT I was increased to 130% of young controls. The strong association between age and reduced anesthetic requirements for four disparate inhalation agents on the one hand, and age and altered PMCA and PLMT I activity on the other, provides strong support to the underlying hypothesis of this work.

7.6.6 Studies with Eosin, A Putatively Specific Inhibitor of PMCA

Recent studies show that eosin (tetrabromofluorescein) is a potent, reversible inhibitor of human erythrocytic PMCA and bovine cardiac sarcolemma PMCA.[46] Inhibition of PMCA by eosin differs from its effect on other P-type pumps in that it does not act by competing with ATP for binding to PMCA. Thus, increasing ATP to physiologic levels does not overcome eosin inhibition of PMCA, as it does with other P-type pumps. The characteristics of eosin inhibition of PMCA (i.e., reversibility and lack of competitiveness with ATP) suggest that eosin is a highly specific inhibitor of this pump. Gatto et al.[46] have shown that in contrast to its effect on sarcolemma PMCA, eosin does not inhibit the Na/Ca exchanger in cardiac myocytes. Therefore, eosin may be a useful tool for studying the role of PMCA in anesthetic action. To explore this possibility, we examined eosin effects on neuronal PMCA in two settings: isolated rat brain SPM and intact rats.

7.6.6.1 Isolated Rat Brain SPM

When low concentrations of eosin were added to the incubation medium containing synaptic plasma membranes, Ca^{2+} transport by PMCA was inhibited in a dose-dependant fashion (ED_{50} = 8 μ*M*), with a maximal inhibition of about 30%, similar to that observed in human erythrocytes and bovine cardiac myocytes. No inhibition of Na^+-K^+-ATPase was observed in SPM.

7.6.6.2 Intact Rats

We also examined the effect of eosin (25 to 150 nmol), injected into the lateral cerebral ventricle of rats, on halothane requirements. For the intraventricular injection of eosin, a metal guiding cannula was chronically implanted, positioned 0.75 mm above the lateral ventricle of male Sprague-Dawley rats. A modified 1 mm microdialysis cannula (dialysate membrane removed, large port sealed with epoxy glue, small port connected to the syringe pump) primed with 20 m*M* eosin in Ringers solution at pH 7.4 or Ringers solution alone was inserted in the guide and perfused at a rate of 0.25 to 0.5 μl/min for 5 to 15 min to deliver 25 to 150 nmol of eosin or an equivalent volume of vehicle intraventricularly. The halothane MED that prevented movement following a noxious stimulus was assessed 30 min later. *In vivo* eosin injection was associated with a significant decrease in halothane MED of 29%, from 1.51 ± 0.09 to 1.07 ± 0.05 vol%. Twenty-four hours after eosin perfusion, MED returned toward baseline.

In summary, we compared brain synaptic PMCA pumping and PLMT I methylating activities with anesthetic requirements for several volatile and gaseous anesthetics in the following diverse groups. Diabetic rats, spontaneously hypertensive rats, and aged rats were observed to have diminished PMCA and increased PLMT I activities and to require significantly lower than expected partial pressures of a wide range of IA for prevention of a response to pain. These processes were reversed when diabetic rats were treated with insulin. Zucker rats, normoglycemic but with high brain insulin levels, and therefore with enhanced synaptic PMCA pumping, showed significantly higher than expected IA requirements. They also showed decreased synaptic phospholipid methylation. The specific PMCA inhibitor, eosin, produced a significant reduction in anesthetic requirement for halothane after injection in the lateral cerebral ventricle. As a specific inhibitor of PMCA in brain SPM, eosin appears to be a useful tool for determining relationships between site-specific effects of PMCA inhibition and the production of anesthesia by IA.

7.7 EFFECT OF INHALATION ANESTHETICS ON PMCA: ELECTROPHYSIOLOGIC AND MICROFLUOROMETRIC STUDIES OF ISOLATED NEURONAL CELL CULTURES

We have carried out studies utilizing whole-cell electrophysiologic techniques and confocal microfluorimetry (at 37°C) in cultured mouse neurons to look for possible anesthetic-induced changes in calcium dynamics.[47] Embryonal cortical and spinal cord neurons and adult DRG cells all expressed PMCA, as determined by immunocytochemistry. Microapplication of halothane (delivered concentrations <0.3 m*M*) or ED_{50} eosin (8 μ*M*), or halothane/eosin combined to cortical and spinal cord neurons undergoing spontaneous burst firing, demonstrated altered electrical activity. Both halothane and eosin caused a reduction in the amplitude of burst-associated action potentials (APs) and a decrease in their frequency when applied to cultured embryonal cortical neurons. Similar alterations of AP firing were observed in cultured spinal cord neurons. The most dramatic consequence of application of halothane/eosin combined was prolongation of the repolarization phase of burst firing by two- to fourfold. All effects were reversible with washout. The effects of halothane/eosin were also examined on adult DRG neurons with electrophysiologic characteristics similar to nociceptive C fibers. Both halothane and eosin altered APs produced by direct current and step depolarization. AP latency was increased, the rate of rise of APs was decreased, repolarization was delayed, and ultimately AP blockade occurred. These agents also prolonged the repolarization phases of capsaicin-mediated Ca^{2+}-dependent depolarization in DRG neurons. Cortical and spinal neurons were also examined by confocal microfluorimetry of single cells loaded with the calcium indicator dye, Fluoro-3-AM. Application of *N*-methyl-D-aspartate (NMDA) resulted in an influx of Ca^{2+} followed by a rapid decline in calcium concentration to baseline. Halothane prolonged clearance of Ca^{2+} from neurons threefold, a response reversed by halothane washout (Figure 7.9). The occurrence and reversibility of these effects at 37°C is relevant because of the sensitivity of the anesthetic response to temperature. In view of the association between neurotransmitter release and changes in the intracellular Ca^{2+} concentration, inhibition of PMCA may, in concert with other selective IA actions, play a functional role in production of the anesthetic state.

7.8 FUTURE DIRECTIONS OF RESEARCH

The observations described in this chapter add support to the hypothesis that PMCA is a functional anesthetic target. The ultimate goal of our current experimental work is to produce mice that manifest diminished expression of PMCA. Examining the anesthetic response of mice with underexpression of a key PMCA isomer will provide the definitive test of our hypothesis.

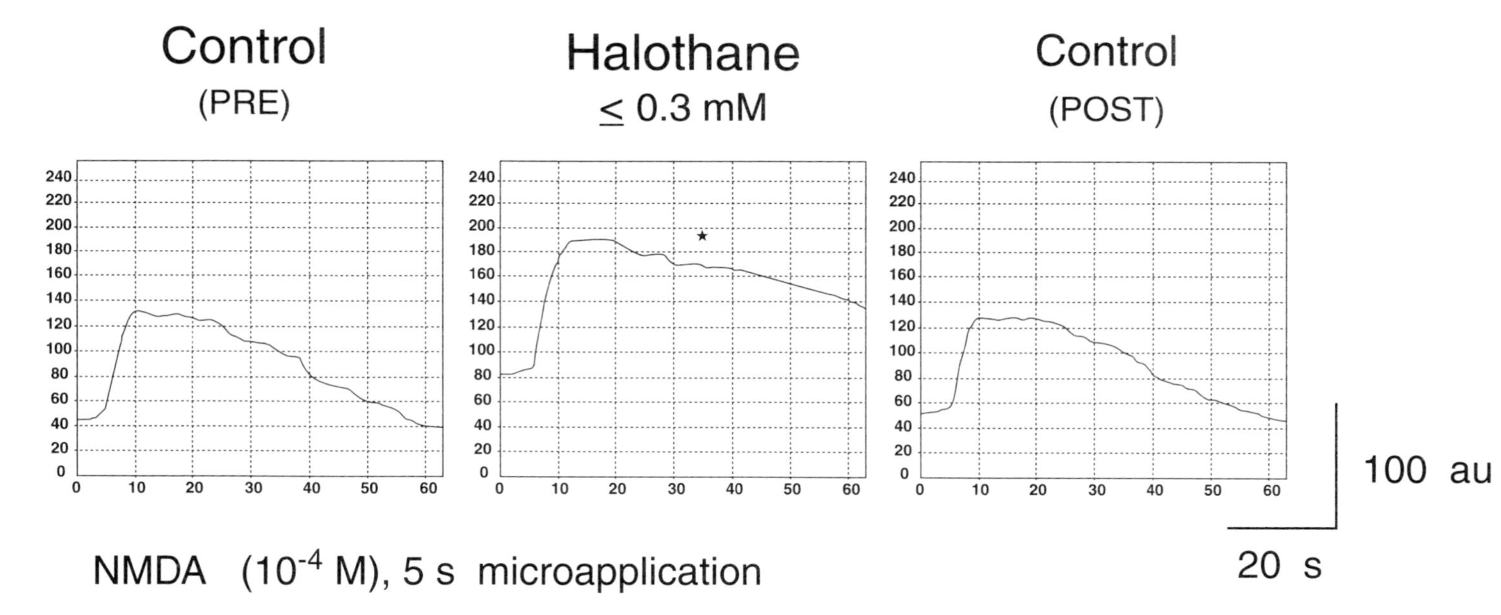

FIGURE 7.9 Effect of halothane on intracellular Ca^{2+} dynamics in cortical neurons.

ACKNOWLEDGMENTS

The authors are grateful for the active participation of Drs. J. L. Horn, A. Wamil, M. Collinge, and G. Singh, and Ms. V. Janson and Mr. W. T. Franks for their essential contributions to the studies described in this chapter. Supported by grant GM 46401 from National Institutes of Health.

REFERENCES

1. Franks, N. P., Lieb, W. R., Molecular and cellular mechanisms of general anaesthesia, *Nature*, 367, 607, 1994..
2. Miller, R. J., Calcium signaling in neurons, *Trends Neurosci*, 11, 415, 1988.
3. Miller, R. J., The control of neuronal Ca^{2+} homeostasis, *Prog Neurobiol*, 37, 255, 1991.
4. Simpson, P. B., Challiss, R. A. J., Nahorski, R., Neuronal Ca^{2+} stores: Activation and function, *Trends Neurosci*, 18, 299, 1995.
5. Thayer, S. A., Miller, R. J., Regulation of the intracellular free calcium concentration in single rat dorsal root ganglion neurones *in vitro*, *J Physiol*, 425, 85, 1990.
6. Tsien, R. W., Tsien, R. Y., Calcium channels, stores and oscillations, *Ann Rev Cell Biol*, 6, 715, 1990.
7. Benham, C. D., Evans, M. L., McBain, C. J., Ca^{2+} efflux mechanisms following depolarization evoked calcium transients in cultured rat sensory neurones, *J Physiol*, 455, 567, 1992.
8. Bleakman, D., Roback, J. D., Wainer, B. H., Miller, R. J., Harrison, N. L., Calcium homeostasis in rat septal neurons in tissue culture, *Brain Res*, 600, 257, 1993.
9. Werth, J. L., Usachev, Y. M., Thayer, S. A., Modulation of calcium efflux from cultured rat dorsal root ganglion neurons, *J Neurosci*, 16, 1008, 1996.
10. Greeb, J., Shull, G. E., Molecular cloning of a third isoform of the calmodulin-sensitive plasma membrane Ca^{2+}-transporting ATPase that is expressed predominantly in brain and skeletal muscle, *J Biol Chem*, 264, 18569, 1989.
11. Verma, A. K., Filoteo, A. G., Stanford, D. R., Wieben, E. D., Penniston, J. T., Strehler, E. E., Fischer, R., Heim, R., Vogel, G., Mathews, S., et al., Complete primary structure of a human plasma membrane Ca^{2+} pump, *J Biol Chem*, 263, 14152, 1988.
12. Carafoli, E., Calcium pump of the plasma membrane, *Physiol Rev*, 71, 129, 1991.
13. Carafoli, E., Biogenesis. Plasma membrane calcium ATPase: 15 years of work on the purified enzyme, *FASEB J*, 8, 993, 1994.
14. Penniston, J. T., Enyedi, A., Plasma membrane Ca^{2+} pump — Recent developments, *Cell Physiol Biochem*, 4, 148, 1994.
15. Wang, K. K. W., Villalobo, A., Roufagalis, B. C., The plasma membrane calcium pump: A multiregulated transporter, *Trends Cell Biol*, 2, 46, 1992.
16. Kuo, T. H., Wang, K. K. W., Carlock, L., Diglio, C., Tsang, W., Phorbol ester induces both gene expression and phosphorylation of the plasma membrane Ca^{2+} pump, *J Biol Chem*, 266, 2520, 1991.
17. Krnjevic, K., Cellular mechanisms of anesthesia, in *Molecular and Cellular Mechanisms of Alcohol and Anesthetics*, Rubin, E., Miller, K. W., Roth, S. H., Eds., NY Acad. Sci., New York, 1991, 1–16.
18. Nicoll, R. A., Madison, D. V., General anesthetics hyperpolarize neurons in the vertebrate central nervous system, *Science*, 217, 1055, 1982.

19. Haydon, D. A., Urban, B. W., The effects of some inhalation anesthetics on the sodium current of the squid giant axon, *J Physiol*, 341, 429, 1983.
20. Moody, E. J., Lewin, A. H., DeCosta, B. R., Rice, K. C., Skolnick, P., Site-specific acylation of GABA-gated Cl channels: Effects on ^{36}Cl-uptake, *Eur J Pharmacol*, 206, 113, 1991.
21. Terrar, D. A., Victory, J. G., Isoflurane depresses membrane currents associated with contraction in myocytes isolated from guinea-pig ventricle, *Anesthesiology*, 69, 742, 1988.
22. Kress, H. G., Tas, P. W., Effects of volatile anaesthetics on second messenger Ca^{2+} in neurones and non-muscular cells, *Br J Anaesth*, 71, 47, 1993.
23. Kress, H. G., Effects of general anaesthetics on second messenger systems, *Eur J Anaesth*, 12, 83, 1995.
24. Kosk-Kosicka, D., Roszczynska, G., Inhibition of plasma membrane Ca^{2+}-ATPase activity by volatile anesthetics, *Anesthesiology*, 79, 774, 1993.
25. Franks, J. J., Horn, J. L., Janicki, P. K., Singh, G., Halothane, isoflurane, xenon, and nitrous oxide inhibit Ca^{2+}-ATPase pump activity in rat brain synaptic plasma membranes, *Anesthesiology*, 82, 108, 1995.
26. Stahl, W. L., Eakin, T. J., Owens, J. W., Jr., Breininger, J. F., Filuk, P. E., Plasma membrane Ca^{2+}-ATPase isoforms: Distribution of mRNAs in rat brain by *in situ* hybridization, *Mol Brain Res*, 16, 223, 1992.
27. Stahl, W. L., Eakin, T. J., Anderson, W. R., Owens, J. W., Jr., Breininger, J. F., Localization of mRNA coding for plasma membrane Ca-ATPase isoforms in rat brain by *in situ* hybridization, *Ann NY Acad Sci*, 671, 433, 1992.
28. Stahl, W. L., Keeton, T. P., Eakin, T. J., The plasma membrane Ca^{2+}-ATPase mRNA isoform PMCA 4 is expressed at high levels in neurons of rat piriform cortex and neocortex, *Neurosci Lett*, 178, 267, 1994.
29. Zacharias, D. A., Dalrymple, S. J., Strehler, E. E.,Transcript distribution of plasma membrane Ca^{2+} pump isoforms and splice variants in the human brain, *Mol Brain Res*, 28, 263, 1995.
30. Rampil, I. J., Mason, P., Singh, H., Anesthetic potency (MAC) is independent of forebrain structures in the rat, *Anesthesiology*, 78, 707, 1993.
31. Roizen, M. F., Newfield, P., Eger, E. I., Hosobuchi, Y., Adams, J. E., Lamb, S., Reduced anesthetic requirements after electrical stimulation of periaqueductal gray matter, *Anesthesiology*, 62, 120, 1985.
32. Angel, A., Central neuronal pathways and the process of anaesthesia, *Br J Anaesth*, 71, 148, 1993.
33. Missiaen, L., Raeymaekers, L., Wuytack, F., Vrolix, M., DeSmedt, H., Casteels, R., Phosphoinositide-protein interactions of the plasma membrane Ca^{2+}-transporting ATPase, *Biochem J*, 263, 687, 1989.
34. Franks, J. J., Sastry, B. V. R., Surber, M. J., England, R. E., Halothane and isoflurane alter phospholipid transmethylation in rat brain synaptosomes, *Anesthesiology*, 73, 984, 1990.
35. Horn, J. L., Janicki, P. K., Franks, J. J., Nitrous oxide and xenon enhance phospholipid-n-methylation in rat brain synaptic plasma membranes, *Life Sci*, 56, PL455, 1995.
36. Panagia, V., Taira, Y., Ganguly, P. K., Tung, S., Dhalla, N. S., Alterations in phospholipid N-methylation of cardiac subcellular membranes due to experimentally induced diabetes in rats, *J Clin Invest*, 86, 777, 1990.
37. Ando, T., Fujimoto, K., Mayahara, H., Miyajima, H., Ogawa, K., A new one-step method for the histochemistry and cytochemistry of Ca^{2+}-ATPase activity, *Acta Histochem Cytochem*, 14, 705, 1981.

38. Franks, J. J., Janicki, P. K., Horn, J. L., Singh, G., Sastry, B. V. R., Surber, M. J., Janson, V. E., Franks, W. T., Catlin, R. W., Johnson, R. F., Inhibition of synaptic plasma membrane Ca^{2+}-ATPase by inhalation anesthetics and its association with mechanisms of general anesthesia, *Prog Anesth Mech,* 3, 262, 1995.
39. Franks, J. J., Horn, J. L., Janicki, P. K., Singh, G., Stable inhibition of brain synaptic plasma membrane Ca^{2+}-ATPase in rats anesthetized with halothane, *Anesthesiology,* 82, 118, 1995.
40. Janicki, P. K., Horn, J. L., Singh, G., Franks, W.T., Franks, J. J., Diminished brain synaptic plasma membrane Ca^{2+}-ATPase activity in rats with streptozocin-induced diabetes: Association with reduced anesthetic requirements, *Life Sci*, 55, PL359, 1994.
41. Singh, G., Janicki, P. K., Horn, J. L., Janson, V. E., Franks, J. J., Inhibition of plasma membrane Ca^{2+}-ATPase pump activity in cultured C6 glioma cells by halothane and xenon, *Life Sci*, 56, PL219, 1995.
42. Janicki, P. K., Horn, J. L., Singh, G., Janson, V. E., Franks, W. T., Franks, J. J., Reduced anesthetic requirements, diminished brain plasma membrane Ca^{2+}-ATPase pumping, and enhanced brain synaptic plasma membrane phospholipid methylation in diabetic rats: Effects of insulin, *Life Sci*, 56, PL357, 1995.
43. Horn, J. L., Janicki, P. K., Franks, J. J, Diminished brain synaptic plasma membrane Ca^{2+}-ATPase activity in spontaneously hypertensive rats: Association with reduced anesthetic requirements, *Life Sci,* 56, PL427, 1995.
44. Janicki, P. K., Horn, J. L., Singh, G., Franks, W. T., Janson, V. E., Franks, J. J., Increased anesthetic requirements for isoflurane, halothane, enflurane and desflurane in obese Zucker rats are associated with insulin-induced stimulation of plasma membrane Ca^{2+}-ATPase, *Life Sci*, 59, PL269, 1996.
45. Horn, J. L., Janicki, P. K., Wamil, A. W., Singh, G., Franks, J. J., Reduced anesthetic requirements in aged rats: Association with alteration of brain synaptic plasma membrane Ca^{2+}-ATPase pump and phospholipid methyl transferase I activities, *Life Sci,* 59, PL263, 1996.
46. Gatto, C., Hale, C. C., Xu, W., Milanick, M. A., Eosin, a potent inhibitor of the plasma membrane Ca pump does not inhibit the cardiac Na-Ca exchanger, *Biochemistry,* 34, 965, 1995.
47. Wamil, A. W., Franks, J. J., Janicki, P. K, Horn, J. L., Franks, W. T., Halothane alters electrical activity and calcium dynamics in cultured mouse cortical, spinal cord, and dorsal root ganglion neurons, *Neurosci Lett*, 216, 93, 1996.

8 Anesthetic Modification of Neuronal Sodium and Potassium Channels

Daniel S. Duch and Tatyana N. Vysotskaya

CONTENTS

8.1 INTRODUCTION

Voltage-gated sodium and potassium channels are best known as the primary mediators of the propagating action potential, the neuronal mechanism by which electrical signals are transmitted from one end of an axon to the other. Although these ionic

0-8493-8555-5/01/$0.00+$.50

channels were the first to be recognized, described, and studied,[1] the role that sodium and potassium channels play in anesthesia is still controversial. Early studies, mostly with nonmammalian neuronal preparations, indicated that clinical concentrations of anesthetics caused only a slight modification of voltage-gated channel functions compared with anesthetic effects on ligand-gated receptors. This led many researchers in the field to consider it unlikely that anesthetic modification of sodium and potassium channels contributed to the clinical state of anesthesia.[2-4] In the past decade, however, there has been increasing evidence of a previously unsuspected diversity in sodium and potassium channel structure, function, and pharmacology. This new understanding of channel variability has called into question the applicability of the earlier results to anesthetic modification of mammalian central nervous system (CNS) sodium and potassium channels. Further supporting a reassessment of the role of these channels in anesthesia, recent studies examining mammalian neuronal channels have provided evidence that these channels are indeed much more sensitive to anesthetic modification than was previously believed.

The purpose of this review is to examine the recent experimental data reported with sodium and potassium channels and compare them with earlier results. The first section presents a general description of the molecular functions of these ionic channels. The second discusses and compares the molecular modifications of different sodium channel subtypes by anesthetics and how these molecular modifications correlate with experimental observations at the cellular level. The final section presents a similar examination of anesthetic interactions with potassium channels at the molecular and cellular levels.

8.2 VOLTAGE-GATED ION CHANNELS

The first quantitative description of the permeability changes that occur during the propagation of an action potential along a nerve cell membrane was presented by Hodgkin, Huxley, and Katz in a series of publications that form the basis for the present understanding of electrical signaling in excitable tissues.[5-8] As depicted in Figure 8.1, the action potential (dashed line) is a transient reversal of the negative resting transmembrane potential caused by a flow of sodium ions into the axon (i.e., an increase in g_{Na} [sodium conductance]). This increase in g_{Na} makes the membrane potential more positive (i.e., depolarizes the membrane). The membrane is repolarized to its negative resting potential by a subsequent increase in potassium conductance (g_K).

Qualitatively, Hodgkin and Huxley described the nerve as having separate pathways for the passage of sodium and potassium through the membrane. When the nerve is in its "resting" state, both channels are closed. Upon depolarization of the nerve past threshold, a voltage-sensitive "gate" opens the sodium channel, allowing sodium to enter the cell and further depolarizing the membrane. After a delay, two other responses occur: (1) the sodium channel "inactivates" as an inactivation gate again closes the channel, and (2) potassium channels activate, allowing potassium ions to leave the cell. These latter two responses return the nerve membrane to its resting potential. The importance of potassium channels in repolarizing

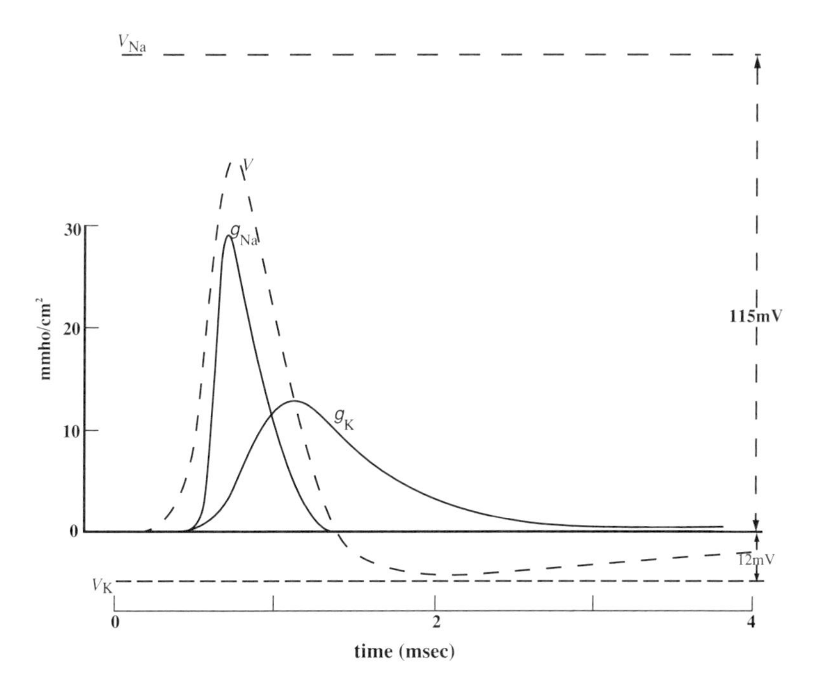

FIGURE 8.1 Theoretical action potential and conductance changes. V, membrane potential; g_{Na}, sodium conductance; g_K, potassium conductance. (From Hodgkin, A. L. *Proc. R. Soc. (Lond.)* 148:1, 1958. With permission.)

the nerve during an action potential has been found to vary from species to species and cell to cell.

Fitting their data to the simplest analysis, Hodgkin and Huxley[5-8] compared the potassium pathway to a channel controlled by a single activation gate with four charged particles that respond to changes in membrane potential. When the membrane was depolarized, all four particles occupied a certain position, which caused the gate to open and allowed potassium to flow out of the cell. Upon repolarization of the membrane, the gate closed when any one of the particles left the "open" position. The conductance of potassium was therefore proportional to the fourth power of a variable that obeys a first order equation (n^4). This explained the fact that during excitation, potassium conductances increased with a delay but fell more rapidly upon repolarization.

In a similar way, sodium conductance was compared to a pathway controlled by two gates. Quantitatively, the gating process was described by the following equation for sodium current:

$$I_{Na} = g_{Na}m^3h(E - E_{Na})$$

where I_{Na} is the sodium current, g_{Na} is the conductance expected when the maximum number of sodium channels are open, m^3h represents the probability that a pathway is open at a particular potential and time, and $E - E_{Na}$ is the electrochemical driving force for sodium ions. In this model, m^3 was the factor representing the sigmoid time course of activation, which was controlled by three gating particles; and h, representing inactivation, was the probability that a channel was not inactivated.

Although specific details of the Hodgkin and Huxley model have been shown to be incorrect, their work provided the molecular model for most of the biophysical studies of voltage-gated ionic channels conducted during the past four decades. As will be discussed later, many different types of potassium channels have since been found, some of which have an inactivation gate. In contrast, all voltage-gated sodium channels have both activation and inactivation gates.

Based on current information, a hypothetical view of a voltage-gated channel is shown in Figure 8.2. These channels have three functionally defined properties: (1) they are aqueous pores in the membrane; (2) they are selective (e.g., sodium channels allow sodium ions to pass more readily than other ions and exclude anions); and (3) they are gated, with one or more voltage-controlled gates. Structurally, they have three distinct biochemical domains:[9] the protein that comprises the channel, covalently attached carbohydrate groups, and strongly or covalently attached fatty acids that interface with the membrane lipids. All three domains present potential sites for anesthetic interactions.

8.3 ANESTHETIC MODIFICATION OF SODIUM CHANNELS

8.3.1 VARIABILITY OF SODIUM CHANNEL STRUCTURE AND FUNCTION

A simplified gating scheme for sodium channels is depicted here:

$$R \underset{\beta_1}{\overset{\alpha_1}{\rightleftarrows}} O \underset{\beta_2}{\overset{\alpha_2}{\rightleftarrows}} I$$

Actual sodium channel gating is much more complex.[1] As discussed above, a strongly voltage-dependent activation process quickly opens (O) the resting channel (R) in response to significant depolarizations from the resting membrane potential, while an inactivation mechanism (I) subsequently closes conducting channels, preventing further influx of sodium. A graphical depiction of the voltage dependence of sodium channel activation, inactivation, and their time constants in squid giant axon is given in Figure 8.3. The steady-state h_∞ curve depicts the fraction of channels that are not inactivated. As can be seen, the number of inactive channels depends on membrane potential.

Although the transitions between the three channel states are similar for all examined sodium channels in different preparations, the voltage dependence and kinetics of the transitions between resting, open, and inactivating states vary widely among tissues.[1] Traditional biophysical preparations such as squid giant axons and

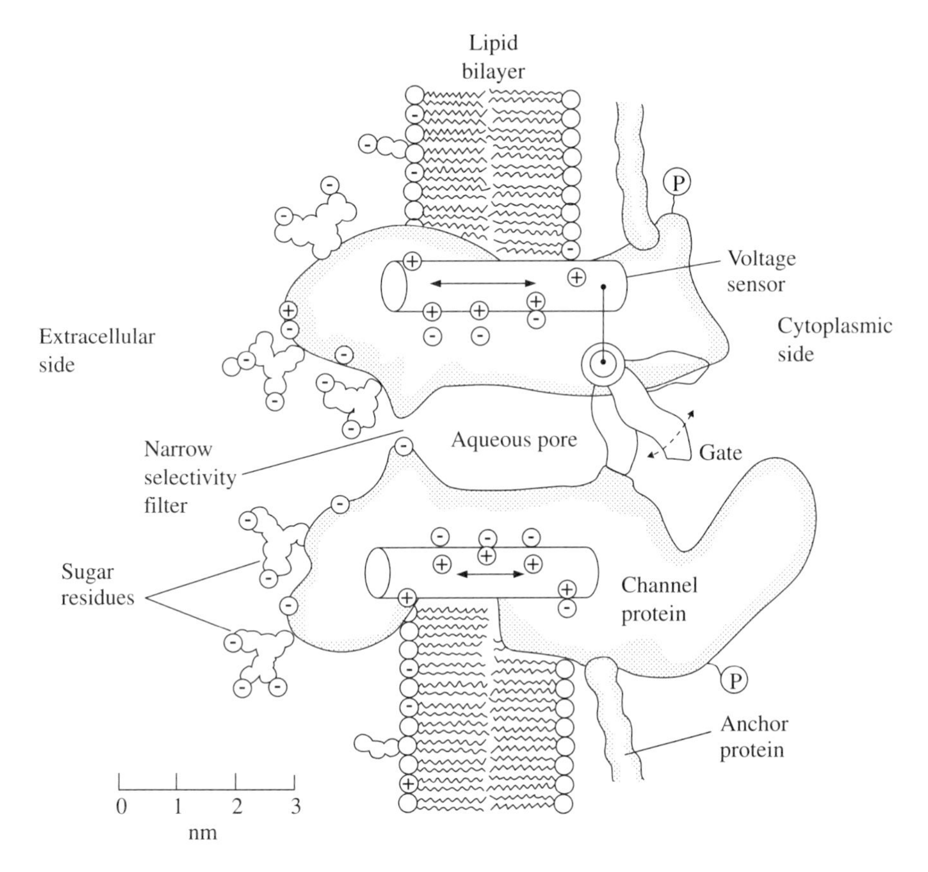

FIGURE 8.2 Model of a voltage-gated channel. Voltage-gated channels have three functionally defined properties: (1) they are aqueous pores, (2) they are selective, and (3) their conductance pathways are opened and closed by "gates" that open and shut dependent on membrane potential. These ionic channels are comprised of protein, which makes up the physical structure of the pore and channel; covalently attached carbohydrate groups, which usually impart a net negative charge to the outside of the channel; and strongly associated fatty acids. (From Hille, B. *Ionic Channels of Excitable Membranes,* Sinauer Assoc., Sunderland, MA, 1992.)

frog nodes of Ranvier have sodium channels with markedly different properties from channels examined in the mammalian CNS.[1] More specifically, sodium channels in mammalian CNS have short openings and generally seem to have slower activation and faster inactivation.[1,10-13] At the other extreme, sodium channels in GH_3 cells have long open times and slow, voltage-dependent inactivation.[1,14,15] Intermediate kinetics have been found in mammalian heart.[1,13,14] Pharmacologically, many substances interact with only one or another of these states, and this variability in channel activation and inactivation properties may contribute to the experimentally observed diversity in the pharmacologic properties of channels from different tissues.[10]

This functional and pharmacologic sodium channel diversity is matched by structural differences among sodium channel isoforms. All sodium channels have in

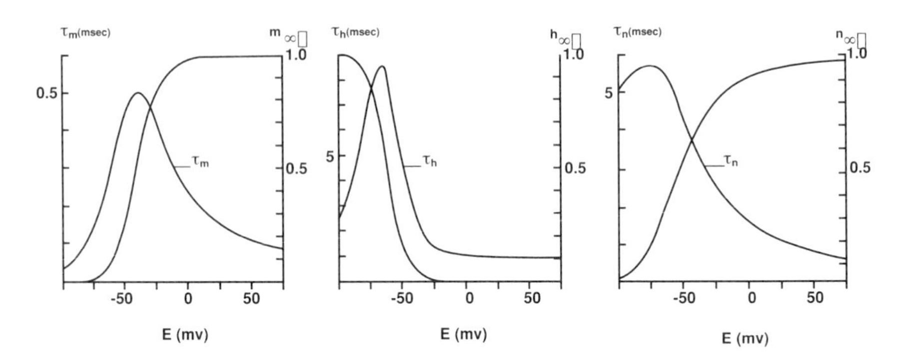

FIGURE 8.3 The steady-state values and time constants for sodium and potassium channels, as a function of membrane potential and time. Abbreviations are as listed in the text. (From Hille, B. *Prog. Biophys. Mol. Biol.* 21:3, 1970. With permission.)

common a large α subunit that, by itself, has the structures needed to carry out all the electrophysiologic functions and pharmacologic interactions of sodium channels.[10,17] A large number of distinct sodium channel α subunits have been found. Within rat brain, five full-length rat α subunit sodium channel cDNAs have been published so far, and it has been estimated that there are at least ten sodium channel genes in the rat.[18] The discovery of these molecular subtypes has "surpassed our understanding of the unique roles played by these channels."[18]

In addition to the α subunit, sodium channels have either none, one or two smaller β subunits, dependent on tissue expression. The functional role of these subunits remains unclear. The β_1 subunit is necessary to obtain normal sodium channel inactivation and pharmacology when expressed with the α subunit in some cells, but not in Chinese Hamster Ovary (CHO) cells.[19-22] On the other hand, the β_2 subunit has not been found to alter channel functions whether or not it is present.[19]

Posttranslational modifications such as phosphorylation[23-26] and glycosylation[27,28] also contribute to functional sodium channel diversity.[10,17,22] In spite of the increasing evidence for the structural, functional, and pharmacologic diversity of sodium channels, the correlation between sodium channel structural and functional differences is still not understood.

Further, electrophysiologic experiments with rat dorsal root ganglion (DRG) cells have provided evidence that sodium channels with distinct pharmacologic and electrophysiologic properties can exist not only in different cells of the same tissue, but even within the same cell.[29,30] This structural diversity can cause functional differences between cells. For example, DRG neurons must be capable of following stimulation frequencies varying between two and several hundred hertz.[31] This has been achieved partly through the use of at least two different types of neurons: one that can allow frequencies of several hundred hertz and one that can follow only very low frequencies and exhibit broader action potentials of increased duration.[29,31] Experiments have indicated that these functional differences correlate with differential expression of sodium channel subtypes in the various neuronal cells.[29] Pharmacologically, these sodium channel subtypes are also quite distinct,[29] one being TTX sensitive and the other TTX insensitive.

In summary, it is increasingly recognized that, first, there is a large diversity in sodium channel structure and pharmacology that is still poorly understood. This diversity occurs not only in sodium channel primary structure, but also in the posttranslational modifications (carbohydrate groups, associated fatty acids, phosphorylation, etc.) of sodium channels. Second, differences in the electrophysiologic and pharmacologic responses of sodium channels can lead to differences in cellular electrophysiologic and pharmacologic responses. Finally, sodium channels have other functions than just the conduction of axonal action potentials. Sodium channels in presynaptic nerve terminals and those in dendrites may affect synaptic responses as well as neurotransmitter release.

8.3.2 Anesthetic Modification of Nonmammalian Sodium Channels

Most investigators have made use of the Hodgkin–Huxley parameters to understand the changes that occur in voltage-gated sodium channels as a result of anesthetic modification. A detailed description of these parameters and their application to understanding the pharmacologic modifications of sodium channels has been given by Elliott and Haydon,[32] including a discussion of the viability of this model in light of more recent experimental evidence on sodium channel properties. Briefly, all sodium channel modifiers alter channel function by one or more of four mechanisms. They may (1) modify channel activation gating, (2) modify channel inactivation gating, (3) block or plug the channel pore, and/or (4) alter the structure and properties of the channel pore (i.e., conductance and selectivity) without blocking it. Modification of the gating parameters may also be separated into steady-state effects (i.e., increases in the number of inactivated channels at the resting membrane potential or decreases in the fraction of resting channels that can open) and kinetic effects (i.e., changes in the time constants of activation and inactivation).[32]

Macroscopically, sodium current can be reduced by depolarizing (positive) shifts in steady-state activation (m_∞ curves) or by hyperpolarizing (negative) shifts in inactivation (h_∞ curves).[32] Either shift may decrease the fraction of sodium channels that can open in response to membrane depolarization (i.e., the channels available for opening). This decrease, if large enough, can alter action potential initiation and propagation. Likewise, making activation slower (increasing the time constants for activation) or inactivation faster (decreasing the time constants for inactivation) will also reduce the observed sodium current.[31] At the cellular level, the voltage dependence and time constants of the activation and inactivation transitions define or affect such critical neuronal properties as spike threshold, speed of action potential propagation, frequency and latency of action potential firing, etc.[1]

Within this Hodgkin–Huxley framework, perhaps the most extensive and detailed examination of anesthetic modification of voltage-gated sodium channels was conducted by Haydon and his co-workers in a series of papers that compared the molecular interactions of different chemical classes of anesthetic substances on sodium and potassium channels in the squid giant axon,[32-38] the same preparation used by Hodgkin and Huxley. The results of Haydon and Urban[36,38] with chloroform depict many of the main anesthetic modifications of squid axon sodium channels found in these and

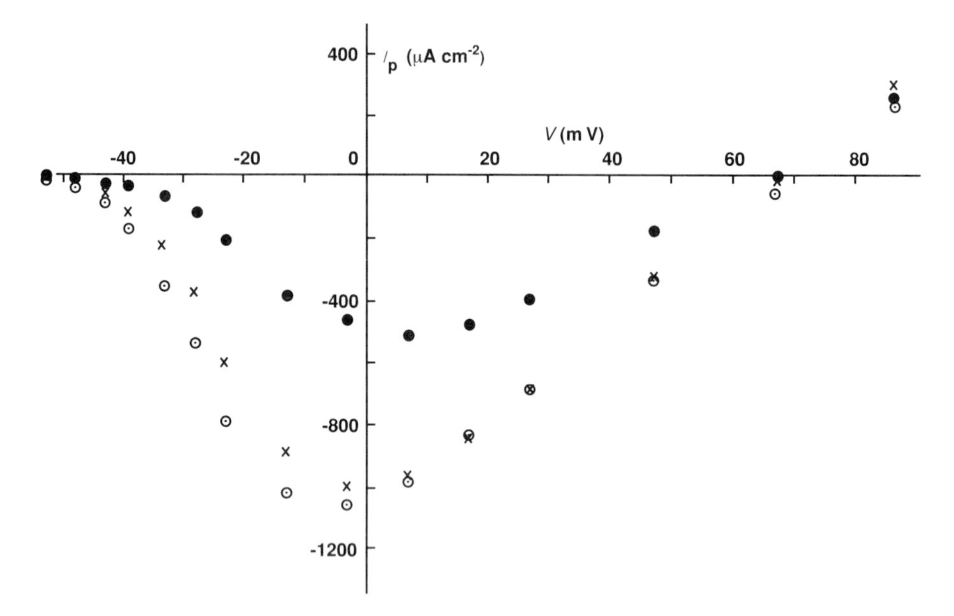

FIGURE 8.4 Chloroform decreases peak current in the squid giant axon. Open circles, before exposure to 5 m*M* chloroform; filled circles, during exposure to chloroform; X, washout of chloroform. (From Haydon, D. A. and Urban, B. W. *J. Physiol. (Lond.)* 341:429, 1983. With permission.)

subsequent studies (Figure 8.4). In these experiments, the membrane was first hyperpolarized to potentials where all channel inactivation is removed before the I-V (current-voltage) relationship was determined (see Figure 8.3, middle panel). Under such conditions, changes in the peak sodium current subsequent to anesthetic addition reflect modification(s) of either the resting or open channel state(s).

As shown in Figure 8.4, Haydon and Urban found that (1) the peak sodium current was reduced; (2) the potential at which the current reached a maximum was shifted in the positive (depolarized) direction, indicating a shift in the voltage dependence of activation (m_∞); and (3) the reversal potential was unaffected, indicating that channel selectivity was not significantly altered at these anesthetic concentrations. Using a different pulse protocol as described by Kimura and Meves,[39] Haydon and Urban next examined the effect of anesthetics on inactivation, determining h_∞ curves (Figure 8.5). It can be seen that all examined volatile anesthetics shifted the h_∞ curves to more hyperpolarized (negative) potentials. These investigators also reported that anesthetics decreased the peak time constants for activation and inactivation. As described, these channel modifications should lead to a reduction in sodium current. Fitting the data to the Hodgkin–Huxley formulation, Haydon and Urban indeed found that the voltage-independent conductance was decreased by these anesthetics (see Ref. 31 for a more detailed explanation). These results, as well as earlier studies with other classes of anesthetic agents,[32-38] thus indicated that anesthetics altered channel activation, inactivation, and macroscopic conductance, thereby altering the channel closed, open, and resting states. However, none of the

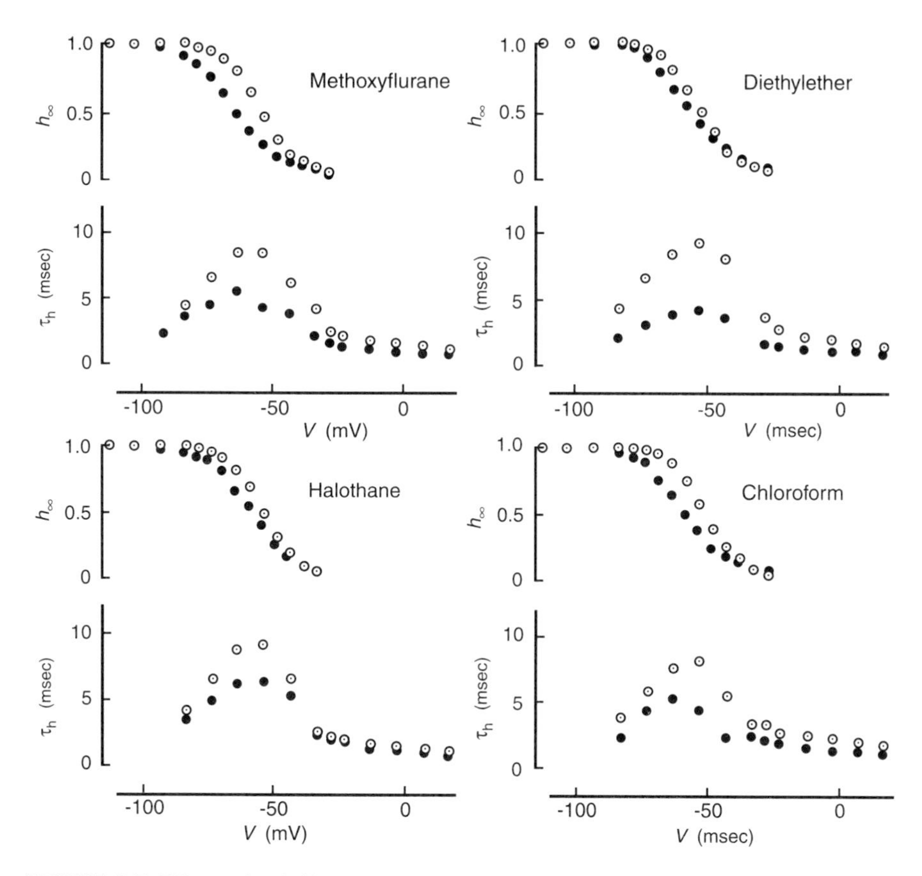

FIGURE 8.5 Effects of volatile anesthetics on steady-state inactivation (h_∞) and inactivation time constant (τ_h). Concentrations of anesthetics were methoxyflurane, 3 m*M*; diethylether, 100 m*M*; halothane, 4 m*M*; and chloroform, 5 m*M*. Open circles, before exposure to anesthetics; filled circles, after about 20 min exposure to anesthetics. (From Haydon, D. A. and Urban, B. W. *J. Physiol. (Lond.)* 341:429, 1983. With permission.)

anesthetic agents appeared to alter the structure of the pore (i.e., channel selectivity was unaffected).

Comparing different classes of anesthetics, Haydon and co-workers did find some differences in their actions.[32,40] The channel modifications caused by volatile anesthetics were intermediate between those found with nonpolar hydrocarbons and polar surface active substances such as the alcohols. The hydrocarbons produced much larger hyperpolarizing shifts in inactivation than found in the volatile anesthetic experiments, whereas alcohols had very little effect on channel activation. Hydrocarbons also shifted channel activation to more negative (hyperpolarized) potentials in contrast to the depolarizing shift in activation observed with the volatile anesthetics. Nonetheless, in all cases, these effects were obtained at concentrations well above the minimum alveolar concentration (MAC) (for example, see Table 8.1).

Other groups examined anesthetic modifications of sodium channels in different neuronal preparations. Kendig et al.,[41] examining the effects of ether on sodium

TABLE 8.1
Anesthetic Block of Mammalian and Non-Mammalian Na+ Channels

	Halothane (MAC = 0.22 m*M*)			Ether (MAC + 9.3 m*M*)			
Cell Examined	**Conc. (m*M*)**	**↓ I_{peak}**	**h_∞ (shift mV)**	**Conc. (m*M*)**	**↓ I_{peak}**	**h_∞ (shift mV)**	**Ref.**
Squid axon	4.70	0.56	–1.8	100	0.61	–3.6	35
Frog node of ranvier	XXX	XXX	XXX	200	0.5	–10.0	41
Crayfish giant axon	32.0	0.6	XXX	100	0.2	–5.0	42
GH_3 pituitary cells	2.6	0.5	–11.0	XXX	XXX	XXX	43
CNaII A-1	1.0	0.58	–33.0	100	0.55	–40.0	51

Note: XXX, no data available.

channel block in frog nodes of Ranvier, found similar hyperpolarizing shifts in the steady-state inactivation curves; however, once again these shifts occurred at relatively high anesthetic concentrations (Table 8.1). Bean et al.[42] examined the effects of ether and halothane on sodium channel gating in the crayfish giant axon. They also found a similar effect of these anesthetics on inactivation (i.e., a hyperpolarizing shift in the steady-state inactivation and a speeding up of inactivation kinetics); however, in contrast to the work of Haydon and Urban, no effects were noted on channel activation. Once again, significant anesthetic modifications occurred only with relatively high concentrations of anesthetic.

Model cell systems have also been used to examine and compare anesthetic effects on various ion channel families. Herrington et al.[43] examined the effects of halothane on ionic channels in pituitary GH_3 cells and found that the anesthetic suppressed peak sodium current and shifted steady-state inactivation to hyperpolarized potentials, similar to the squid axon channels examined by Haydon and Urban. Once again these changes occurred only at relatively high anesthetic concentrations, although the h_∞ shift was markedly more pronounced than in earlier studies. Pancrazio et al.[44] found that enflurane (1.7 m*M*) had little or no effect on sodium currents in bovine adrenal chromaffin cells with no change in channel properties. The results from these studies are summarized and compared in Table 8.1. Together, these experiments on different preparations indicate that anesthetic modification of sodium channels depends not only on the anesthetic agent being studied, but also on the particular cell and sodium channel isoform being examined.

8.3.3 IV Anesthetic Modification of Sodium Channels from Mammalian Neurons

For technical reasons, it was not feasible to record sodium currents from mammalian neurons until the late 1980s.[2] The first detailed examination of anesthetic modification of mammalian CNS sodium channels was conducted by Frenkel, Urban, and colleagues, who used planar lipid bilayer technology to investigate IV (intravenous) anesthetic modification of sodium channels from human brain.[45-49] These experiments used batrachotoxin (BTX), a sodium channel modifier, to remove inactivation and allow the channels to be recorded in planar bilayers, potentially interfering with anesthetic modification of the channel inactivation. Under the steady-state conditions in these studies, all investigated anesthetic compounds had two major effects on channel properties: (1) a voltage-independent reduction of the fractional channel open time and (2) a hyperpolarizing shift in channel activation (opposite to that found in the squid axon studies). Pentobarbital and propofol significantly altered sodium channel function in the clinical concentration range, but ketamine and midazolam did not. However, even propofol and pentobarbital blocked only about 10% of sodium channel current at clinical concentrations. Thus these results were, in general, similar to earlier results with nonmammalian sodium channels.

To investigate anesthetic interactions with brain sodium channels in the absence of BTX, Rehberg et al.[50] examined and compared pentobarbital modification of sodium channels from rat brain and rat muscle stably transfected into CHO cells. With intact inactivation, pentobarbital caused both a voltage-independent block of sodium current at hyperpolarized potentials (indicating an interaction with either the resting or open state of the channel, or both) and, more significantly, a hyperpolarizing shift in channel inactivation. The summed result of these modifications was that pentobarbital blocked about 50% of sodium current at concentrations that were three- to fivefold higher than clinical concentrations. No significant differences were found in the anesthetic interactions with either the muscle or brain sodium channel isoforms. Thus, with intact inactivation, pentobarbital was more effective at blocking sodium channels than in experiments with altered inactivation.

8.3.4 Volatile Anesthetics Significantly Suppress Mammalian Neuronal Sodium Channels at Clinical Concentrations

The first electrophysiologic examination of volatile anesthetic modification of mammalian CNS sodium channels also was conducted by Rehberg et al.,[51] again using type IIA sodium channels from rat brain stably transfected into CHO cells. This study focused on the two main anesthetic effects already discussed: block of peak sodium current and the hyperpolarizing shift in channel inactivation. Volatile anesthetics blocked peak sodium currents at concentrations that averaged about threefold higher than MAC values (Figure 8.6), producing a much larger block than reported in other experimental systems. More important, all of the examined volatile anesthetics gave a much larger hyperpolarizing shift in the steady-state inactivation curve than had been found previously (Figure 8.7). Because of the magnitude of this shift, anesthetic block of sodium current was markedly voltage dependent. As indicated in Figure 8.8, at a membrane holding potential of –120 mV, IC_{50} values were about three times greater than MAC values. However, at –60 mV (close to the resting potential of nerves) the IC_{50} for anesthetic block of sodium currents overlapped with clinical concentrations. In other words, clinical concentrations of volatile anesthetics blocked approximately half of the sodium channel current at normal resting membrane potentials. Thus, in contrast to the previous reports with nonmammalian and/or nonneuronal sodium channels, the rat brain IIA sodium channels were found to be much more sensitive to clinical concentrations of volatile anesthetics (Table 8.2).

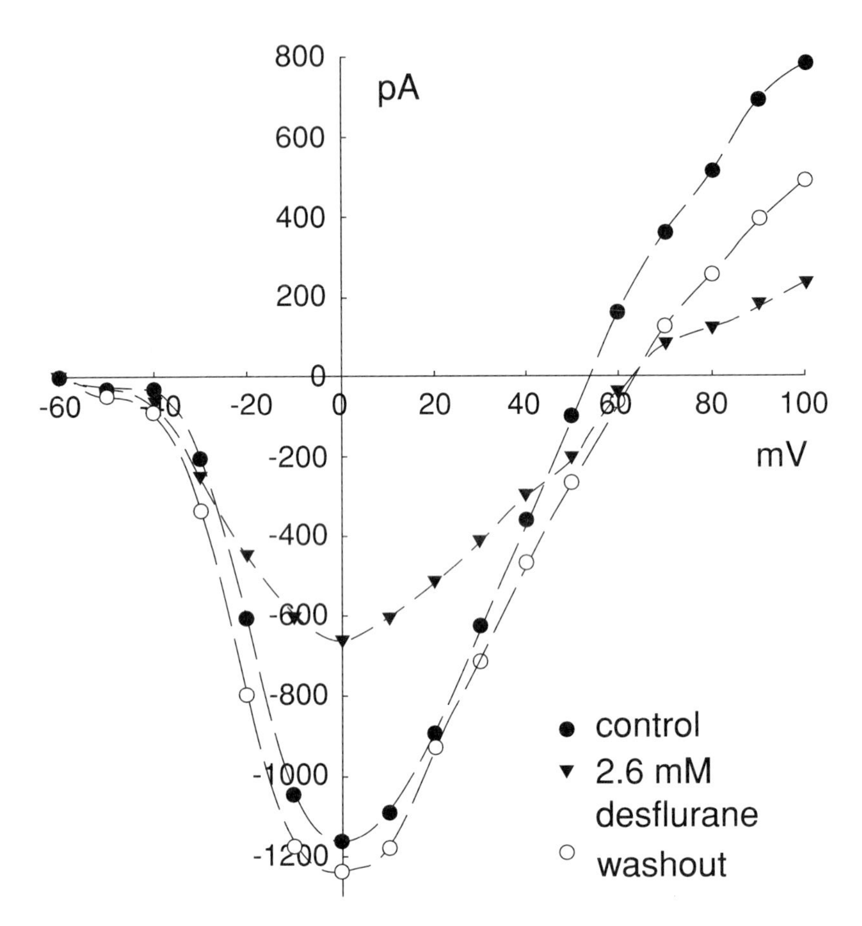

FIGURE 8.6 Effect of desflurane on whole-cell sodium currents in a CNaIIA-1 cell. Traces are before (left panel), during (upper right panel) superfusion with 2.6 m*M* desflurane, and after washout (lower right panel) with desflurane-free extracellular solution. A sustained outward current at positive potentials was sometimes observed after prolonged perfusion with high anesthetic concentrations (lower right panel). Calibration bars are 200 pA and 0.9 ms. Peak current–voltage relationship for the current traces shown in left panel. (From Rehberg, B., Xiao, Y.-H. and Duch, D. S. *Anesthesiology* 84:1223, 1996. With permission.)

A comparison of anesthetic block of channels from the different experimental preparations of sodium channels for halothane and ether is given in Table 8.1. Both peak current reduction and the shift of steady-state inactivation are significantly larger for the rat brain sodium channels than for the other preparations.

8.3.5 Anesthetic Block of Sodium Currents in Acutely Dissociated Neurons

The question that remains is whether these anesthetic interactions, which occur in a model system, also occur in neurons. Preliminary experiments examining anesthetic modification of sodium currents in neurons isolated from rat DRG indicate

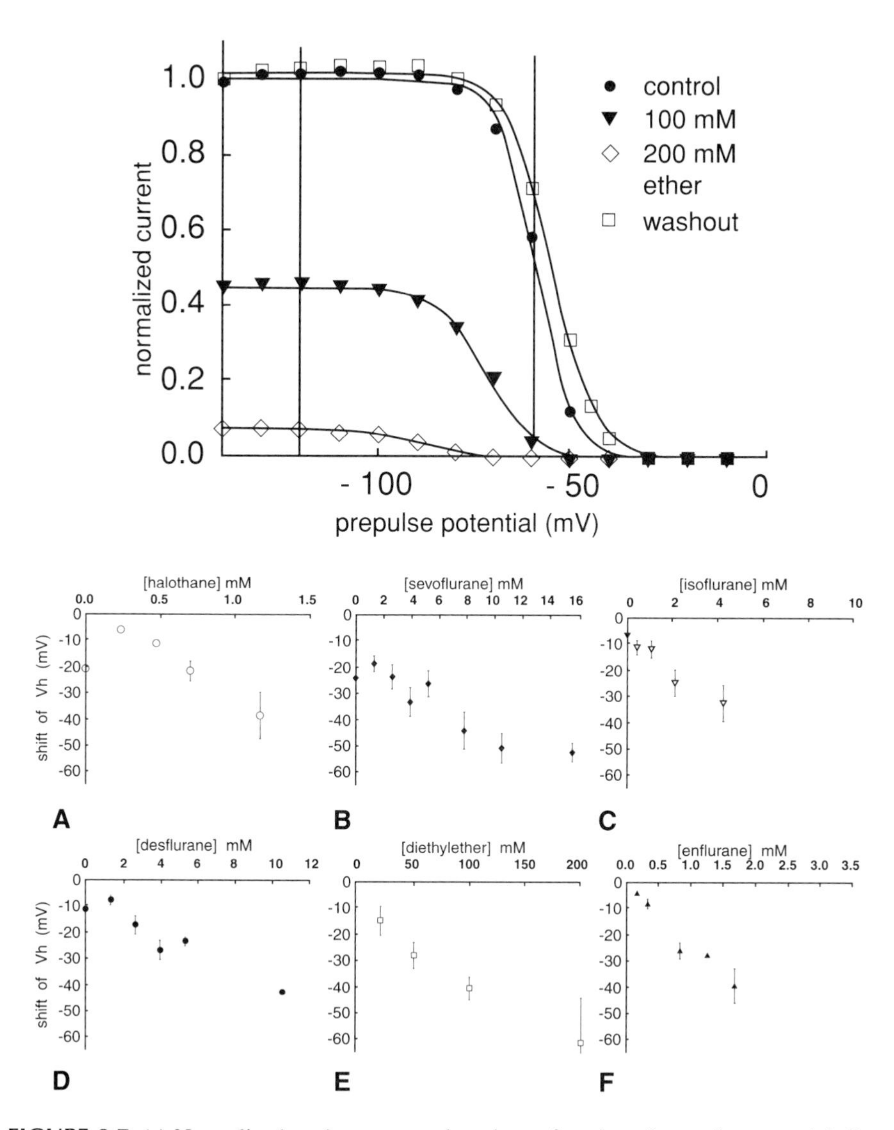

FIGURE 8.7 (a) Normalized peak currents, plotted as a function of prepulse potential. Currents were normalized to the maximum current before addition of the anesthetic. The midpoint of channel inactivation shifted from –58.8 mV (control) to –72.0 and –89.1 mV during perfusion with 100 m*M* and 200 m*M* diethylether, respectively. After washout it returned to a value of –55.0 mV. (From Rehberg, B., Xiao, Y.-H. and Duch, D. S. *Anesthesiology* 84:1223, 1996. With permission.) (b) Volatile anesthetics cause a concentration-dependent shift of the midpoint of steady-state sodium channel inactivation. Data was obtained from experiments as in Figure 8.7a. (A) Halothane (averages of five to six cells), (B) sevoflurane (averages of three to nine cells), (C) isoflurane (averages of three to ten cells; data point at 8.5 m*M* single cell), (D) desflurane (averages of six to nine cells; data point at 10.6 m*M* from two cells), (E) diethylether (averages of three to ten cells), and (F) enflurane (averages of three to four cells; data point at 1.26 m*M* from a single cell). Values for washout are shown at 0 m*M*. (From Rehberg, B., Xiao, Y.-H. and Duch, D. S. *Anesthesiology* 84:1223, 1996. With permission.)

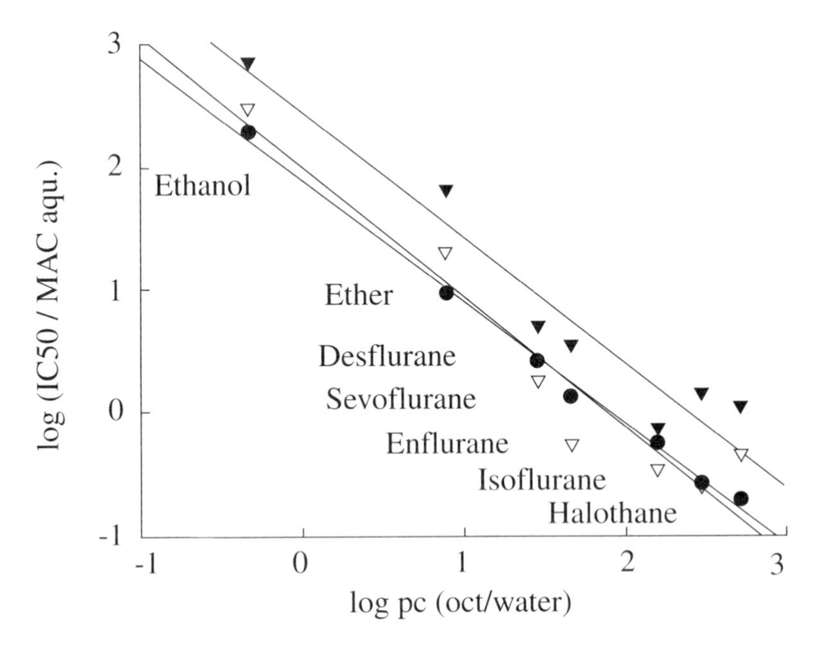

FIGURE 8.8 Linear correlation of the logarithm of the octanol/water partition coefficients and the logarithm of the IC_{50} values for sodium current block at two different prepulse potentials (solid triangles, –120 mV; open triangles, –60 mV). Human MAC values as aqueous concentrations are plotted on the same scale (filled circles). Lines are linear regression fits to the data. (From Rehberg, B., Xiao, Y.-H. and Duch, D. S. *Anesthesiology* 84:1223, 1996. With permission.)

TABLE 8.2
MAC and IC_{50} Values for CNaIIA-1 Cells

Anesthetic	Aqueous MAC (μM)	IC_{50} at –60 mV (μM)
Desflurane	2700	1790
Diethylether	9300	1990
Enflurane	580	340
Halothane	220	460
Isoflurane	280	250
Sevoflurane	1400	560

Source: From Rehberg, B., Xiao, Y.-H. and Duch, D. S. *Anesthesiology* 84:1223, 1996. With permission.

that these sodium channels are also significantly blocked by volatile anesthetics at clinical concentrations.[52,53] Primary cultures of spinal and trigeminal ganglionic neurons from adult rat were obtained after enzymatic dissociation with trypsin and/or collagenase. Cells were used two hours after dissociation and during the next 3 days of cultivation. Control measurements of the cells elicited normal action potentials and stable ionic currents.

Sodium currents were measured using the whole-cell patch-clamp configuration. Sodium currents and anesthetic suppression were measured with depolarizing test potentials from a hyperpolarizing holding potential (–80 mV). These initial studies used relatively high anesthetic concentrations. Similar to results with CNS channels, halothane (2 m*M* and less) reversibly decreased about 50% of peak sodium current. The effect of isoflurane was also similar to the results in CHO cells (0.5 m*M* caused 30% block of peak current), as were results with pentobarbital, which blocked sodium current in the same concentration range as CNS channels (≤1.8 m*M*). As noted above, anesthetic concentrations that block peak currents under these experimental conditions are about five- to tenfold higher than clinical concentrations and are comparable to anesthetic potency in squid axons.[51]

As also observed in CHO cells, however, both anesthetics significantly shifted steady-state inactivation to more negative potentials, resulting in a voltage-dependent block of sodium current. The two effects combined to completely block sodium current at resting membrane potentials (~ –60 mV) at this anesthetic concentration. The halothane concentration that blocked 50% of peak current also caused a –20 mV shift in the midpoint of the steady-state inactivation, similar to that found when CNS channels in CHO cells are modified with volatile anesthetics but different from other sodium channels and pentobarbital action. All of these results are within the range of variability of the previous results with rat CNS sodium channels in CHO cells, and the similarity in anesthetic shift of inactivation supports the hypothesis that mammalian neuronal sodium channels *in situ* are also significantly blocked by clinical concentrations of volatile anesthetics.

8.3.6 Correlation of Molecular, Cellular, and Clinical Anesthetic Actions

In considering the clinical significance of anesthetic effects on ion channels, it is important to determine whether the resulting changes impact functions at higher levels of CNS integration such as the cellular level or, ultimately, cognitive function. One of the fundamental problems encountered in any such assessment is that the quantitative association between changes in the molecular function of ionic channels and functional changes at the cellular level is inherently nonlinear. For example, it has been reported that less than a 2% change in the open probability of resting sodium channels can cause myotonia,[54] less than a 0.1% change in the function of these same channels underlies the toxicity of insecticides,[55] and activation of less than 1% of adenosine triphosphate (ATP)-sensitive potassium channels results in the observed shortening of the action potential duration of ventricular myocytes during early hypoxia.[56] This nonlinear relationship between ion channel function and cellular function means that subtle anesthetic alterations of ion channel properties can have much larger effects at the cellular level, and, conversely, large molecular effects may not result in significant changes in cell function. Moreover, the amount of molecular change required to alter cell function varies among cells, dependent on such parameters as ionic channel densities, nerve fiber diameters, physiologic frequencies of stimulation, and channel subtypes present.[57-59] Thus, whether anesthetic actions on ionic channels result in significant cellular or cognitive changes depends on the particular make up of the neuron or neuronal network responsible for cognitive function.

Given these considerations, it is not surprising that results at the cellular level have been mixed, as has been found also with other sensitive molecular anesthetic targets.[60-63] The results obtained depend not only on the functions examined but also on the neuron or neuronal system used. Action potential propagation in myelinated axons of the CNS appears to be relatively insensitive to anesthetic modification.[2,64] Other studies, however, have found alterations in the axonal properties of rat and human CNS neurons that are consistent with a change in sodium channel function.[65,66]

In addition to the generation of action potentials, there is increasing evidence that sodium channels are employed for more subtle, subthreshold signal processing in neurons.[17,18] It has long been known, for example, that sodium channels in neuronal cell bodies and axon initial segments determine the threshold for action potential generation and influence the frequency of neuronal firing.[67-70] Additionally, it has been found that sustained sodium currents in some projection neurons may affect action potential duration and repetitive firing.[71,72] Sodium channels are also present in nerve terminals,[73,74] as well as having been found more recently in the dendrites of cerebellar Purkinje cells and of pyramidal cells in both hippocampus and cortex.[76-79] The sodium channels in these dendrites are capable of generating dendritic action potentials, and their density and distribution in dendrites, as well as in presynaptic membranes, will modify synaptic responses[18] and potentially influence the amount of neurotransmitter released from presynaptic nerve endings.[17] Catterall[17] has suggested that these latter, integrative functions of sodium channels are more likely targets for neuromodulation than the sodium channels that mediate the axonal action potentials. If so, then they may also be more likely targets for anesthetic modification than the axonal sodium channels.

Examining these neuronal properties, the threshold for action potential firing has been found to increase, consistent with anesthetic modification of sodium channels.[58,66,80,81] Anesthetic block of sodium channels also has been found to directly inhibit glutamate release from synaptosomes,[82,83] and it may contribute to the anesthetic depression of glutamate release recently reported in hippocampus.[84]

Several hypotheses of anesthetic action have been proposed that include vital or contributory roles for sodium channels,[41,58,80,85-88] stressing the changes that occur in neuronal firing patterns rather than a simple block of action potentials. These hypotheses generally involve the failure of impulse propagation of action potentials at such sites as neuronal bifurcations.[88] Although sodium channels play a vital role in most of these hypotheses, it is evident that other ionic channels also are essential in anesthetic action. Thus, anesthesia is generally attributed to a collective effect on many distinct channels, with recent experiments providing strong evidence that voltage-gated sodium channels also play a role.

8.4 ANESTHETIC MODIFICATION OF POTASSIUM CHANNELS

8.4.1 ANESTHETIC MODIFICATION OF THE DELAYED RECTIFIER POTASSIUM CHANNELS

In describing the actions of potassium channels, it is useful to begin with a review of the Hodgkin–Huxley parameters that describe the potassium channel of squid axon, the "delayed rectifier" channel (K_V). This is the first and perhaps best studied of potassium channel families. As described above, the delayed rectifier potassium channels are responsible for the repolarization of the action potential, opening with a delay after the membrane is depolarized and returning it to its hyperpolarized resting potential. In contrast to the sodium channel of the squid axon, however, this potassium channel has no inactivation gate. A good description of these currents and their modification by anesthetics is given by Elliott and Haydon[32] in the same context as for the sodium channel.

Quantitatively, the steady-state potassium current is described by the following equation:

$$I_K = g_K\, n^4\, (E - E_K)$$

where I_K is the potassium current, g_K is the potassium current expected when the maximum number of potassium channels are open, n^4 represents the probability that a pathway (potassium channel) is open at a particular potential and time, and $E - E_K$ represents the driving force for potassium ions. In the Hodgkin–Huxley model, n^4 was the factor controlling the time course of potassium activation and deactivation (closing). The activation gate of the channel was modeled as being controlled by four gating particles. These parameters allowed a quantitative description[32] of pharmacologic effects on the steady-state channel properties by measuring potassium conductance as well as shifts in the voltage dependence of the n_∞ curve representing channel activation (see Figure 8.3, last panel). Additionally, kinetic effects could be examined by measuring τ_n, the time constant of channel activation. Thus, pharmacologic modification of these currents could be described by changes in three parameters: g_K, n_∞, and τ_n.

Once again the most thorough examination of the effects of various classes of anesthetics on potassium channels was conducted by Dennis Haydon and his colleagues.[37,38,89,90] In these experiments, it was found that, similar to the effects on sodium currents, anesthetics altered all Hodgkin–Huxley parameters to varying degrees. A comparison of the effects of isoflurane on sodium and potassium currents in squid axon is shown in Figure 8.9. It can be seen that the potassium (outward) current is more greatly suppressed than the sodium (inward) current by this anesthetic. These experiments indicated that halothane, isoflurane, diethylether, methoxyflurane, and enflurane all caused a greater reduction in peak potassium current than in peak sodium current,[32] thereby having a greater effect on potassium channels than on sodium channels. Additionally, halothane and methoxyflurane both caused a two- to threefold greater depolarizing shift in the steady-state activation of potassium

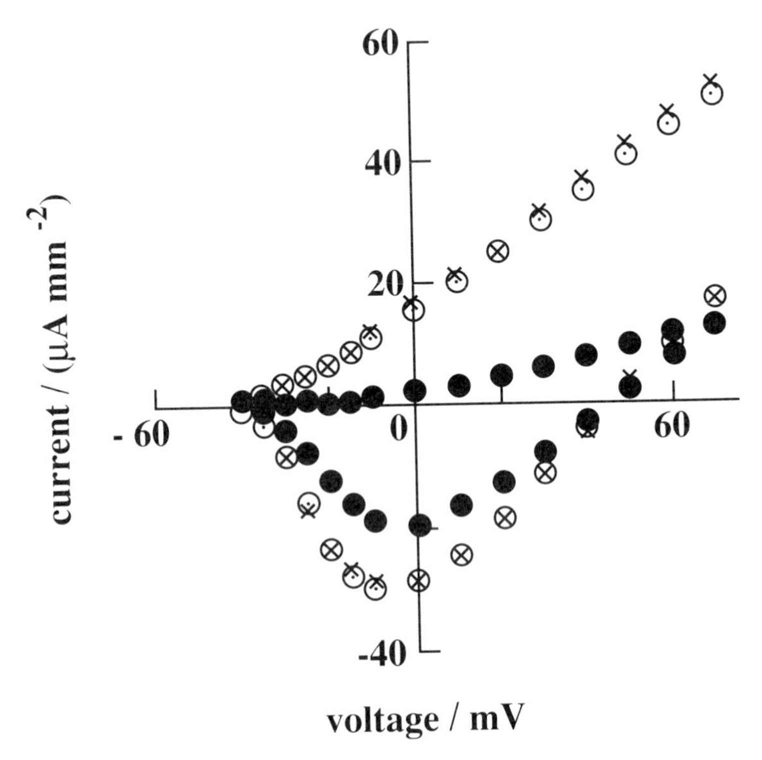

FIGURE 8.9 Current–voltage relationships for a squid giant axon. Peak inward (sodium) currents and steady-state outward (potassium) currents before (open symbols), during (filled symbols), and after (x) exposure to 2.55 m*M* isoflurane. (From Urban, B. W. and Haydon, D. A. *Proc. R. Soc. (Lond.)* 231:13, 1987. With permission.)

current compared with sodium activation, whereas there was no difference in the shifts caused by diethylether with either channel type. Changes in the activation time constants were similar for both potassium and sodium channels.

Overall, however, it was found that anesthetic sensitivity for the delayed rectifier potassium channels was relatively low compared with clinical anesthetic concentrations. This overall finding was supported by other researchers. Bean et al.[42] examined the effects of ether on delayed rectifier channels in frog node of Ranvier and found no change in the voltage dependence of steady-state potassium currents, but activation was faster in the presence of ether. Like the channels in squid axon, this change occurred at high (40 to 480 m*M* ether) anesthetic concentrations. More recently, Benoit[91] found that ketamine, etomidate, alphaxalone, and alphadolone acetate depressed sodium and potassium currents in frog myelinated nerve fibers, although the latter three anesthetics were more effective in blocking potassium currents than sodium. These effects occurred at anesthetic concentrations that were 25- to 250-fold greater than clinical levels.

Other recent studies have also examined the effects of anesthetics on delayed rectifier potassium channels[92-95] (see Table 8.3). It can be seen that in most cases

TABLE 8.3
Anesthetic Modification of Potassium Channels

Channel Type	Anesthetics Conc.	Effect	Species/Tissue	Physiologic Function Affected	Method	Year (Ref.)
K(V) K(V-indep)	Volatile anesthetics: isoflurane, enflurane 0.4–0.8 m*M*	Did not affect K_V conductance and inhibited 40–80% K (v-indep)	Squid axon	Depolarization of resting potential and reduction of resting K conductance	Voltage-clamp	1991[92]
K(V) K(I)	IV anesthetics: ketamine, propofol, methohexital	Ketamine — decreased only K(I); propofol — decreased K(V); methohexital — no effect on either	Single guinea pig ventricular myocyte	IV anesthetics had distinct spectrum of activity on K currents	Whole cell recordings	1993[93]
K(V) K(I)	Volatile anesthetics: sevoflurane	K_V was markedly depressed, but K(I) was only slightly affected by 0.35 m*M* sevoflurane	Isolated guinea pig ventricular papillary muscle; 37C	Depression of K_V appears to underlie the increase of action potential duration	Whole cell volt.-clamp, microelectrode, isomet-ric force and cooling contractures measur.	1996[94]
K(V) K(I)	IV anesthetics: propofol, thiopentone	1. 10 μ*M* propofol decreased K_V to 78%, no effect on K(I); 2. 10 μ*M* thiopentone decreased K_V to 78%, but 100 μ*M* thiopentone decreased K_V to 27% and K(I) to 67%	Isolated guinea pig ventricular myocyte	Biphasic effect on APD were caused by depression of K(V) and concomitant suppression of K(A) at higher con; cardiodepression	Whole cell current and voltage clamp	1996[95]
K(A) K(I)	IV anesthetics: etomidate, propofol midazolam	1. 60 μ*M*: etomidate 8%, propofol 9%, midazolam 23%-decrease K(A) 2. 60 μ*M*: etomidate 20%, propofol 0%, midazolam 14% decrease K(I)	Canine myocardial cells	Partly antagonized effect of decreased ICa	Whole cell voltage-clamp	1996[98]
K(A)	Volatile anesthetics: isoflurane	Dysfunction of I(A) conductance	*Drosophila melanogaster*	Movement of *Shaker* flies when heated	Indirect experiment: number of flies moving	1991[99]

TABLE 8.3 *(continued)*
Anesthetic Modification of Potassium Channels

Channel Type	Anesthetics Conc.	Effect	Species/Tissue	Physiologic Function Affected	Method	Year (Ref.)
K(I)	Ethanol 40–200 m*M*	Enhanced K(I): increased maximal slope conductance, shifted voltage range for activation in depolarizing direction; effects were concentration dependent and fully reversible	Rat locus ceruleus neurons	Ethanol also reduced the amplitude of after hyperpolarization. Shift K equilibrium	Voltage-clamp in brain slice	1994[100]
K(I)	Barbiturates: methohexital, pentobarbital	Reversibly inhibit current in concentration-dependent manner with IC_{50} 145 μ*M* methohexital, 218 μ*M* pentobarbital; inhibition was voltage dependent	Rat basophilic granulocyte cell line (RBL-1)		Whole cell voltage-clamp	1996[101]
K(Ca)	Volatile anesthetics: halothane 0.4–0.9 m*M*	1. 18–35% (nondefined K current) IK inhibition 2. 17–30% IK(Ca) inhibition 3. Decreased open state probability, open time, and frequency of opening; did not alter amplitude	Dog cerebral arterial muscle	Concluded that K channels do not mediate anesthetic induced vasodilation	Pach — whole cell and single channel	1995[102]
K(Ca)	Volatile anesthetics: halothane 0.2–1.0 m*M*	1. Decreased open state probability without changing channel conductance	Rat skeletal muscle, in bilayers	Halothane altered channel gating kinetics	Voltage-clamp in planar bilayers	1993[103]
K(Ca)	Volatile anesthetics: halothane, isoflurane	Decreased open state probability of channels	Bovine aortic endothelial cells		Pach — whole cell and fura-2; single channel inside-out patch	1996[104]

K baseline K(S)	Volatile anesthetics: halothane, isoflurane	1. Slow hyperpolarization of neuron 2. Block spontaneous firing conductance 3. Activation increases open probability of S channels 4. Halothane inhibited small noninactiving K channels	Aplysia neurons		Nerve-electrical activity, single channel patch clamp	1996[105]
K(ATP)	Barbiturates, halothane, isoflurane	Barbiturate — directly interact with K(ATP) volatile anesthetic-modulated part of channels	Ventricular myocytes	Myocardial protection	Whole cell recordings	1995[106]
K(ATP)	Isoflurane	Induce activation of K(ATP)	Dog heart	Improved recovery of contractive function of postischemic-vasodilation during ischemia	Indirect experiment, measurements of hemodynamics	1996[107]
K(ATP)	Halothane 0.62 m*M*	Little or no relaxation of smooth muscle caused by induced changes in K(ATP)	Canine tracheal smooth muscle	None	Organ bath system	1996[108]
K(ATP)	Isoflurane	Inhibit channel activity without change in single-channel conductance; decreased burst duration and increased interburst duration	Rabbit ventricle myocyte	Isoflurane mediated cardioprotective effect via K(ATP)	Patch-clamp	1996[109]

K(V) — Delayed rectifier
K(V-indep) — Voltage independent
K(I) — Inward rectifier
K(A) — Fast transient K channels
K(Ca) — Calcium activated K channels
K(S) — Second messenger-modulated K channels
K(ATP) — ATP-modulated K channels

the effects of anesthetics on the delayed rectifier sodium currents are relatively small at clinical concentrations. However, as with sodium channels, it should be noted that these studies have not directly examined the effects of anesthetics on delayed rectifiers in mammalian neurons.

8.4.2 Other Potassium Channels

Since the experiments of Hodgkin and Huxley[5-8] many other types of potassium channels have been found, and it is now established that potassium channels are a heterogeneous family of membrane ionic channels. Potassium channels have been found in virtually all cells that mediate action potentials, whether by sodium or other ions.[1,96] Thus, although the squid axon has been a useful neuronal model, both its resting potential and action potential are mediated by a relatively restricted set of ionic conductances. Many neurons have a more complicated mix of potassium channels. In addition to returning the membrane potential to its resting levels after an action potential, these ionic channels have such diverse roles as setting the resting membrane potential, controlling action potential discharge patterns, and terminating burst activities. In general, all neuronal potassium currents are inhibitory,[96] counteracting the inward sodium and calcium currents that depolarize the cells. Therefore, these channels play an important role in regulating neuronal excitability.[96] Potassium channels also have been found in nonexcitable cells.

Three main classes of potassium channels have been described and reviewed.[96] The *voltage-dependent potassium channels* include the delayed rectifier channel (K_V) modeled by Hodgkin and Huxley and described above, as well as *fast transient channels* (K_A), which open on depolarization and have fast inactivation, in contrast to the K_V channels, which do not inactivate. These voltage-dependent channels help define the latency to first spike, regulate neuronal firing rate, and help in action potential repolarization. The second class of potassium channels is *inward rectifiers* (K_I), also known as anomalous rectifiers. These channels increase potassium entry when the membrane is hyperpolarized and provide the resting potassium conductance in some cells. Additionally, they dominate the outward current at potentials close to rest, help regulate firing frequency, and are involved with cardiac pacemaker activity. The third class of potassium channels is *calcium-activated potassium channels* (K_{Ca}), which help to regulate calcium entry, action potential repolarization, firing rate, spike frequency adaptation, and burst termination. Additionally, *neurotransmitter and second-messenger-regulated potassium channels* have been described, which can effect increases in neurotransmitter release and presynaptic facilitation. These include the "M" channels which are turned off by acetylcholine acting on muscarinic receptors as well as several peptide neurotransmitters. "S" channels are outward rectifiers with a weak voltage dependence. They are closed by serotonin action on the cell and are present in aplysia sensory neurons.

By their activation and blockade by neurotransmitters, potassium channels participate in both inhibitory and excitatory slow transmission at the level of the postsynaptic cell body. Likewise, localization of presynaptic potassium channels allows the possibility of controlling transmitter release by modulating terminal excitability (and thereby Ca^{2+} influx) via both intrinsic channel properties and extrinsic humoral influence. *ATP-sensitive potassium channels* (KATP), which are blocked by ATP, are present in cardiac cells as well as pancreatic B cells.[96]

The main function of these diverse potassium channels in the mammalian CNS may be viewed as controlling neuronal excitability and determining discharge patterns of neurons. Potassium channels therefore play key roles in neuronal function, making them plausible molecular sites of anesthetic action. A further summation of K^+ channels and various isoforms in these families has been presented by Halliwell.[97]

8.4.3 Anesthetic Modification of Other Neuronal Potassium Channels

Aside from the delayed rectifier potassium channel, the interactions of anesthetics with several other types of potassium channels have been examined (summarized in Table 8.3), mostly in nonneuronal preparations. In examining the interactions with the other voltage-dependent potassium channel, the fast transient current (K_A), it was found that several intravenous agents decreased K_A in canine myocardia, partly antagonizing the effect of a decreased I_{Ca} caused by the anesthetics.[98] In *Drosophila Shaker* mutants, K_A was found to play a significant role in isoflurane anesthesia of these flies.[99] In this latter study, the behavioral insensitivity (i.e., movement) of the flies to isoflurane was correlated with the degree to which the potassium conductance through various K_A mutants was impaired, suggesting that K_A channels play a role in volatile anesthetic action.

As was found with sodium channels, the anesthetic sensitivity of potassium channels appears to vary among tissues and cell types. For example, inward rectifiers (K_I) of various muscle preparations were not greatly affected by either intravenous agents or sevoflurane.[95,98] In rat neurons, on the other hand, ethanol (although not strictly an anesthetic) did cause significant changes in channel conductance and the voltage dependence of activation, whereas in a granulocyte cell line model system clinical levels of barbiturates significantly inhibited K_I in a voltage-dependent manner[100,101] (Table 8.3).

K_{Ca} channels are activated by internal calcium and are widely distributed in secretory and excitable cells.[96] Again, most studies with these channels have been with nonneuronal cells.[102-104] As summarized in Table 8.3, it was generally found that isoflurane and halothane significantly decreased the open state probability of these channels as well as channel open time and the frequency of channel opening, but they did not affect channel amplitude. Although these effects were significant at clinical concentrations, one group concluded that these potassium channels did not mediate anesthetic-induced vasodilation.[102]

Anesthetic effects on the neurotransmitter and *second-messenger-modulated K channels* (K(S)) are also described in Table 8.3. K_S channels have been reported in aplysia neurons, and it has been found that halothane and isoflurane increase the open probability of these channels.[105]

ATP-modulated potassium channels have been implicated in providing myocardial protection during anesthesia[106-109] (see Ref. 106 for a review). It has generally been found that these channels are sensitive to barbiturates as well as halothane and isoflurane.[106,107,109] On the other hand, in canine tracheal smooth muscle halothane had little effect on relaxation of smooth muscle.[108]

Another potassium channel, $K_V2.1$, is a rat brain potassium channel that mediates a slowly inactivating current on membrane depolarization.[110] Kulkarni and co-workers examined the inhibitory effects of ketamine and halothane on this channel using site-directed mutagenesis. They found that, although both anesthetics inhibited potassium currents by reducing peak current amplitude through these cells, their mechanisms of action were different. Halothane peak current reduction was voltage dependent whereas ketamine action was not. Additionally, halothane accelerated the time constant of current inactivation but ketamine had only a minimal effect on this parameter. Deletions of the C terminus of the channel resulted in decreased sensitivity of the channel to modification by both anesthetics, indicating that this part of the channel influences the actions of both anesthetics.

8.4.4 Anesthetic Modification of Cellular Potassium Currents

As can be seen from Table 8.3, many laboratories have proposed or examined potassium channels as molecular targets of anesthetics. Complicating an understanding of the role of these channels in anesthesia, however, several channel types are present in most cells and their individual contributions to total ionic current are often difficult to distinguish.[1] Additionally, not all potassium channel subtypes have yet been described.[96] Thus, for most cells it can be known only that a current is carried by potassium ions, but not what types of channels are present. Therefore, several laboratories have looked at the effects of anesthetics on general or nonspecific potassium currents in several cell types, especially central neurons. Nicoll and Madison,[111] using intracellular recordings, found that volatile anesthetics hyperpolarized rat hippocampal neurons, probably due to an increase in potassium conductance. Berg-Johnsen and Langmoen[66,112] found similar effects in rat hippocampus and cerebral cortex as well as human neocortex. These results suggested that the anesthetic-induced membrane hyperpolarization in these neurons resulted at least partly from an increase in potassium conductance. In aplysia, neurons have baseline inactivating potassium channels as well as outwardly rectifying baseline channels with a weak voltage-dependent potassium current.[105] The potassium current in these cells also was found to be turned on or enhanced by volatile anesthetics.[113] This enhanced current occurred at clinical anesthetic concentrations and may be mediated by K_S channels. Thus, although the potassium channels involved in these anesthetic interactions are not specifically known, they do point to an important role for potassium channels in anesthesia.

8.5 SUMMARY

Voltage-gated sodium and potassium channels generally have been regarded as being relatively insensitive to anesthetic modification due primarily to initial studies on classical electrophysiologic models. Recent experimental evidence from the fields of molecular biology and electrophysiology, however, has indicated the existence of a previously unsuspected diversity of channel isoforms having distinct pharmacologic and electrophysiologic properties. Many cells express several of these distinct channel isoforms, and the cellular functions and pharmacology of these diverse channels are still poorly understood. More recent molecular studies with sodium and potassium channels from mammalian neuronal preparations have indicated that the voltage-gated channels in these cells are more sensitive to volatile and other anesthetics than was previously indicated by earlier studies. Results obtained at the cellular level have given mixed support for these channels, as has also been found with other ionic channels that have been found sensitive at the molecular level, such as the $GABA_A$ chloride channel.[3] This recent experimental evidence suggests that the role sodium and potassium channels play in general anesthesia should be reexamined.

REFERENCES

1. Hille, B. *Ionic Channels of Excitable Membranes*, Sinauer Assoc., Sunderland, MA, 1992.
2. Krnjevic, K. Cellular mechanisms of anesthesia. *Ann. NY Acad. Sci.* 625:1, 1991.
3. Franks, N. P. and Lieb, W. R. Molecular and cellular mechanisms of general anaesthesia. *Nature* 367:607, 1994.
4. Richards C. D. The synaptic basis of general anaesthesia. *Eur. J. Anaesth.* 12:5, 1995.
5. Hodgkin, A. L., Huxley, A. F. and Katz, B. Measurements of current-voltage relations in the membrane of the giant axon of *Loligo*. *J. Physiol. (Lond.)* 116:424, 1952.
6. Hodgkin, A. L. and Huxley, A. F. Currents carried by sodium and potassium ions through the membrane of the giant axon of *Loligo*. *J. Physiol. (Lond.)* 116:449, 1952.
7. Hodgkin, A. L. and Huxley, A. F. The components of membrane conductance in the giant axon of *Loligo*. *J. Physiol. (Lond.)* 116:473, 1952.
8. Hodgkin, A. L. and Huxley, A. F. The dual effect of membrane potential on sodium conductance in the giant axon of *Loligo*. *J. Physiol. (Lond.)* 116:497, 1952.
9. Levinson, S. R., Duch, D. S., Urban, B. W. and Recio-Pinto, E. The sodium channel from *Electrophorus electricus*. *Ann. NY Acad. Sci.* 479:162, 1986.
10. Catterall, W. A. Structure and function of voltage-gated ion channels. *Annu. Rev. Biochem.* 64:493, 1995.
11. Aldrich, R. W., Corey, D. P. and Stevens, C. F. A reinterpretation of mammalian sodium channel gating based on single channel recording. *Nature* 306:436, 1983.
12. Barres, B. A., Chun, L. L. Y. and Corey, D. P. Glial and neuronal forms of the voltage-dependent sodium channel: Characteristics and cell-type distribution. *Neuron* 2:1375, 1989.
13. Kirsch, G. E. and Brown, A. M. Kinetic properties of sodium channels in rat heart and rat brain. *J. Gen. Physiol.* 93:85, 1989.
14. Horn, R. and Vandenberg, C. A. Statistical properties of single sodium channels. *J. Gen. Physiol.* 84:505, 1984.
15. Vandenberg, C. A. and Horn, R. Inactivation viewed through single sodium channels. *J. Gen. Physiol.* 84:535, 1984.
16. Berman, M. F., Camardo, R. B., Robinson, R. B. and Siegelbaum, S. A. Single sodium channels from canine ventricular myocytes: Voltage dependence and relative rates of activation and inactivation. *J. Physiol. (Lond.)* 415:505, 1989.
17. Catterall W. A. Cellular and molecular biology of voltage-gated sodium channels. *Phys. Rev.* 2:S15, 1992.

18. Schaller, K. L., Krzemien, D. M., Yarowsky, P. J., Krueger, B. K. and Caldwell, J. H. A novel, abundant sodium channel expressed in neurons and glia. *J. Neurosci.* 15:3231, 1995.
19. Isom, L. L., DeJongh, K. S. and Catterall, W. A. Auxiliary subunits of voltage-gated ion channels. *Neuron* 12:1183, 1994.
20. Isom, L. I., Scheuer, T., Brownstein, A. B., Ragsdale, D. A. and Catterall, W. A. Functional co-expression of the B1 and type IIA alpha subunits of sodium channels in a mammalian cell line. *J. Biol. Chem.* 270:3306, 1995.
21. Ragsdale, D. S., Scheuer, T. and Catterall, W. A. Frequency and voltage-dependent inhibition of type IIA Na+ channels, expressed in a mammalian cell line, by local anesthetic, antiarrhythmic, and anticonvulsant drugs. *Mol. Pharm.* 40:756, 1991.
22. Bonhaus, D. W., Herman, R. C., Brown, C. M., Cao, Z., Chang, L.-F., Loury, D. N., Sze, P., Zhang, L. and Hunter, J. C. The β1 sodium channel subunit modifies the interactions of neurotoxins and local anesthetics with rat brain IIA α sodium channel in isolated membranes but not in intact cells. *Neuropharmacology* 35:605, 1996.
23. Lotan, I., Dascal, N., Naor, Z. and Boton, R. Modulation of vertebrate brain Na+ and K+ channels by subtypes of protein kinase C. *FEBS Lett.* 267:25, 1990.
24. Dascal, N. and Lotan, I. Activation of protein kinase C alters voltage dependence of Na+ channel. *Neuron* 6:165, 1991.
25. Schreibmayer, W., Dascal, N., Lotan, I., Wallner, M. and Weigl, L., Molecular mechanism of protein kinase C modulation of sodium channel alpha-subunits expressed in *Xenopus* oocytes. *FEBS Lett.* 291:341, 1991.
26. Li, M., West, J. W., Lai, Y., Scheuer, T. and Catterall, W. A., Functional modulation of brain sodium channels by cAMP-dependent phosphorylation. *Neuron* 8:1151, 1992.
27. Recio-Pinto, E., Thornhill, W. B., Duch, D. S., Levinson, S. R. and Urban, B. W. Neuraminidase treatment modifies the function of electroplax sodium channels in planar lipid bilayers. *Neuron* 5:675, 1990.
28. Bennett, E., Urcan, M. S., Tinkle, S. S., Koszowski, A. G. and Levinson, S. R. Contribution of sialic acid to the voltage-dependence of sodium channel gating: A possible electrostatic mechanism. *J. Gen. Physiol.* 109:327, 1997.
29. Campbell, D. T. Large and small vertebrate sensory neurons express different Na and K channel subtypes. *PNAS* 89:9569, 1992.
30. Campbell, D. T. Single-channel current/voltage relationships of two kinds of Na+ channel in vertebrate sensory neurons. *Pflug. Arch.* 423:492, 1993.
31. Mandel, G. Tissue specific expression of the voltage-sensitive sodium channel. *J. Membrane Biol.* 125:193, 1992.
32. Elliott, J. R. and Haydon, D. A. The actions of neutral anaesthetics on ion conductances of nerve membranes. *Biochim. Biophys. Acta.* 988:257, 1989.
33. Haydon, D. A. and Urban, B. W. The action of hydrocarbons and carbon tetrachloride on the sodium current of the squid giant axon. *J. Physiol. (Lond.)* 338:435, 1983.
34. Haydon, D. A. and Urban, B. W. The action of alcohols and other non-ionic surface active substances on the sodium current of the squid giant axon. *J. Physiol. (Lond.)* 341:411, 1983.
35. Haydon, D. A. and Urban, B. W. The effects of some inhalation anaesthetics on the sodium current of the squid giant axon. *J. Physiol. (Lond.)* 341:429, 1983.
36. Elliott, J. R., Haydon, D. A. and Hendry, B. M. Anaesthetic action of esters and ketones: Evidence for an interaction with the sodium channel protein in squid axons. *J. Physiol. (Lond.)* 354:407, 1984.

37. Haydon, D. A. and Urban, B. W. The action of hydrophobic, polar, and some inhalation anaesthetic substances on the potassium current of the squid giant axon. *J. Physiol. (Lond.)* 373:311, 1986.
38. Urban, B. W. and Haydon, D. A. The action of halogenated ethers on the ionic currents of the squid giant axon. *Proc. R. Soc. (Lond.)* 231:13, 1987.
39. Kimura, J. E. and Meves, H. The effect of temperature on the asymmetrical charge movement in squid giant axons. *J. Physiol. (Lond.)* 289:479, 1979.
40. Urban, B. W. Differential effects of gaseous and volatile anaesthetics on sodium and potassium channels. *Br. J. Anaesth.* 71:25, 1993.
41. Kendig, J. J., Courtney, K. R. and Cohen, E. S. Anesthetics: Molecular correlates of voltage- and frequency-dependent sodium channel block in nerve. *J. Pharm. Exp. Ther.* 210:446, 1979.
42. Bean, B. B., Shrager, P. and Goldstein, D. A. Modification of sodium and potassium channel gating kinetics by ether and halothane. *J. Gen. Physiol.* 77:233, 1981.
43. Herrington, J., Stern, R. C., Evers, A. S. and Lingle, C. J. Halothane inhibits two components of calcium current in clonal (GH_3) pituitary cells. *J. Neurosci.* 11:2226, 1991.
44. Pancrazio, J. J., Park, W. K. and Lynch, C., III, Effects of enflurane on the voltage-gated membrane currents of bovine adrenal chromaffin cells. *Neurosci. Lett.* 146:147, 1992.
45. Frenkel, C., Duch, D. S., Recio-Pinto, E. and Urban, B. W. Pentobarbital suppresses human brain sodium channels. *Mol. Brain Res.* 6:211, 1989.
46. Frenkel, C., Duch, D. S. and Urban, B. W. Molecular actions of pentobarbital isomers on sodium channels from human brain cortex. *Anesthesiology* 72:640, 1990.
47. Frenkel, C. and Urban, B. W. A molecular target site for propofol: Voltage-clamp studies on human CNS sodium channels in bilayers. *Eur. J. Pharmacol.* 208: 75, 1991.
48. Frenkel, C. and Urban, B. W. Molecular actions of racemic ketamine on single sodium channels from human brain cortex. *Br. J. Anaesth.* 69:292, 1992.
49. Frenkel, C., Duch, D. S. and Urban B. W. Effects of i.v. anaesthetics on human brain sodium channels. *Br. J. Anaesth.* 71:15, 1993.
50. Rehberg, B., Bennett, E., Xiao, Y.-H., Levinson, S. R. and Duch, D. S. Voltage- and frequency-dependent pentobarbital suppression of rat brain and muscle sodium channels expressed in a mammalian cell line. *Mol. Pharmacol.* 48:89, 1995.
51. Rehberg, B., Xiao, Y.-H. and Duch, D. S. Central nervous system sodium channels are significantly suppressed at clinical concentrations of volatile anesthetics. *Anesthesiology* 84:1223, 1996.
52. Vysotskaya, T. N. and Duch, D. S. Sodium channels in peripheral neurons are blocked by halothane and pentobarbital. *Biophys. J.* 72:A362, 1997.
53. Vysotskaya, T. N. and Duch, D. S. Isoflurane and halothane block sodium channels in DRG neurons. *Soc. Neurosci. Abstr.* 23:1472, 1997.
54. Cannon, S. C., Brown, R. H. and Corey, D. P. Theoretical reconstruction of myotonia and paralysis caused by incomplete inactivation of sodium channels. *Biophys. J.* 65:270, 1993.
55. Narahashi, T. Nerve membrane Na+ channels as targets of insecticides. *Trends in Pharmacol. Sci.* 13:236, 1992.
56. Weiss, J. N., Venkatesh, N. and Lamp, S. T. ATP-sensitive K+ channels and cellular K+ loss in hypoxic and ischaemic mammalian ventricle. *J. Physiol.* 447:649, 1992.
57. Mandel, G. Tissue specific expression of the voltage-sensitive sodium channel. *J. Membr. Biol.* 125:193, 1992.

58. Butterworth J. F., Raymond S. A. and Roscoe R. F. Effects of halothane and enflurane on firing threshold of frog myelinated axons. *J. Physiol.* 411:493, 1989.
59. Raymond, S. A. Subblocking concentrations of local anesthetics: Effects on impulse generation and conduction in single myelinated sciatic nerve axons in frog. *Anesth. Analg.* 75:906, 1992.
60. Collins, J. G., Kendig, J. J. and Mason, P. Anesthetic actions within the spinal cord: Contributions to the state of general anesthesia. *TINS* 18:549, 1995.
61. Study, R. E. Highlights: Halothane blocks synaptic excitation of inhibitory interneurons. *Anesthesiology* 85:29a, 1996.
62. Perouansky, M., Kirson, E. D. and Yaari, Y. Halothane blocks synaptic excitation of inhibitory interneurons. *Anesthesiology* 85:1431, 1996.
63. Franks, N. P. and Lieb, W. R. Molecular and cellular mechanisms of general anaesthesia. *Nature* 367:607, 1994.
64. Richards, C. D. and White, A. E. The actions of volatile anaesthetics on synaptic transmission in the dentate gyrus. *J. Physiol. (Lond.)* 252:241, 1975.
65. MacIver, M. B. and Roth, S. H. Inhalation anesthetics exhibit pathway specific and differential actions on hippocampal synaptic responses *in vitro*. *Br. J. Anaesth.* 60:680, 1988.
66. Berg-Johnson, J. and Langmoen, I. A. Mechanisms concerned in the direct effect of isoflurane on rat hippocampal and human neocortical neurones. *Brain Res.* 507:28, 1990.
67. Araki, T. and Terzuolo, C. A. Membrane currents in spinal motoneurons associated with the action potential and synaptic activity. *J. Neurophysiol.* 25:772, 1962.
68. Barrett, J. N. and Crill, W. E. Voltage-clamp of motor neuron somata: Properties of the fast inward current. *J. Physiol. (Lond.)* 304:231, 1980.
69. Furshpan, E. J. and Furukawa, T. Intracellular and extracellular responses of the several regions of the Mauthner cell of goldfish. *J. Neurophysiol.* 25:732, 1962.
70. Smith, T. G., Jr. Sites of action potential generation in cultured vertebrate neurons. *Brain Res.* 288:381, 1983.
71. Stafstrom, C. E., Schwindt, P. C. and Crill, W. E. Negative slope conductance due to a persistent subthreshold sodium current in cat neocortical neurons in vitro. *Brain Res.* 236:221, 1982.
72. Stafstrom, C. E., Schwindt, P. C., Chubb, M. C. and Crill, W. E. Properties of persistent sodium conductance and calcium conductance of layer V neurons from cat sensorimotor cortex in vitro. *J. Physiol. (Lond.)* 53:163, 1985.
73. Abita, J.-P., Chicheportiche, R., Schweitz, H. and Lazdunski, M. Effects of neurotoxins (veratridine, sea anemone toxin, tetrodotoxin) on transmitter accumulation and release by nerve terminals in vitro. *Biochemistry* 16:1838, 1977.
74. Krueger, B. K. and Blaustein, M. P. Sodium channels in presynaptic nerve terminals *J. Gen. Physiol.* 76:287, 1980.
75. Tamkun, M. M. and Catterall, W. A. Ion flux studies of voltage-sensitive sodium channels in synaptic nerve-ending particles. *Mol. Pharmacol.* 19:78, 1981.
76. Regehr, W. G., Konnerth, A. and Armstrong, C. Sodium action potentials in the dendrites of cerebellar Purkinje cells. *PNAS* 89:5492, 1992.
77. Regehr, W. G., Kehoe, J., Ascher, P. and Armstrong, C. Synaptically triggered action potentials in dendrites. *Neuron* 11:145, 1993.
78. Huguenard, J. R., Hamill, O. P. and Prince, D. A. Sodium channels in dendrites of rat cortical pyramidal neurons. *PNAS* 86:2473, 1989.
79. Stuart, G. J. and Sakmann, B. Active propagation of somatic action potentials into neocortical pyramidal cell dendrites. *Nature* 367:69, 1994.

80. Fujiwara, N., Higashi, H., Nishi, S., Shimoji, K., Sugita, S. and Yoshimura, M. Changes in spontaneous firing patterns of rat hippocampal neurones induced by volatile anaesthetics. *J. Physiol.* 402:155, 1988.
81. Yoshimura, M., Higashi, H., Fujita, S. and Shimoji, K. Selective depression of hippocampalinhibitory postsynaptic potentials and spontaneous firing by volatile anesthetics. *Brain Res.* 340:363, 1985.
82. Schlame, M. and Hemmings, H. C. Inhibition by volatile anesthetics of endogenous glutamate release from synaptosomes by a presynaptic mechanism. *Anesthesiology* 82:1406, 1995.
83. Ratnakumari, L. and Hemmings, H. C. Inhibition by propofol of [^{3}H]-batrachotoxinin-A 20-α-benzoate binding to voltage-dependent sodium channels in rat cortical synaptosomes. *Br. J. Pharmacol.* 119:1498, 1996.
84. MacIver, M. B., Mikulec, B. A., Amagasu, S. M. and Monroe, F. A. Volatile anesthetics depress glutamate transmission via presynaptic actions. *Anesthesiology* 85:823, 1996.
85. Urban, B. W. Modifications of excitable membranes by volatile and gaseous anesthetics. In: *Effects of Anesthesia, Clinical Physiology Series.* Covino, B. G., Fozzard, H. A., Rehder, K. and Strichartz, G., (Eds.), Am. Physiol. Soc., Bethesda, MA, 1985, p. 13.
86. Berg-Johnsen, J. and Langmoen, I. A. The effect of isoflurane on unmyelinated and myelinated fibres in the rat brain. *Acta Physiol. Scand.* 127:87, 1986.
87. Raymond, S. A. A new hypothesis for the neural basis of anesthesia. *Soc. Neurosci. Abstr.* 11:1109, 1985.
88. Kendig, J. J. and Grossman, Y. Homogeneous and branching axons: Differing responses to anesthetics and pressure. In: *Molecular and Cellular Mechanisms of Anesthetics.* Roth, S. H. and Miller, K. W. (Eds.), Plenum Press, New York, 1986, pp. 333–353.
89. Haydon, D. A. and Kimura, J. E. Some effects of n-pentane on the sodium and potassium currents of the squid giant axon. *J. Physiol. (Lond.)* 312:57, 1981.
90. Haydon, D. A., Requena, J. and Urban, B. W. Some effects of aliphatic hydrocarbons on the electrical capacity and ionic currents of the squid giant axon membrane. *J. Physiol. (Lond.)* 309:229, 1980.
91. Benoit, E. Effects of intravenous anaesthetics on nerve axons. *Eur. J. Anaesthesiol.* 12:59, 1995.
92. Hendry, B. M. and Haydon, D. A. Effects of anesthetics and convulsants on the resting potassium conductance in squid nerve. *Ann. NY Acad. Sci.* 625:355, 1991.
93. Baum, V. C. Distinctive effects of three intravenous anesthetics on the inward rectifier (IK1) and the delayed rectifier (IK) potassium currents in myocardium: Implications for the mechanism of action. *Anesth. Analg.* 76:18, 1993.
94. Park, W. K., Pancrazio, J. J., Suh, C. K., and Lynch, C., III, Myocardial depressant effects of sevoflurane. Mechanical and electrophysical actions in vitro. *Anesthesiology* 84:1166, 1996.
95. Sakai, F., Hiraoka, M. and Amaha, K. Comparative action of propofol and thiopentone on membranes of isolated guinea pig ventricular myocytes. *Br. J. Anaesth.*, 77:508, 1996.
96. Rudy, B. Diversity and ubiquity of K channels. *Neuroscience* 25:729, 1988.
97. Halliwell, J. V. K+ channels in central nervous system. In: *Potassium Channels: Structure, Classification, Function and Therapeutic Potential.* Cook, N. S. (Ed.), Ellis Horwood Ltd., London, 1990, pp. 339.

98. Buljubasic, N., Marijic, J., Berczi, V., Supan, D. F., Kampine, J. P. and Bosnjak, Z. J. Differential effects of etomidate, propofol, and midazolam on calcium and potassium channel currents in canine myocardial cells. *Anesthesiology* 85:1092, 1996.
99. Tinkerberg, J. A., Segal, I. S., Guo, T. Z. and Maze, M. Analysis of anesthetic action on potassium channels of Shaker mutant of Drosophila. *Ann. NY Acad . Sci.* 625:532, 1991.
100. Osmanovic, S. S. and Shefner, S. A. Ethanol enhances inward rectification in rat locus ceruleus neurons by increasing the extracellular potassium concentration. *J. Pharmacol. Exp. Ther.* 271:334, 1994.
101. Gibbons, S. J., Nunez-Fernandez, R., Maze, G. and Harrisson, N. L. Inhibition of a fast rectifying potassium conductance by barbiturates. *Anesth. Analg.* 82:1242, 1996.
102. Eskinder, H., Gebremedhin, D., Lee, J. G., Rusch, N. J., Supan, F. D., Kampine, J. P. and Bosnjak, Z. J. Halothane and isoflurane decrease the open state probability of K+ channels in dog cerebral arterial muscle cells. *Anesthesiology* 82:479, 1995.
103. Beeler, T. and Gable, K. Effect of the general anesthetic halothane on the activity of the transverse tubule Ca(2+)-activated K+ channel. *FEBS Lett.* 331:207, 1993.
104. Simoneau, C., Thuringer, D., Cai, S., Garneau, L., Blaise, G. and Sauve, R. Effects of halothane and isoflurane on bradykinin-evoked Ca2+ influx in bovine aortic endothelial cells. *Anesthesiology* 85:366, 1996.
105. Winegar, B. D., Owen, D. F., Yost, C. S., Forsayeth, J. R. and Mayeri, E. Volatile general anaesthetics produce hyperpolarization of Aplysia neurons by activation of a discrete population of baseline potassium channels. *Anesthesiology* 85:889, 1996.
106. Cason, B. A., Gordon, H. J., Avery, E. G., IV and Hickey, R. F. The role of ATP sensitive in myocardial protection. *J. Card. Surg.* 10(4 Suppl):441, 1995.
107. Kersten, J. R., Schmeiling, T. J., Hettrick, D. A., Pagel, P. S., Gross, G. J. and Waritier, D. C. Mechanism of myocardial protection by isoflurane. Role of adenosine triphosphate-regulated potassium (KATP) channels. *Anesthesiology* 85:794, 1996.
108. Fukusima, T., Hirasaki, A., Jones, K. A. and Warner, D. O. Halothane and potassium channels in airway smooth muscle. *Br. J. Anaesth.* 76:847, 1996.
109. Han, J., Kim, E., Ho, W. K. and Earm, Y. E. Effects of volatile anesthetic isoflurane on ATP-sensitive K+ channels in rabbit ventricular myocytes. *Biochem. Biophys. Res. Commun.* 229:852, 1996.
110. Kulkarni, R. S., Zorn, L. J., Anantharam, V., Bayley, H. and Treistman, S. N. Inhibitory effects of ketamine and halothane on recombinant potassium channels from mammalian brain. *Anesthesiology* 84:900, 1996.
111. Nicoll, R. A. and Madison, D. V. General anesthetics hyperpolarize neurons in the vertebrate central nervous system. *Science* 217:1055, 1982.
112. Berg-Johnsen, J. and Langmoen, I. A. Mechanisms concerned in direct effect of isoflurane on rat hippocampal and human neocortical neurons. *Brain Res.* 507:28, 1990.
113. Franks, N. P. and Lieb, W. R. An anaesthetic-activated channel. *Alcohol. Suppl.* 1:197, 1991.
114. Hodgkin, A. L. The Croonian Lecture: Ionic movements and electrical activity in giant nerve fibres. *Proc. R. Soc. (Lond.)* 148:1, 1958.
115. Hille, B. Ionic channels in nerve membranes. *Prog. Biophys. Mol. Biol.* 21:3, 1970.

9 Volatile Anesthetic Effects at Excitatory Amino Acid Receptors

Jo Ellen Dildy-Mayfield and R. Adron Harris

CONTENTS

9.1 OVERVIEW OF EXCITATORY AMINO ACID (EAA) RECEPTORS

The EAA receptors may be divided into NMDA (*N*-methyl-D-aspartate) and non-NMDA receptors, and the latter may be further divided into AMPA (α-amino-3-hydroxy-5-methyl-4-isoxazole propionic acid) and kainate receptors. These are all ligand-gated cation channels, although the channel properties are distinct, as discussed below. There is also a group of metabotropic glutamate receptors that are coupled with guanosine triphosphate (GTP)-activated proteins, rather than directly to ion channels. Our review will focus mainly on the ligand-gated ion channels. Distinctions between these receptors were originally based on the agonists that activate the channels (e.g., NMDA vs. AMPA), but molecular cloning and sequencing have demonstrated that the protein sequence of these receptor subunits also corresponds to the pharmacologic divisions. The molecular biology of these receptors has been reviewed in detail[1-3] and will only be briefly summarized here.

The NMDA receptors are composed of two groups of subunits, NR1 and NR2. NR1 subunits are encoded in a single gene, but eight splice variants have been identified.[3] There are four NR2 subunits, each encoded in a separate gene. The

0-8493-8555-5/01/$0.00+$.50

stoichiometry of native receptors is not known but likely contains both NR1 and NR2 subunits. The NR1 subunit is widely distributed in brain; the NR2 subunits have more discrete neuroanatomical localization, leading to the idea that NMDA receptors in different brain regions have different subunit compositions.[2] NMDA receptors are unusual in that they require glycine as a coagonist for channel activation. In addition, they display a voltage-dependent block by Mg^{2+} and a large single-channel conductance with high calcium permeability. These unique properties may underlie the importance of NMDA receptors in neuronal plasticity and learning and memory.

Although much research attention has been given to the function of NMDA receptors, most fast excitatory neurotransmission is carried out by AMPA receptors. They are found at the majority of excitatory synapses and have a low Ca^{2+} permeability. Kainate receptors are also widespread in brain but in lower density than AMPA receptors. AMPA receptors are encoded in four genes, GluR1 through 4, and kainate receptors are encoded by three genes termed GluR5 through 7 as well as two genes for poorly understood kainate binding protein, KA1 and 2.[1,3] The subunit stoichiometry of these receptors is not defined. Although GluR1 and GluR3 form functional homomeric receptors, it is likely that native AMPA receptors also contain GluR2. For kainate receptors, homomeric GluR6 subunits are functional, but it is possible that native receptors contain mixtures of GluR5 through 7 as well as KA1 and 2. Phylogenetically, all of the EAA receptors appear to be related, and sequence homologies allow the subunits to be grouped into NMDA subunits (NR), AMPA (GluR1–4), kainate (GluR5–7), and kainate binding proteins (KA1–2).[1] As discussed in this review, these subunits can be expressed in mammalian cells or *Xenopus* oocytes, and these systems have been used to study effects of anesthetic agents.[4]

In addition to the well-established EAA receptors discussed above, there is some evidence for novel receptors that do not fit into these families. Yuzaki et al.[5] reported that aspartate activates a calcium conductance in cerebellar Purkinje cells in mutant mice lacking functional NMDA receptors. This action of aspartate was sensitive to both NMDA and non-NMDA receptor antagonists and appears to be distinct from all other EAA receptors. In addition, a group of glutamate binding proteins has been characterized biochemically and found to bind NMDA and other ligands known to act on the NMDA receptor complex (e.g., dizocilpine).[6] These proteins are distinct from the NR, KA, and GluR gene families and may prove to be novel NMDA receptors.[7] Thus far, there is little functional characterization of these proteins, although a recent report shows upregulation of these proteins in mouse brain by chronic alcohol treatment.[8]

9.2 ANESTHETIC–EAA INTERACTIONS: *IN VIVO* STUDIES

Tests of EAA receptor agonists and antagonists *in vivo* are critical to determining the significance of the findings from simpler preparations that are discussed below. The rationale for the *in vivo* studies is that if anesthetic action on an EAA receptor (e.g., inhibition of NMDA action) observed *in vitro* is important for anesthesia, then receptor agonists and antagonists should have predictable effects on anesthesia. For example, if inhibition of NMDA action causes anesthesia, then NMDA receptor antagonists should produce anesthesia and, at lower doses, potentiate actions of volatile anesthetics. Conversely, NMDA (or glycine site) agonists should reduce the action of volatile anesthetics. These behavioral studies are necessary but not sufficient to implicate a given receptor in anesthesia. They are not sufficient because the anesthetic and the receptor antagonist could be acting at different sites yet have a physiologic integration that results in potentiation of anesthetic action. Thus, the *in vivo* and *in vitro* approaches are complementary, and information from both are required to define sites of anesthetic action. A potentially powerful approach that would combine *in vivo* and *in vitro* approaches is production of mutant animals with changes in a specific receptor (e.g., NMDA) or receptor subunit that changes the anesthetic sensitivity of the receptor in vitro without changing the neurotransmitter action. These animals could then be tested for anesthetic sensitivity *in vivo*.

Because there is evidence that volatile anesthetics inhibit glutaminergic transmission (see below), a first question is whether behavioral effects of anesthetics can be mimicked by receptor antagonists of NMDA, AMPA, or kainate receptors. Indeed, ketamine, CGS19755, and phencyclidine, which block NMDA receptors, can be used as general anesthetic agents.[9,10] However, ketamine or PCP anesthesia is characterized as a "dissociative" state rather than complete loss of consciousness and is quite different from the anesthetic state produced by volatile anesthetics. Riluzole is a new inhibitor of glutaminergic neurotransmission that apparently has both pre- and postsynaptic actions that have yet to be precisely defined.[11] Riluzole produces loss of righting reflex in rats,[12] but it is not clear if this state most closely resembles that produced by ketamine or volatile anesthetics.

Another *in vivo* approach is to ask if glutamate receptor agonists or antagonists alter the potency of volatile anesthetics. A number of different NMDA antagonists (e.g., dizocilpine, CPP, CGS 19755) all potentiate (decreased minimum alveolar concentration, MAC) isoflurane anesthesia in rats.[13-16] In addition, an antagonist of the glycine recognition site on the NMDA receptor (ACEA-1021) also decreases MAC for halothane in rats.[17] Thus, these two studies show that inhibition of NMDA receptor function enhances the action of volatile anesthetics. Another study showed that the AMPA receptor antagonist, NBQX, decreases MAC for halothane in rats.[18] Furthermore, riluzole, an inhibitor of glutaminergic neurotransmission, albeit by an unknown mechanism, also potentiates the action of halothane.[12] Finally, McFarlane et al.[19] tested the combined actions of an AMPA antagonist (NBQX) and an NMDA antagonist (CGS 19755) on halothane MAC. Consistent with other studies from this group, each antagonist decreased MAC, and the combination was at least additive and perhaps synergistic.

An important point about these studies of anesthetic sensitivity is that different tests often are used to define anesthesia. Many studies simply use duration or production of loss of righting reflex (sleep time) as anesthesia, whereas others use a response to a noxious stimulus such as tail clamp[17-19] or paw pinch as the criterion. It is likely that these different behaviors rely on different neuronal pathways and may give conflicting information about the mechanism of anesthetic action.

Taken together, these *in vivo* studies provide support for the idea that glutaminergic neurotransmission at both NMDA and AMPA receptors is important for the actions of volatile anesthetics. This is based on the ability of receptor antagonists to decrease the MAC of volatile anesthetics, and, as noted above, this may reflect a physiologic interaction rather than a direct effect of anesthetics at these receptors. However, it is important to note that the receptor antagonists do not mimic the actions of volatile anesthetics. Thus, reduction of glutaminergic neurotransmission does not appear to be sufficient to account for the *in vivo* actions of volatile anesthetics but may contribute to these actions. There are few *in vivo* studies of interactions between anesthetics and drugs that affect metabotropic glutamate receptors (see below), and this area deserves additional attention.

9.3 ANESTHETIC–EAA INTERACTIONS: NEURONAL STUDIES

Most of these studies used either brain slices or the isolated spinal cord and recorded electrical activity after stimulation of glutaminergic pathways or application of glutamate receptor agonists. Because these are *in vitro* studies, it is important to consider what concentrations of anesthetic correspond to MAC *in vivo*. As noted by Franks and Lieb,[20] anesthetic concentrations that are four to ten times MAC are often used *in vitro* and produce effects on many proteins, including ion channels, but these actions are unlikely to have any relevance to anesthesia. For this discussion, we will use the values calculated by Franks and Lieb,[20] which give concentrations corresponding to 1 MAC of 0.25 m*M* for halothane and 0.3 m*M* for isoflurane.

There are few detailed studies of effects of volatile anesthetics on neuronal glutamate responses, and most of the work is rather recent. In their review of the literature, Pocock and Richards[21] noted that effects of volatile anesthetics on glutamate receptors remained to be defined. In fact, one of the early studies postulated actions on glutamate mechanisms that have proven accurate.[22] These workers suggested that volatile anesthetics depress excitatory neurotransmission and that this action of halothane is likely due to inhibition of glutamate release rather than an inhibition of glutamate receptor sensitivity, although other volatiles (ether, methoxyflurane) may affect the receptors. There is now ample evidence that volatile anesthetics inhibit excitatory neurotransmission and, as discussed below, at least part of this action is due to inhibition of neurotransmitter release.

The delineation of NMDA and non-NMDA (AMPA, kainate) ionotropic glutamate receptors has led to efforts to determine if one of these types of receptors is more sensitive than the other.[23] In spinal cord, a cyclobutane anesthetic (but not the nonanesthetic congener) inhibited NMDA and non-NMDA pathways.[24] In agreement with this result, a patch-clamp study of hippocampal CA1 neurons found that both NMDA and non-NMDA excitatory postsynaptic currents (EPSCs) were inhibited by halothane (about 1.5 to 10 MAC), and the inhibition was equal for both receptors.[25] However, halothane did not affect currents produced by both applications of glutaminergic agonists, in agreement with the earlier study of Richards and Smaje.[22] One study of the hippocampal CA1 region using population spike analysis found that NMDA-receptor-mediated responses were more sensitive to halothane than the non-NMDA responses, and another study of hippocampal CA1 activity suggests that enflurane inhibits non-NMDA but not NMDA responses.[26] Thus, all three possibilities (nonselectivity or NMDA or non-NMDA selectivity) are represented in studies of CA1 activity. In all of these studies, the NMDA and non-NMDA responses were distinguished by pharmacologic isolation (receptor antagonists). It is possible that some of the conflicting findings are due to a lack of selectivity or specificity of these antagonists. The receptor selectivity is an important issue because if the anesthetics act solely by inhibiting glutamate release, then there should be no receptor specificity.

In contrast to several studies discussed above, Carla and Moroni[27] found that volatile anesthetics inhibited the excitatory responses of cortical slices to applications of NMDA and AMPA. Chloroform, halothane, isoflurane, and ether inhibited the action of maximal concentrations of NMDA or AMPA, and the inhibition was generally greater for AMPA than for NMDA. For example, 1 m*M* halothane inhibited the AMPA response by 45% with no effect on the NMDA response, and 0.5 m*M* isoflurane inhibited the AMPA response by 35% and the NMDA response by 20%. This is generally consistent with a study of NMDA or quisqualate-stimulated increases in intracellular calcium in cultured hippocampal neurons.[28] These workers found that isoflurane inhibited the action of both glutamate agonists, but quisqualate was somewhat more sensitive than NMDA. In contrast, intracellular recording of NMDA and quisqualate responses in hippocampal slices found that isoflurane and halothane did not alter NMDA responses but increased glutamate and quisqualate responses.[29] Although it is difficult to reconcile these results, it should be noted that volatile anesthetics have been shown to enhance the function of recombinant kainate receptors (GluR6) even though they inhibit NMDA and AMPA receptor function (discussed below). In addition, activation of kainate receptors reduces glutamate release in hippocampus;[30] thus, enhancement of kainate receptor function by anesthetics could indirectly inhibit the function of NMDA and AMPA receptors by reducing glutamate release.

In summary, there is clear evidence that volatile anesthetics inhibit excitatory neurotransmission in many brain regions, including spinal cord, hippocampus, and cortex, and this can be detected at concentrations corresponding to 1 to 2 MAC. The mechanism of this inhibition is less clear. Direct application of glutamate agonists in the presence of 1 to 2 MAC concentrations of volatile anesthetics has given highly variable results. Most studies have found little or no effect on NMDA receptors and weak effects on non-NMDA receptors, but other studies have shown clear inhibitory (and facilitory) effects on these receptors. A problem of these studies is the mixed population of receptors (of unknown subunit composition) and the reliance on "specific" agonists or antagonists to pharmacologically dissect the different receptors. An alternative approach is the use of recombinant receptors, as discussed below.

9.4 ANESTHETIC–NMDA RECEPTOR INTERACTIONS: *IN VITRO* STUDIES

Biochemical approaches, including ion flux and receptor binding, have been used to study interactions between anesthetics and NMDA receptors. At or below 1 MAC, the volatile anesthetics enflurane and halothane inhibited NMDA-stimulated but not basal calcium uptake in rat brain microvesicles and reduced the agonist-induced desensitization of uptake observed when NMDA is preincubated.[31] The NMDA receptor coagonist, glycine, which is a positive modulator of NMDA receptor function, reduced inhibition of uptake in the presence of enflurane and halothane.

Anesthetic concentrations of halothane, enflurane, and diethylether also inhibited NMDA-mediated increases in intracellular calcium concentration in mouse hippocampal microsacs.[32] Enflurane did not alter resting intracellular calcium concentrations, and only large concentrations of halothane and diethylether produced increases. Thus, the ability of volatile anesthetics to inhibit NMDA responses does not appear to result indirectly from increased resting calcium concentrations.

A series of volatile anesthetics inhibited glutamate-stimulated MK-801 binding to the NMDA channel in a noncompetitive manner, and this inhibition was reversed by glycine.[33] The inhibition occurred within concentration ranges achieved during surgical anesthesia. The site of volatile anesthetic action is not likely within the channel because the inhibition was specific for agonist but not basal MK-801 binding and was reduced by glycine, which does not act within the cation channel. Binding of CGS 19755 to the glutamate site of the NMDA receptor was also inhibited by these volatile anesthetics, although in general the anesthetics were more potent inhibitors of glutamate-stimulated MK-801 binding. Anesthetic inhibitory potency did not correlate with their lipophilicity profiles. A second site that positively modulates NMDA receptor function is known as the polyamine site. Spermidine, a polyamine agonist, reversed the inhibition of glutamate-stimulated MK-801 binding produced by volatile anesthetics,[34] which was similar to the above report in which glycine reversed the inhibition of binding. These results provide evidence for anesthetic action at the NMDA receptor complex, which may encompass multiple sites of the receptor–channel activation mechanism. Whereas the agonist site or the channel pore were apparently not direct sites of volatile anesthetic action, the channel activation process including positive, allosteric sites may be involved. These biochemical studies support the NMDA receptor complex as a potential site of volatile anesthetic action.

Studies of NMDA receptors expressed from brain mRNAs or cloned receptor subunits also provide evidence for modulation of NMDA receptors by volatile anesthetics. Enflurane (1.8 m*M*) inhibited NMDA-stimulated currents in *Xenopus* oocytes expressing mouse brain mRNA by approximately 30%.[35] The percent inhibition was similar in the presence of maximal and minimal concentrations of NMDA or different concentrations of the NMDA coagonist, glycine, thus indicating that the agonist or coagonist binding sites on the NMDA complex are not sites of action for enflurane. *Xenopus* oocyte expression studies offer the advantages of studying direct effects of volatile anesthetics on receptor function and comparing different glutamate receptors in a common cellular environment. This study provides evidence that volatile anesthetics directly, although weakly, inhibit NMDA receptor function.

9.5 ANESTHETIC–AMPA–KAINATE RECEPTOR INTERACTIONS: *IN VITRO* STUDIES

Biochemical studies also indicate interactions of volatile anesthetics with AMPA and kainate receptors. For example, halothane and isoflurane at 1 MAC inhibited kainate-induced glutamate release from cerebellar granule cells, but halothane had no effect on glutamate reuptake.[36] Halothane inhibition was not dependent on kainate concentration, indicating that halothane does not compete with kainate at its binding site on the receptor. In contrast to halothane and isoflurane, 1 MAC enflurane had no effect on kainate-induced glutamate release, and none of the inhalation anesthetics tested significantly affected NMDA-induced glutamate release.[36] These results provide evidence for receptor specificity in the mechanisms of volatile anesthetics.

However, enflurane (1.8 m*M*) inhibited both AMPA- and kainate-stimulated currents in *Xenopus* oocytes expressing mouse brain mRNA by approximately 30%.[35] As noted above, this study reported a similar magnitude of inhibition for NMDA currents in these oocytes. The percent inhibition was not different in the presence of maximal and minimal concentrations of AMPA or kainate, suggesting that enflurane does not compete for these agonist binding sites. This site of action also was ruled out for enflurane at NMDA receptors. Of the glutamate agonists, only kainate produced measurable currents in oocytes expressing post-mortem human brain mRNA. Human kainate channels were inhibited by enflurane to a similar degree as mouse kainate responses. Thus, no species or glutamate agonist differences were observed with respect to enflurane action. However, the agonists AMPA and kainate can both stimulate AMPA subtypes of glutamate receptors. It is only this subtype that would be activated in oocytes expressing brain mRNA because the kainate subtype is a fast desensitizing receptor whose responses would not likely have been measured in the absence of a desensitization blocker.

A study of volatile anesthetics on defined receptor subunits expressed in *Xenopus* oocytes revealed opposite actions at AMPA versus kainate receptor subtypes.[37] Although kainate can activate AMPA receptors, the glutamate receptor subunits (GluR1–4) are termed AMPA receptors because AMPA acts as a high-affinity agonist at these receptors.[38] The volatile anesthetics enflurane, isoflurane, halothane, and 1-chloro-1,2,2-trifluorocyclobutane either had no effect or inhibited kainate-induced currents in oocytes expressing AMPA receptors (GluR1, GluR3, GluR2+3). The glutamate receptor subunits (GluR5–7) are kainate-selective receptors based on high-affinity kainate binding and insensitivity to AMPA. Surprisingly, all volatile anesthetics tested potentiated kainate-induced currents in ooyctes expressing the kainate-selective subunit GluR6.[37] Compared to the percent inhibition of AMPA receptors, the percent potentiation of kainate receptor function by volatile anesthetics was much greater. For example, enflurane, halothane, or 1-chloro-1,2,2-trifluorocyclobutane at 1 MAC only inhibited AMPA receptors by 15 to 20% yet potentiated kainate receptor function by 30 to 70%.[37] The percent inhibition of GluR3 (AMPA) and potentiation of GluR6 (kainate) receptor function by halothane approximately doubled when halothane alone was preexposed for several minutes.[37] This enhanced, time-dependent sensitivity may be explained by more indirect intracellular actions. For example, Dildy-Mayfield et al.[37] showed that halothane directly (i.e., in the absence of preexposure) affects GluR channel function, and the enhanced effects after minutes of halothane preexposure might be due to other indirect mechanisms. Two nonanesthetics, structurally similar to the anesthetic 1-chloro-1,2,2-trifluorocyclobutane but without *in vivo* anesthetic action, had no effect on AMPA or kainate subunits,[37] indicating that lipophilicity does not predict anesthetic potency at glutamate receptors.

In contrast to volatile anesthetics, anesthetic concentrations of barbiturates[37] and ethanol[39] inhibited GluR6 channels. Ethanol also inhibited the function of AMPA glutamate receptors,[39] whereas the barbiturates phenobarbital and pentobarbital produced only slight inhibition.[37] GluR6 is thus a unique glutamate receptor whose function is enhanced rather than inhibited by volatile anesthetics specifically.

Because the function of GluR6, an excitatory receptor, is enhanced by volatile anesthetics, this receptor might not appear to be a physiologic candidate in the mechanisms of anesthesia. However, increased stimulation of excitatory receptors could in turn produce increased activation of inhibitory receptors, which would then be consistent with anesthetic action. One might also question the relevance of homomeric GluR6 expression studies to GluRs expressed *in vivo*. However, there is evidence for the existence of pure kainate-type channels resembling the GluR6 subtype in hippocampal neurons.[40] The high-affinity kainate-selective receptor subunits (KA-1 and KA-2) do not form functional channels alone but do combine with GluR5 or GluR6 kainate receptors in functional heteromeric assemblies.[41,42] Halothane also potentiated GluR6+KA-2 channel responses expressed in *Xenopus* oocytes.[37] Given the *in vivo* evidence for homomeric GluR6-like channels or coassembly of GluR6 with either KA-1 or KA-2 receptors but not with AMPA receptors,[43] the potentiating effects of volatile anesthetics on homo- and heteromeric GluR6 assemblies expressed in oocytes may be physiologically relevant. Further support for this comes from a study showing that isoflurane and halothane enhance the glutamate (but not NMDA) response in hippocampal CA1 slices.[29] GluR6 mRNA and protein are expressed in the hippocampus, including the CA1 region,[44,45] which may explain the unexpected enhancement by volatile anesthetics.

9.6 ANESTHETICS AND GLUTAMATE RELEASE

In addition to postsynaptic sites of action at glutamate receptors, there is biochemical (as well as electrophysiologic — see above) evidence for presynaptic effects of volatile anesthetics on glutamate release. For example, at clinically relevant concentrations of 1 or 2 MAC, halothane, enflurane, and isoflurane each inhibited calcium-dependent glutamate release.[46] In contrast, anesthetic concentrations of pentobarbital failed to inhibit glutamate release. Both calcium-independent and -dependent glutamate release were inhibited by isoflurane.[47] A recent study by Miao et al.[48] of synaptosomes suggests that decreased K^+ depolarization–induced glutamate release by 1 to 2 MAC isoflurane, enflurane, or halothane may be due to depressed intracellular concentrations of calcium. This study proposed that volatile anesthetics do not affect intrasynaptosomal calcium-dependent exocytosis mechanisms, and that the decreased intracellular calcium concentration was sufficient to account for inhibition of glutamate release. Lower intracellular calcium levels could result from decreased calcium entry via voltage-dependent calcium channels. Although not reviewed here, there is some evidence for volatile anesthetic action on neuronal calcium channels, which could account for inhibition of neurotransmitter (glutamate) release.

Halothane and enflurane at 1 MAC also decreased hypoxia-evoked release of glutamate.[49] Isoflurane at 0.5 MAC reduced glutamate accumulation in a rat model of forebrain ischemia,[50] although an increase was observed in a rabbit model of global ischemia.[51] Despite evidence for reduced glutamate release, volatile anesthetics have failed to demonstrate protective effects in hypoxic models.

Finally, presynaptic inhibition of glutamate release by riluzole reduced halothane MAC in the rat by approximately 50%.[12] Riluzole does not bind to any known glutamate receptors but inhibits both the presynaptic release of glutamate and some postsynaptic effects.[11] Riluzole alone induced loss of righting reflex in rats.[12] Thus, this novel glutamatergic agent not only exerts anesthetic action but also enhances anesthesia induced by the volatile anesthetic halothane.

At clinically relevant concentrations, isoflurane and halothane failed to alter glutamate uptake in rat synaptosomes,[52] indicating that uptake processes are not responsible for the volatile anesthetic inhibition of glutamate release cited above. It appears likely that volatile anesthetics may modulate the release process itself. Inhibition of glutamate release, the major excitatory neurotransmitter in the mammalian central nervous system, is a mechanism consistent with anesthesia. In the absence of decreased postsynaptic glutamate sensitivity, decreased glutamate excitatory postsynaptic potentials by volatile anesthetics could be due to a presynaptic inhibition of calcium entry through voltage-gated calcium channels, which in turn decreases glutamate release. This mechanism may combine with postsynaptic effects on glutamate receptors in addition to actions at other receptors to induce anesthesia.

9.7 ANESTHETICS AND METABOTROPIC GLUTAMATE RECEPTORS

Metabotropic glutamate receptors are coupled to signal transduction pathways by activation of GTP binding proteins whereas ionotropic glutamate receptors gate ion channels directly. As for the ionotropic glutamate receptors discussed above, there are multiple subtypes of metabotropic glutamate receptors. To date, only one study has examined the role of metabotropic glutamate receptors in anesthesia. In this study, agonists for selective metabotropic glutamate receptors delayed recovery from halothane anesthesia in rats.[53] One agonist, DCG-IV, which selectively activates mGluR2 and mGluR3 subtypes, was the most potent of the agonists tested. These receptors are negatively coupled to adenylyl cyclase, suggesting that adenylyl cyclase activity may be involved in recovery from halothane anesthesia. DCG-IV also was reported previously to be the most potent agonist in reducing monosynaptic excitation in isolated rat spinal cord.[54] Thus, some types of metabotropic glutamate receptors may have depressant actions in the rat.

9.8 CONCLUSIONS AND FUTURE DIRECTIONS

The findings reviewed above raise several questions, including, Which of the EAA receptors is important for anesthesia? as well as the more general question, How can we determine if an ion channel or other protein is the site of action of volatile anesthetics? To address the latter question, several necessary criteria include:

1. Channel function must be affected at 1 MAC of several anesthetics.
2. The structurally related nonanesthetic, transitional, and anesthetic compounds should have effects on the function consistent with their *in vivo* pharmacology.
3. Other activators/inhibitors of the channel should either produce anesthesia or change the MAC of known anesthetics *in vivo*.
4. Transgenic animals with a mutated channel that is resistant to anesthetics *in vitro* should show an increased MAC *in vivo*.

These four criteria have not been fulfilled for any of proposed sites of anesthetic action. However, data are not sufficient to reject NMDA, AMPA, or kainate receptors as potential sites of volatile anesthetics, although the glutamate receptors can be discarded as sites of action for long-chain alcohols.[37] One problem for the EAA receptors is Criterion 1 above, because, as discussed above, the volatile anesthetics have very small effects on channel function at concentrations corresponding to 1 MAC. However, Criterion 2 is fulfilled by NMDA, AMPA, and kainate receptors, and Criterion 3 also provides good pharmacologic support for the importance of these receptors in anesthesia. It is important to remember that although the study of these receptors in isolated expression systems may be used to address Criterion 1, it may not adequately address the sensitivity of excitatory synaptic transmission to volatile anesthetics. In fact, this appears to be the case as electrophysiologic studies of brain and spinal cord suggest a sensitivity of excitatory neurotransmission far greater than would be expected from studies of EAA receptors in isolated cells. This may reflect actions of anesthetics on presynaptic processes, particularly glutamate release. Pursuit of Criterion 4, expression of anesthetic-resistant receptors in mice, should provide an opportunity to study the role of specific postsynaptic receptors as well as presynaptic processes in electrophysiologic actions of anesthetics.

Still to be addressed are the questions of whether all volatile anesthetics act by a single, identical mechanism (unitary theory) and whether a specific "receptor" can be defined for these agents. Answers to both questions require more molecular information about the sites of action of volatile anesthetics on ligand-gated ion channels and more physiologic information about the relevant targets *in vivo*. Although the molecular and physiologic represent different poles of reductionism, the approaches may be interdependent, because mutagenesis of specific sites on ligand-gated ion channels *in vitro* and *in vivo* is the strategy most likely to answer both of these questions.

ACKNOWLEDGMENTS

Supported by funds from the Department of Veterans Affairs, NIH grants GM47818 and AA06399, and the UCSF Anesthesia Research Foundation. We thank Dr. Ted Eger for support and advice.

REFERENCES

1. Seeburg, P. H., The TiNS/TiPS lecture: The molecular biology of mammalian glutamate receptor channels, *Trends Neurosci.*, 16, 359, 1993.
2. Schoepfer, R., Monyer, H., Sommer, B., Wisden, W., Sprengel, R., Kuner, T., Lomeli, H., Herb, A., Kohler, M., Burnashev, N., Gunther, W., Ruppersberg, P., and Seeburg, P., Molecular biology of glutamate receptors, *Prog. Neurobiol.*, 42, 353, 1994.
3. Hollmann, M. and Heinemann, S., Cloned glutamate receptors, *Annu. Rev. Neurosci.*, 17, 31, 1994.
4. Harris, R. A., Mihic, S. J., Dildy-Mayfield, J. E., and Machu, T. K., Actions of anesthetics on ligand-gated ion channels: Role of receptor subunit composition, *FASEB J.*, 9, 1454, 1995.
5. Yuzaki, M., Forrest, D., Curran, T., and Connor, J. A., Selective activation of calcium permeability by aspartate in Purkinje cells, *Science*, 273, 1112, 1996.
6. Kumar, K. N., Babcock, K. K., Johnson, P. S., Chen, X., Eggeman, K. T., and Michaelis, E. K., Purification and pharmacological and immunochemical characterization of synaptic membrane proteins with ligand-binding properties of N-methyl-D-aspartate receptors, *J. Biol. Chem.*, 269, 27384, 1994.
7. Kumar, K. N., Babcock, K. K., Johnson, P. S., Chen, X., Ahmad, M., and Michaelis, E. K., Cloning of the cDNA for a brain glycine-, glutamate- and thienylcyclohexylpiperidine-binding protein, *Biochem. Biophys. Res. Commun.*, 216, 390, 1995.
8. Hoffman, P. L., Bhave, S. V., Kumar, K. N., Iorio, K. R., Snell, L. D., Tabakoff, B., and Michaelis, E. K., The 71 kDa glutamate-binding protein is increased in cerebellar granule cells after chronic ethanol treatment, *Mol. Brain Res.*, 39, 167, 1996.
9. France, C. P., Winger, G. D., and Woods, J. H., Analgesic, anesthetic, and respiratory effects of the competitive N-methyl-D-aspartate (NMDA) antagonist CGS 19755 in rhesus monkeys, *Brain Res.*, 526, 355, 1990.
10. Irifune, M., Shimizu, T., Nomoto, M., and Fukuda, T., Ketamine-induced anesthesia involves the N-methyl-D-aspartate receptor-channel complex in mice, *Brain Res.*, 596, 1, 1992.
11. Dobono, M.-W., Le Guern, J., Canton, T., Doble, A., and Pradier, L., Inhibition by riluzole of electrophysiological responses mediated by rat kainate and NMDA receptors expressed in Xenopus oocytes, *Eur. J. Pharmacol.*, 235, 283, 1993.
12. Mantz, J., Cheramy, A., Thierry, A.-M., Glowinski, J., and Desmonts, J.-M., Anesthetic properties of riluzole (54274 RP), a new inhibitor of glutamate transmission, *Anesthesiology*, 76, 844, 1992.
13. Kuroda, Y., Strebel, S., Rafferty, C., and Bullock, R., Neuroprotective doses of N-methyl-D- aspartate receptor antagonists profoundly reduce the minimum alveolar anesthetic concentration for isoflurane in rats, *Anesth. Analg.*, 77, 795, 1993.
14. Daniell, L. C., The noncompetitive N-methyl-D-aspartate antagonists, MK-801, phencyclidine and ketamine, increase the potency of general anesthetics, *Pharmacol. Biochem. Behav.*, 36, 111, 1990.
15. Daniell, L. C., Effect of CGS 19755, a competitive N-methyl-D-aspartate antagonist, on general anesthetic potency, *Pharmacol. Biochem. Behav.*, 40, 767, 1991.
16. Scheller, M., Zornow, M., Fleischer, J., Shearman, G., and Greber, T., The noncompetitive N-methyl-D-aspartate receptor antagonist, MK-801 profoundly reduces volatile anesthetic requirements in rabbits, *Neuropharmacology*, 28, 677, 1989.
17. McFarlane, C., Warner, D. S., Nader, A., and Dexter, F., Glycine receptor antagonism: Effects of ACEA-1021 on the minimum alveolar concentration for halothane in the rat, *Anesthesiology*, 82, 963, 1995.

18. McFarlane, C., Warner, D. S., Todd, M. M., and Nordholm, L., AMPA receptor competitive antagonism reduces halothane MAC in rats, *Anesthesiology*, 77, 1165, 1992.
19. McFarlane, C., Warner, D. S., and Dexter, F., Interactions between NMDA and AMPA glutamate receptor antagonists during halothane anesthesia in the rat, *Neuropharmacology*, 34, 659, 1995.
20. Franks, N. P. and Lieb, W. R., Molecular and cellular mechanisms of general anaesthesia, *Nature*, 367, 607, 1994.
21. Pocock, G. and Richards, C. D., Cellular mechanisms in general anaesthesia, *Br. J. Anaesth.*, 66, 116, 1991.
22. Richards, C. D. and Smaje, J. C., Anaesthetics depress the sensitivity of cortical neurones to L-glutamate, *Br. J. Pharmacol.*, 58, 347, 1976.
23. Collins, J. G., Kendig, J. J., and Maso, P., Anesthetic actions within the spinal cord: Contributions to the state of general anesthesia, *Trends Neurosci.*, 18, 549, 1995.
24. Kendig, J. J., Kodde, A., Gibbs, L. M., Ionescu, P., and Eger, E. I., II, Correlates of anesthetic properties in isolated spinal cord: Cyclobutanes, *Eur. J. Pharmacol.*, 264, 427, 1994.
25. Perouansky, M., Baranov, D., Salman, M., and Yaari, Y., Effects of halothane on glutamate receptor-mediated excitatory postsynaptic currents, *Anesthesiology*, 83, 109, 1995.
26. MacIver, M. B. and Kendig, J. J., Enflurane-induced burst discharge of hippocampal CA1 neurones is blocked by the NMDA receptor antagonist APV, *Br. J. Anaesth.*, 63, 296, 1989.
27. Carla, V. and Moroni, F., General anaesthetics inhibit the responses induced by glutamate receptor agonists in the mouse cortex, *Neurosci. Lett.*, 146, 21, 1992.
28. Puil, E., El-Beheiry, H., and Baimbridge, K. G., Anesthetic effects on glutamate-stimulated increase in intraneuronal calcium, *J. Pharm. Exp. Ther.*, 255, 955, 1990.
29. Berg-Johnsen, J. and Langmoen, I. A., The effect of isoflurane on excitatory synaptic transmission in the rat hippocampus, *Acta. Anaesthesiol. Scand.*, 36, 350, 1992.
30. Chittajallu, R., Vignes, M., Dev, K. K., Barnes, J. M., Collingridge, G. L., and Henley, J. M., Regulation of glutamate release by presynaptic kainate receptors in the hippocampus, *Nature*, 379, 78, 1996.
31. Aronstam, R. S., Martin, D. C., and Dennison, R. L., Volatile anesthetics inhibit NMDA-stimulated 45Ca uptake by rat brain microvesicles, *Neurochem. Res.*, 19, 1515, 1994.
32. Daniell, L. C., Effect of volatile general anesthetics and n-alcohols on glutamate-stimulated increases in calcium ion flux in hippocampal membrane vesicles, *Pharmacology*, 50, 154, 1995.
33. Martin, D. C., Plagenhoef, M., Abraham, J., Dennison, R. L., and Aronstam, R. S., Volatile anesthetics and glutamate activation of N-Methyl-D-Aspartate receptors, *Biochem. Pharmacol.*, 49, 809, 1995.
34. Martin, D. C. and Aronstam, R. S., Spermidine attenuation of volatile anesthetic inhibition of glutamate-stimulated [3H](5D,10S)-(+)-methyl-10,11-dihydro-5H-dibenzo[a,d]cyclohepten-5,10-imine ([3H]MK-801) binding to N-methyl-D-Aspartate (NMDA) receptors in rat brain, *Biochem. Pharmacol.*, 50, 1373, 1995.
35. Lin, L.-H., Chen, L. L., and Harris, R. A., Enflurane inhibits NMDA, AMPA, and kainate- induced currents in Xenopus oocytes expressing mouse and human brain mRNA, *FASEB J.*, 7, 479, 1993.
36. Zhu, S. and Baker, R. C., Effects of inhalation anesthetics on kainate-induced glutamate release from cerebellar granule cells, *Life Sci.*, 58, 1359, 1996.

37. Dildy-Mayfield, J. E., Eger, E. I., II, and Harris, R. A., Anesthetics produce subunit-selective actions on glutamate receptors, *J. Pharmacol. Exp. Ther.*, 276, 1058, 1996.
38. Sommer, B. and Seeburg, P. H., Glutamate receptor channels: Novel properties and new clones, *Trends Pharmacol. Sci.*, 13, 291, 1992.
39. Dildy-Mayfield, J. E. and Harris, R. A., Ethanol inhibits kainate responses of glutamate receptors expressed in Xenopus oocytes: Role of calcium and protein kinase C, *J. Neurosci.*, 15, 3162, 1995.
40. Lerma, J., Paternain, A. V., Naranjo, J. R., and Mellstrom, B., Functional kainate-selective glutamate receptors in cultured hippocampal neurons, *Proc. Natl. Acad. Sci. USA*, 90, 11688, 1993.
41. Herb, A., Burnashev, N., Werner, P., Sakmann, B., Wisden, W., and Seeburg, P. H., The KA-2 subunit of excitatory amino acid receptors shows widespread expression in brain and forms ion channels with distantly related subunits, *Neuron*, 8, 775, 1992.
42. Sakimura, K., Morita, T., Kushiya, E., and Mishina, M., Primary structure and expression of the g2 subunit of the glutamate receptor channel selective for kainate, *Neuron*, 8, 267, 1992.
43. Brose, N., Huntley, G. W., Stern-Bach, Y., Sharma, G., Morrison, J. H., and Heinemann, S. F., Differential assembly of coexpressed glutamate receptor subunits in neurons of rat cerebral cortex, *J. Biol. Chem.*, 269, 16780, 1994.
44. Egebjerg, J., Bettler, B., Hermans-Borgmeyer, I., and Heinemann, S., Cloning of a cDNA for a glutamate receptor subunit activated by kainate but not AMPA, *Nature*, 351, 745, 1991.
45. Good, P. F., Huntley, G. W., Rogers, S. W., Heinemann, S. F., and Morrison, J. H., Organization and quantitative analysis of kainate receptor subunit GluR5-7 immunoreactivity in monkey hippocampus, *Brain Res.*, 624, 347, 1993.
46. Schlame, M. and Hemmings, H. C., Inhibition by volatile anesthetics of endogenous glutamate release from synaptosomes by a presynaptic mechanism, *Anesthesiology*, 82, 1406, 1995.
47. Larsen, M., Grondahl, T. O., Haugstad, T. S., and Langmoen, I. A., The effect of the volatile anesthetic isoflurane on Ca2+-dependent glutamate release from rat cerebral cortex, *Brain Res.*, 663, 335, 1994.
48. Miao, N., Frazer, M. J., and Lynch, C., Volatile anesthetics depress Ca2+ transients and glutamate release in isolated cerebral synaptosomes, *Anesthesiology*, 83, 593, 1995.
49. Bickler, P. E., Buck, L. T., and Feiner, J. R., Volatile and intravenous anesthetics decrease glutamate release from cortical brain slices during anoxia, *Anesthesiology*, 83, 1233, 1995.
50. Patel, P. M., Drummond, J. C., Cole, D. J., and Goskowicz, R. L., Isoflurane reduces ischemia-induced glutamate release in rats subjected to forebrain ischemia, *Anesthesiology*, 82, 996, 1995.
51. Illievich, U. M., Zornow, M. H., Choi, K. T., Strant, M. A., and Scheller, M. S., Effects of hypothermia or anesthetics on hippocampal glutamate and glycine concentrations after repeated transient global cerebral ischemia, *Anesthesiology*, 80, 177, 1994.
52. Nicol, B., Rowbotham, D. J., and Lambert, D. G., Glutamate uptake is not a major target site for anaesthetic agents, *Br. J. Anaesth.*, 75, 61, 1995.
53. Miyamoto, M., Ishida, M., Kwak, S., and Shinozaki, H., Agonists for metabotropic glutamate receptors in the rat delay recovery from halothane anesthesia, *Eur. J. Pharmacol.*, 260, 99, 1994.
54. Ishida, M., Saitoh, T., Shimamoto, K., Ohfune, Y., and Shinozaki, H., A novel metabotropic glutamate receptor agonist: marked depression of monosynaptic excitation in the isolated newborn rat spinal cord, *Br. J. Pharmacol.*, 409, 4469, 1993.

10 Effects of Volatile Anesthetics on $GABA_A$ Receptors: Electrophysiologic Studies

Robert A. Pearce

CONTENTS

The $GABA_A$ receptor is a ligand-gated chloride ionophore that underlies synaptic inhibition in the brain. It has been recognized for many years as the target of barbiturates and benzodiazepines, two important classes of drugs that have found broad clinical utility. More recently, it has been recognized that other structurally diverse drugs with general anesthetic properties, including volatile agents, act on these receptors as well. Furthermore, the concentrations at which these drugs enhance $GABA_A$ receptor function correlate well with their anesthetic potencies.

0-8493-8555-5/01/$0.00+$.50

The combination of these findings, together with the intuitively plausible suggestion that increased inhibition might be expected to produce "anesthesia," has resulted in acceptance of this receptor as an important and relevant anesthetic target.[1,2] This chapter will review some of the basic structural, pharmacologic, and physiologic properties of the $GABA_A$ receptor, describe how synaptic currents and responses to exogenous application of GABA are altered by volatile anesthetic agents, discuss possible mechanisms by which these changes may be produced, and speculate on how alterations of $GABA_A$ receptor function may influence cellular integrative and circuit functions to produce the clinical state of "anesthesia."

10.1 BASIC PROPERTIES: RECEPTOR STRUCTURE, PHARMACOLOGY, AND PHYSIOLOGIC PROPERTIES

The $GABA_A$ receptor is a member of the superfamily of ligand-gated ionophores that includes also the nicotinic acetylcholine receptor (nAChR), the glycine receptor, the $GABA_C$ receptor, and the 5HT3 receptor, and has been the subject of several recent reviews.[3,4] Each of the subunits that make up these heteromultimeric proteins contains four membrane-spanning domains (M1 to M4), with a large hydrophobic N terminal domain that contributes to the binding site for agonist and other allosteric modulators, such as benzodiazepines (Figure 10.1a). There is also a large intracellular loop between the third and fourth transmembrane domains that contains potential phosphorylation sites. By analogy with the nAChR, for which the most complete structural data are present,[5] each receptor is thought to be composed of five separate subunits (Figure 10.1b). This structural arrangement has been supported by electron microscopic observations of the $GABA_A$ receptor itself.[6] To date, 15 different varieties of mammalian subunits have been described, designated $\alpha_{1\text{-}6}$, $\beta_{1\text{-}3}$, $\gamma_{1\text{-}3}$, δ, ε, and π.[4,7,8] In addition, two unique nonmammalian subunits have been described, alternative splice variants of at least the γ_2 and α_6 receptor are known to exist, and two related subunits ρ_1 and ρ_2 are present in the retina and brain. These latter subunits are the only varieties that are known to combine to form functional homomultimeric receptors, which are designated $GABA_C$ receptors.

The subunit composition of native brain receptors is of interest because of its influence on the pharmacologic sensitivity and physiologic properties of the receptor/channel. For example, the α subunit is a major determinant of benzodiazepine selectivity,[4] and the presence of the γ subunit is required for sensitivity to benzodiazepines and relative insensitivity to zinc inhibition of the channel.[9] Deactivation and desensitization kinetics may be influenced by multiple subunits.[10,11] Although the precise subunit composition and structure has not yet been determined with certainty for any native receptors,[12] it is clear from *in situ* hybridization studies that $GABA_A$ receptor subunits are differentially distributed within the brain and that certain subunits tend to be expressed together (reviewed recently by Sieghart[4] and McKernan and Whiting[13]). The stoichiometry by which subunits combine to form functional channels remains a matter of debate, and is an area of active investigation. A recent study using a mutational approach analogous to that employed for the

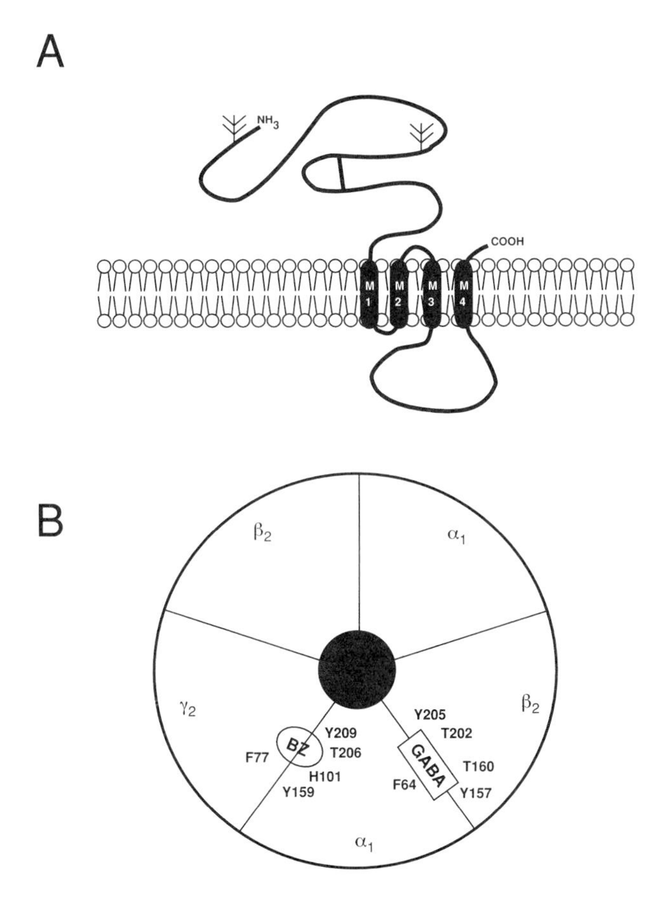

FIGURE 10.1 Structural features of the $GABA_A$ receptor. (A) Each subunit contains four membrane-spanning domains, a large N terminal domain, and a large intracellular loop. (B) Receptors are composed of five subunits arranged around a central pore. Binding sites for the agonist GABA and the modulatory benzodiazepine diazepam are formed by residues on adjacent subunits. (This drawing was kindly provided by Drew Boileau and Cynthia Czajkowski, University of Wisconsin, Department of Physiology.)

nAChR led to the conclusion that two α, two β, and one γ subunit combine to form channels in an exogenous expression system.[14] However, receptors containing only α and β subunits or α and γ subunits do form functional receptors, and it is likely that heterogeneity of at least the α subunit can exist within a single receptor.[15] Thus, the potential exists for a large number of $GABA_A$ receptor subtypes to exist in the brain, and possibly even within individual neurons, although the actual number of distinct subtypes that occur naturally remains unknown.

Electrophysiologic studies of the $GABA_A$ receptor have revealed complex gating and binding characteristics.[3] The dose–response relationship for GABA has a Hill coefficient between one and two, suggesting that at least two molecules of agonist bind to activate the channel. Although binding studies (reviewed in Chapter 11 of this volume) consistently have demonstrated high- and low-affinity GABA binding, it is not clear whether this represents two separate and nonequivalent binding sites, cooperativity between equivalent sites, or changes in apparent affinity for agonist produced by agonist-induced gating steps. Macroscopic responses to varying concentrations of agonist have been used to estimate opening rates and composite closing/unbinding rates, demonstrating that rapid opening rates (on the order of several thousand per millisecond) and relatively slow deactivation (corresponding to long burst durations of hundreds of milliseconds) are characteristics of $GABA_A$ receptors of dorsal root ganglion (DRG) cells, pyramidal neurons, Purkinje cells,[16-18] and likely many other types of neurons. Desensitization (a declining response in the continued presence of a fixed concentration of agonist) is a universal feature of $GABA_A$ receptors, and as methods for producing more rapid solution changes to the surface of cells or membrane patches containing receptors have been developed, it has been recognized that several components may be present, including quite rapid (several milliseconds) components as well as multiple slower phases of complex responses.[18-20] Deactivation following removal of applied agonist, and synaptic currents as well, have multiple exponential decay components. In the case of agonist-induced responses, the multiphasic decay has been postulated to arise from single and double liganded states of the receptor[21] or to be caused by reopening from rapidly equilibrating desensitized states of the receptor.[17]

The complexity observed in macroscopic kinetics is also seen in single-channel recordings. The distribution of open and closed times reveals multiple components in the presence of low concentrations of agonist, suggesting multiple open and closed states of receptors. In recordings obtained from mouse spinal cord neurons[22,23] and chick cerebral neurons[24] in culture, up to five closed and three open states have been observed during steady-state application of low concentrations of GABA. These openings are grouped into bursts, the properties of which are altered by barbiturates[23] and benzodiazepines.[25] A main conductance state of 26 to 30 picosiemens (pS) has been most consistently observed, but additional subconductance levels ranging from 8 to 54 pS are seen also. Indeed, recent studies of native receptors obtained from cells in brain slice preparations have demonstrated a remarkable heterogeneity of single-channel conductances and gating characteristics that may be observed even in a single type of neuron.[26,27]

Permeability of the receptors is greatest for chloride (of the physiologic anions), which for most neurons has a reversal potential close to or only slightly hyperpolarized to the resting potential. Thus, unless a neuron is depolarized by excitatory synaptic inputs or experimental manipulations, activation of these receptors leads to an increase in membrane conductance, but not necessarily hyperpolarization, a phenomenon sometimes termed shunting inhibition. However, these receptors also have a significant bicarbonate permeability, which may be up to 25% of the chloride permeability. Under certain circumstances, such as exogenous application of GABA to small dendritic processes, or pathologic conditions or experimental manipulations that lead to large amounts of synaptic GABA release, bicarbonate permeability may lead to net depolarizing responses.[28]

Synaptic currents mediated by $GABA_A$ receptors display some heterogeneity in kinetic properties, but generally have been found to have rapid rise times (millisecond or less) and exponential decays with one or two components, the time constants of which range from several milliseconds to hundreds of milliseconds.[18,29-31] This heterogeneity has been suggested to arise from the presence of separate populations of physiologically distinct receptors, the kinetics of which may depend on the subunit composition of the underlying receptors.[31-33] In general, the time course rather than the amplitude of synaptic currents has been found to be altered by agents that act at the $GABA_A$ receptor. Benzodiazepines, anticonvulsant and anesthetic barbiturates, other intravenous anesthetics, and as discussed more fully below, volatile anesthetics, have all been found to prolong the time course of synaptic current decay, measured directly[34-40] and indirectly.[41-46]

The diversity and complexity of $GABA_A$ receptor subtypes and interneurons provides evidence that this transmitter system plays varied and important roles in the central nervous system (CNS). This proposition is supported by the variety of clinically useful drugs that are known to affect $GABA_A$ receptors. Although general concepts about the role of inhibition are emerging, our present understanding is relatively rudimentary. Nevertheless, sufficient evidence exists that this receptor system is integral to the behavioral alterations brought about by general anesthetics, that it is a topic of intense study among those interested in anesthetic mechanisms. The following discussion reflects the many levels of inquiry that will ultimately contribute to a fuller understanding of the role of $GABA_A$ receptors in "anesthesia."

10.2 PHYSIOLOGIC STUDIES OF VOLATILE ANESTHETIC EFFECTS

10.2.1 Studies of Intact Tissues and Synaptic Circuits

The earliest indications that volatile anesthetics might enhance GABA-mediated inhibition came from studies of extracellular field potentials recorded *in vivo*. In the spinal cord it was found that spinal presynaptic inhibition, which was thought to involve GABA in its production,[47] was prolonged by a variety of anesthetics.[48] Extending these studies to higher CNS centers, Nicoll[49] found that halothane and other anesthetics prolonged the duration of granule cell postsynaptic inhibition in the olfactory bulb, and Pearce et al.[46] found that halothane, isoflurane, and enflurane prolonged paired-pulse depression of the hippocampal population spike, circuit phenomena thought to be mediated by $GABA_A$ receptors.

More direct evidence that inhibitory synaptic potentials and currents are altered by volatile agents was provided by intracellular recordings in brain slice preparations, where it was shown that halothane prolonged the duration of evoked recurrent inhibitory postsynaptic potentials (IPSPs) in olfactory cortex neurons[50,51] and prolonged spontaneous inhibitory postsynaptic currents (IPSCs) in voltage-clamped hippocampal pyramidal neurons[36,37] (Figure 10.2). However, results using intact tissues have been sometimes contradictory and confusing. Other investigators found that inhibitory postsynaptic potentials were depressed by volatile anesthetics,[52-54] or depressed at low but enhanced at high anesthetic concentrations.[55] Although some of the inconsistency in findings is undoubtedly the result of complex effects on polysynaptic pathways, with changes in responsiveness or excitation of inhibitory neurons affecting the generation of inhibitory potentials,[56,57] depression of responses

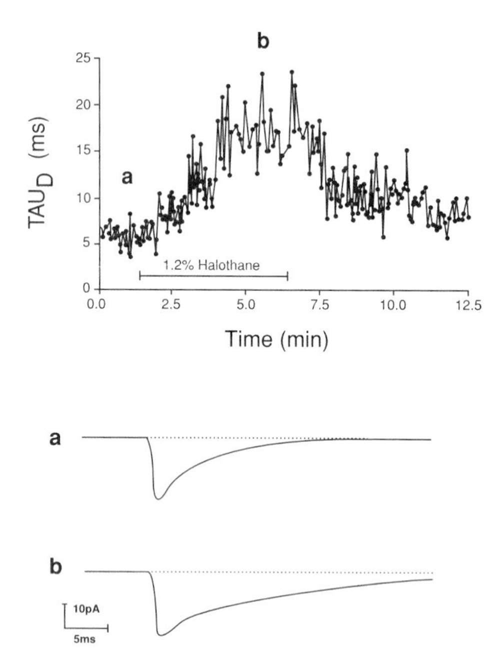

FIGURE 10.2 Plot of decay time constant (TAU_D) of $GABA_A$-mediated spontaneous inhibitory postsynaptic currents (sIPSCs) during recording from a single representative hippocampal CA1 neuron. The bar labeled 1.2% halothane indicates the time of exposure. Trace (a) is an average of 100 individual sIPSCs from control and trace (b) is in the presence of halothane. (From Mody, I., Tanelian, D. L. & Maciver, M. B. *Brain Res.* 538, 319–323, 1991. With permission.)

to exogenous GABA application was observed as well.[53,54] Additional confounding factors in the interpretation of even these experiments with direct agonist application to brain slices may have been indirect responses arising from a depolarizing action of GABA on inhibitory neurons, and desensitization of responses during prolonged application of GABA.

To circumvent the influence of indirect actions produced by anesthetic effects on polysynaptic pathways, yet still observe anesthetic actions on functioning synapses, several recent studies have focused on monosynaptic inhibitory responses, both in synaptically coupled cultured neurons and in pharmacologically isolated synapses within brain slice preparations. These experiments have shown generally an enhancement or prolongation of inhibition by volatile agents, but interesting differences between agents, and differences between effects of individual agents on separate components or types of inhibitory synaptic responses, have been found. In cultured rat hippocampal neurons, enflurane, isoflurane, and halothane were all found to increase the overall duration of IPSCs.[38] The decay of these synaptic currents was described by biexponential functions, with approximately equal proportions of fast and slow components under control conditions. All three anesthetics produced qualitatively similar changes in the IPSC, increasing the amplitude and proportion of the slow component and decreasing the amplitude and proportion of the fast component of decay. Overall, changes in the amplitudes of the two components rather than their time constants appeared to account for the results. Subsequent studies that compared synaptic currents with responses to brief, saturating pulses of GABA to outside-out patches from cultured hippocampal neurons suggested that the fast and slow components may be produced by a single class of receptors, the two components resulting from a combination of channel closing, recovery from desensitization, and agonist unbinding.[17] However, the precise steps that are altered by anesthetics to produce the pattern of changes observed have not been defined.

Experiments with native receptors studied in hippocampal brain slice preparations have provided additional evidence that volatile agents alter $GABA_A$-mediated synaptic inhibition,[37,40] but in this case the two physiologically distinct classes of $GABA_A$ synapses were found to be affected differently by different agents. A long-lasting dendritic inhibitory current that produces the early IPSP in CA1 pyramidal neurons was found to be prolonged by halothane and enflurane, without a significant change in the peak amplitude.[40,58] In contrast, a rapidly decaying inhibitory current that is localized to the somatic region, and is the source of spontaneous and miniature IPSCs, is prolonged by halothane without a change in its amplitude,[37,40] but is blocked to a substantial degree by enflurane[40] (Figure 10.3A). The reduction in amplitude by enflurane could be the result of a direct action on the postsynaptic receptor, or an indirect effect caused by a reduction in synaptic transmitter release (e.g., by reducing presynaptic calcium entry or causing action potential failure by block of sodium channels or activation of potassium channels). A concentration-dependent reduction in the amplitude of spontaneous (action potential dependent) and miniature (action potential independent, in the presence of tetrodotoxin) IPSCs in these neurons by enflurane indicates that the effect is on the postsynaptic receptors themselves.[59] A similar conclusion was reached by Antkowiak and Heck,[60] who also observed a reduction in the amplitude of spontaneous and miniature IPSCs when enflurane was applied to cerebellar Purkinje neurons (Figure 10.3B). Effects of other volatile agents on this preparation were not tested, so it is not known whether these receptors would show agent-specific modulation, as observed in hippocampal neurons. Depression by enflurane of direct agonist-activated currents in mouse cortical and recombinant receptors expressed in *Xenopus* oocytes[61,62] and of the fast component of IPSCs in cultured neurons[38] have been noted as well, supporting the conclusion that this anesthetic is able to block the current or reduce the response directly for many $GABA_A$ receptors.

The differences that have been observed between the effects of two anesthetics, halothane and enflurane, on a single synaptic response, and between the effects of enflurane on two synaptic responses in the same cell, illustrate a heterogeneity in anesthetic responses that has been observed also in intact circuits[63] and expressed receptors,[64] and that may be pharmacologically and behaviorally significant. Evidently, receptor subtype and individual anesthetic properties both are important in determining the synaptic effects of anesthetics. Additional factors that might contribute to overall anesthetic actions on $GABA_A$ receptors include the concentration and time course of neurotransmitter that activates the receptors, the influence of other channels or signaling systems on the state of the receptors, the degree of

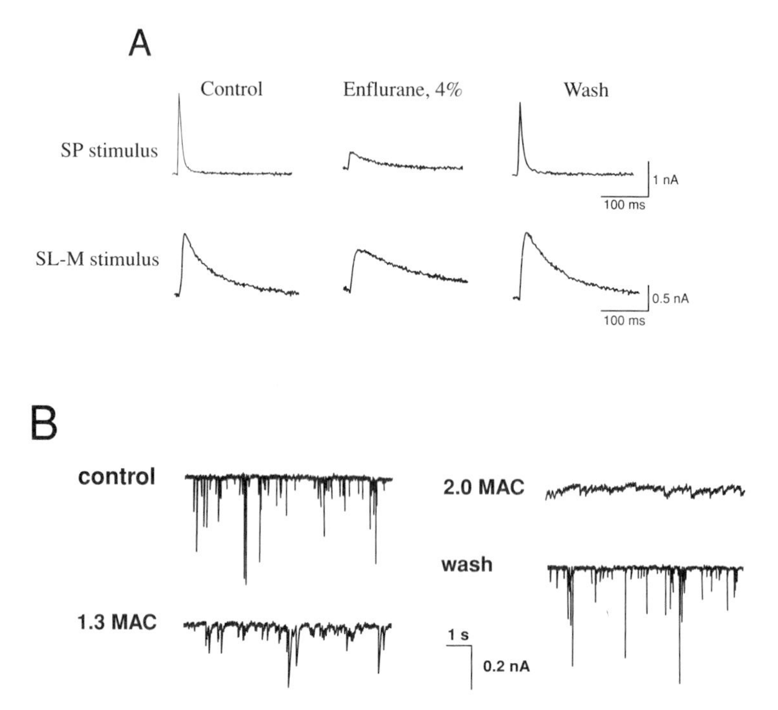

FIGURE 10.3 Effects of enflurane on IPSCs in (A) hippocampal CA1 pyramidal neurons and (B) cerebellar Purkinje cells. (A) Voltage clamp recording of evoked monosynaptic IPSCs in response to electrical stimulation of stratum pyramidale (SP) or stratum lacunosum-moleculare (SL-M). Enflurane (4%) prolonged the decay of $GABA_{A,slow}$ after SL-M stimulation, with little effect on response amplitude, but markedly reduced and prolonged the SP response. (B) Depression of spontaneous IPSCs in cerebellar Purkinje cells by enflurane was dose dependent, with an estimated half-maximal inhibition at 0.9% enflurane. (Part A from Pearce, R. A. *J. Physiol.* 492.3:823–840, 1996. With permission. Part B from Antkowiak, B. & Heck, D. *J. Neurophysiol.* 77:2525–2538, 1997. With permission.)

desensitization produced by ongoing activity, and the degree to which anesthetics directly activate receptors.

10.2.2 Studies of Isolated Cells and Receptors

To study more directly the effect of volatile agents on $GABA_A$ receptors, without the complicating influence of other cellular and circuit properties involved in synaptic transmission, many investigators have turned to the study of receptors in isolated neurons and expressed recombinant receptors. These investigations have generally used the patch-clamp or two-electrode voltage clamp techniques to perform whole-cell recordings, and exogenous application of GABA to whole cells or oocytes to activate receptors. Although the rates of solution exchange and duration of drug application have varied substantially between experimental preparations, results have been relatively consistent in one regard: at low agonist concentrations, volatile agents increase the peak amplitude of the agonist-evoked current (Figure 10.4A). At higher

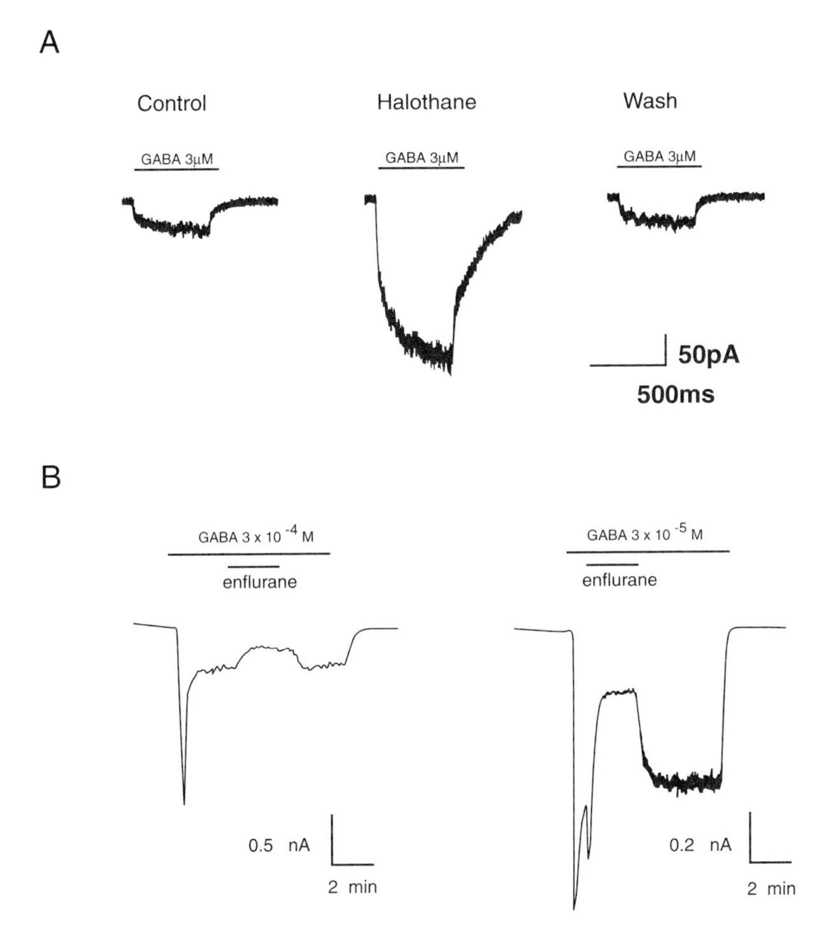

FIGURE 10.4 Effect of volatile agents on GABA-activated currents. (A) The response of recombinant $GABA_A$ receptors ($\alpha_1\beta_2\gamma_{2l}$) expressed in human embryonic kidney cells to application of 3 μ*M* GABA is increased in the presence of halothane (0.43 m*M*). (B) The effects of enflurane (1.89 m*M*) on currents recorded from cultured rat DRG neurons. (Part A from unpublished data from the author's laboratory (X.S. Li and R.A. Pearce). Part B from Nakahiro, M., Yeh, J. Z., Brunner, E. & Narahashi, T. *FASEB J.* 3:1850–1854, 1989. With permission.)

agonist concentrations, a somewhat more variable picture of anesthetic effects emerges, depending on the particular anesthetic, receptor or cell type, and method of drug application.

10.2.2.1 Native Receptors

A series of studies on rat DRG neurons in culture was performed to investigate the effects of halothane (0.86 m*M*), isoflurane (0.96 m*M*), and enflurane (1.89 m*M*) on responses to exogenous GABA application.[65,66] Each of the anesthetics increased by approximately threefold the peak response to a low concentration of bath-applied GABA (3 μ*M* for 3 to 5 min). Effects at higher concentrations of GABA were more

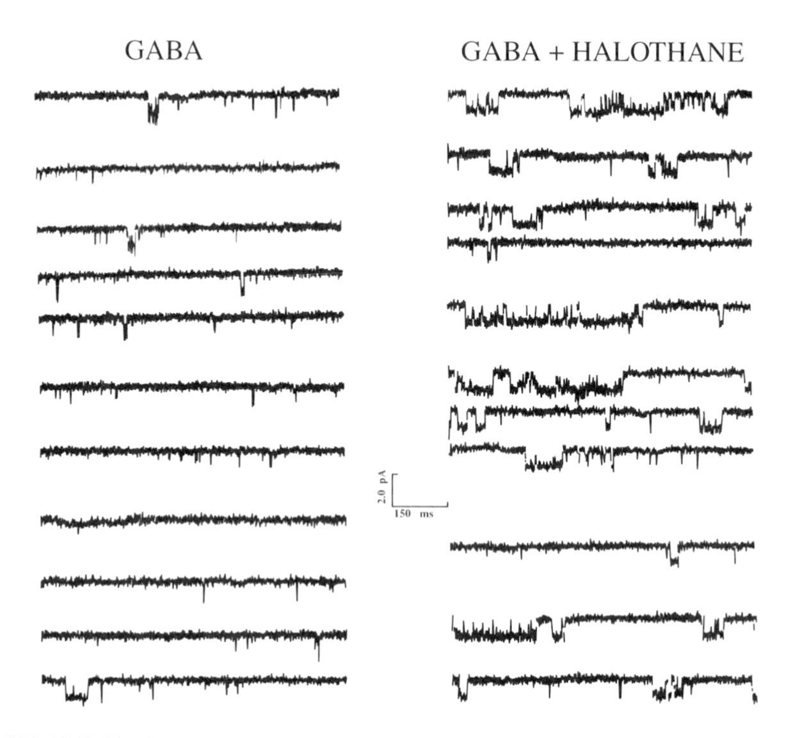

FIGURE 10.5 Single-channel currents recorded from cell-attached patches on acutely dissociated rat DRG neurons. The recording pipette contained 10 μ*M* GABA to activate the receptors. Halothane was applied in the bath at a concentration of 0.86 m*M* (2 MAC). (From Yeh, J. Z., Quandt, F. N., Tanguy, J., Nakahiro, M., Narahashi, T. & Brunner, E. A. *Ann. NY Acad. Sci.* 625:155–173, 1991. With permission.)

complex (Figure 10.4B). Responses to 30 and 300 μ*M* GABA were desensitizing; that is, an early peak current was followed by a declining response (35 s to decline to 70% of the peak) in the continued presence of agonist. The peak response to 30 μ*M* GABA was increased by all three anesthetics, but the rate of decline of the desensitizing component was accelerated by approximately threefold, and the final level of steady-state current was depressed in the presence of all three anesthetics. At higher (300 μ*M*) GABA concentrations, the only additional effect observed was a decrease in the steady-state current level. The investigators interpreted this pattern of activity to represent opposite effects of the anesthetics on two forms of the receptor: augmentation of the nondesensitized form of the receptors and depression of the desensitized form of the receptor.

The effect of halothane on single-channel currents in acutely isolated and cultured DRG neurons was also investigated, using cell-attached and outside-out patches with 3 to 10 μ*M* GABA in the recording pipette[66] (Figure 10.5). No change in the channel conductance (28 pS) was observed. The overall probability of a channel being open was increased from 0.02 to 0.04 in the presence of GABA alone, to 0.11 in the presence of GABA plus 1.7 m*M* halothane. This increase in opening was produced by three distinct changes (Table 10.1): (1) the mean open time was

TABLE 10.1
Alteration of Gating Kinetics of GABA-Activated Channels by Halothane[a]

Parameter	GABA	Halothane
P_{open}	0.04 ± 0.02	0.11 ± 0.05
Burst interval	1.00[b]	0.60
Burst duration (ms)	15.1 (range 5–25)	41.9 (range 10–49)
Mean open time (ms)	1.95 ± 1.50	2.89 ± 2.00
Closed time in a burst (ms)	1.64 ± 0.32	1.72 ± 0.21

[a] Data are from four cell-attached patches recorded from acutely dissociated rat DRG neurons, exposed to 10 μ*M* GABA with or without 4 MAC halothane.

[b] Because the number of channels in a patch was unknown, the change in burst interval was normalized to the control burst interval. Temperature 24 to 25°C.

Source: From Yeh, J. Z., Quandt, F. N., Tanguy, J., Nakahiro, M., Narahashi, T. & Brunner, E. A. *Ann. NY Acad. Sci.* 625:155–173, 1991.

prolonged, as a result of a change in the proportion but not time constants of the fast (1 ms) and slow (7 ms) components of the open time distribution; (2) interburst intervals were reduced by nearly one half; and (3) burst duration was lengthened. This overall pattern of effects was postulated to result from changes in both gating and agonist binding characteristics of the receptors.[66] However, it was noted that the postulated changes could not account for the effects of halothane on macroscopic currents in response to high (desensitizing) concentrations of agonist.

These studies of DRG neurons pointed to an interesting complexity in the actions of volatile agents that depended on the concentration of agonist and the desensitized state of the receptors. However, the temporal resolution of the drug delivery system employed was relatively slow, compared with the rates of activation and desensitization that have been recognized more recently, thus limiting the ability to observe rapidly obtained desensitized states and nondesensitized states induced by high agonist concentrations.[16-19] Studies with more rapid exchange were performed on acutely dissociated neurons of the nucleus of the tractus solitarius and hippocampus[67,68] using the "Y-tube" method for delivery of agonist and anesthetic. These studies again showed that halothane (1 m*M*), enflurane (1 m*M*), and sevoflurane (0.3 to 1 m*M*) increased the peak response at low agonist concentration (1 to 3 μ*M* GABA). Higher concentrations of sevoflurane produced depression rather than enhancement of responses to 1 μ*M* GABA, so that the dose response to sevoflurane showed a peak at 1 m*M* sevoflurane and a steep concentration dependence of depression with higher levels of anesthetic. Halothane caused the GABA dose–response curve to be shifted to the left, without a change in the Hill coefficient (1.5) or in the peak response at high agonist concentrations, whereas sevoflurane caused the responses to high concentrations of GABA (10 μ*M* and greater) to be

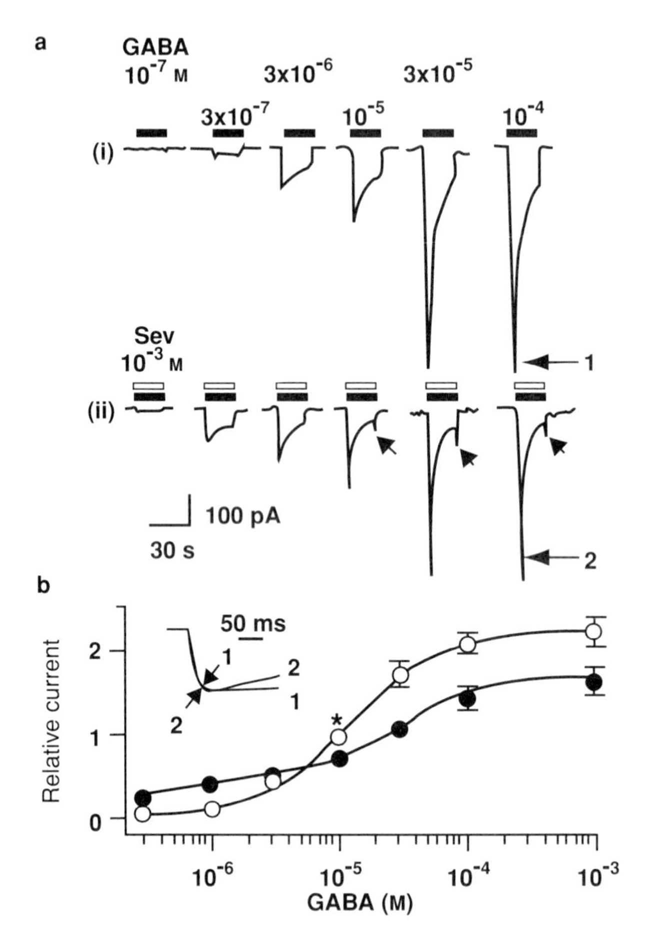

FIGURE 10.6 Effect of sevoflurane on the concentration–response relationship of the GABA-induced chloride current (I_{Cl}) in acutely dissociated rat hippocampal CA1 pyramidal neurons. (a i) I_{Cl} elicited by GABA at various concentrations. (a ii) I_{Cl} elicited by various concentrations of GABA plus 1 m*M* sevoflurane. Arrowheads indicate the "hump" current that appeared with washout of the anesthetic. (b) Concentration–response curves for GABA in the absence (o) or presence (•) of sevoflurane. Amplitudes of the peak current components were normalized to that induced by 10^{-5} *M* GABA alone. Inset to (b) shows the superimposed I_{Cl} induced by 10^{-4} *M* GABA (1) and GABA plus 10^{-3} *M* sevoflurane (2). (From Wu, J., Harata, N. & Akaike, N. *Br. J. Pharmacol.* 119, 1013–1021, 1996. With permission.)

depressed (Figure 10.6). As in studies of DRG neurons,[65] the desensitization was accelerated and the plateau response was depressed.[67] Although the time course of desensitization in these studies was not reported, it appears to be on the order of seconds, faster by approximately tenfold than the previous studies.[65] Despite the more rapid exchange rates that can be achieved using this method, they still do not appear to be fast enough to discern effects of anesthetics on rapid (tens of milliseconds) desensitization, which has been postulated to play an important role in determining the time course of inhibitory synaptic currents.[17]

An additional anesthetic action was apparent when the Y-tube application method was used to deliver pulses of agonist and anesthetic at the same time (coapplication). After coapplication was completed, a transient inward current, which the investigators referred to as a "hump current," was observed after application of GABA, pentobarbital, or muscimol together with halothane (1 m*M*) or enflurane (1 m*M*).[68] A similar transient current also was observed after application of sevoflurane alone at very high concentration (5 mm) or coapplication with GABA[67] (see Figure 10.6). It was suggested that these hump currents might be the result of a direct channel blocking effect of the anesthetics, observed in these experiments because the anesthetic was cleared from the receptors faster than the receptors deactivated after agonist removal.

10.2.2.2 Expressed Receptors

To determine whether volatile anesthetics have subunit-specific requirements or actions, as do several intravenous anesthetics, including steroids, etomidate, and propofol,[64,69-71] a number of investigators have examined their effects on $GABA_A$ receptors of defined subunit composition, expressed either in oocytes or mammalian expression systems. Recombinant receptors composed of human α_1 or α_2 subunits, together with either β_1 or γ_2 subunits or both, displayed qualitatively similar responses to isoflurane, with enhancement of the response to pressure ejection of 100 μM GABA.[72] The γ subunit, which is required for augmentation of $GABA_A$ receptor responses by benzodiazepines, was found not to be necessary for enhancement of responses of $\alpha_1\beta_1$ receptors to enflurane.[64,73] An apparent difference in the potentiation of $\alpha_1\beta_1$ versus $\alpha_1\beta_1\gamma_{2s}$ or $\alpha_1\beta_1\gamma_{2l}$ receptors[73] was subsequently attributed to the difference between these receptors in baseline responses to GABA, which activates $\alpha_1\beta_1$ receptors with an EC_{50} of 6 μM but $\alpha_1\beta_1\gamma_2$ receptors with an EC_{50} of 17 μM.[64] As observed for native receptors, the ability of anesthetics to enhance peak responses was found to decrease as the concentration of GABA increased.[61,62] This result underscores the importance of comparing responses at equipotent concentrations of agonist, and the difficulty in comparing studies performed using different receptors or conditions. Thus, in contrast to findings with intravenous agents, a strong subunit dependence in ability to modulate the response to GABA has not been identified for volatile agents. However, it must be recognized that only a small fraction of the possible combination of subunits has been examined. Also, subunit specificity of action does appear to be present for direct activation of the receptors.[74,75]

Subunit-specific modulation has been described for the structurally related ρ subunit, which forms homomeric $GABA_C$ receptors. These receptors, which are found predominantly in the retina (hence the term rho), are depressed by enflurane and isoflurane,[76] demonstrating that not all members of the $GABA_A$ receptor family are enhanced by volatile agents. Also, there was recently described a novel $GABA_A$ subunit, designated ε, that confers upon $GABA_A$ receptors with which it is expressed a resistance to modulation by intravenous agents.[7] If this turns out to be the case for volatile agents as well, it will represent not only another example of subunit-specific responses of $GABA_A$ receptors to volatile anesthetics, but also the only known $GABA_A$ receptor response that is not modulated by volatile agents.

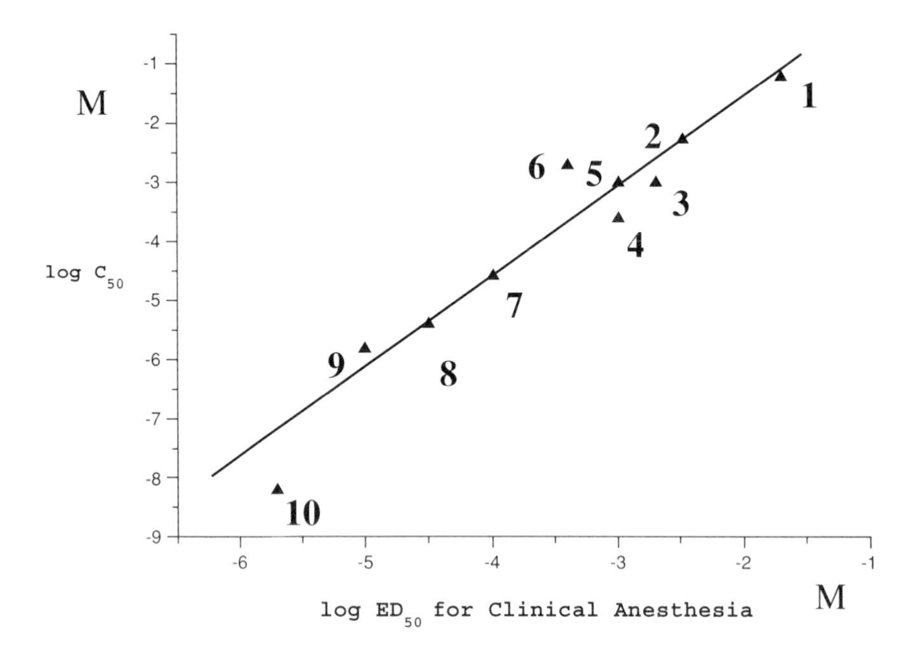

FIGURE 10.7 Drug potency for modulation of $GABA_A$ receptors is highly correlated with *in vivo* anesthetic potency. C_{50}, the concentration of anesthetic required to produce a 50% increase in the curation of GABA-activated currents is plotted on a double logarithmic scale against the concentration (ED_{50}) required for surgical anesthesia. The value of the correlation coefficient r = 0.97. Anesthetics are referred to as follows: 1, urethane; 2, chloroform; 3, trichloroethanol; 4, isoflurane; 5, enflurane; 6, halothane; 7, pentobarbital; 8, methohexital; 9, propofol; and 10, alphaxalone. (From Zimmerman, S. A., Jones, M. V. & Harrison, N. L. *J. Pharmacol. Exp. Ther.* 270:987–991, 1994. With permission.)

The excellent correlation that has been observed between the *in vivo* potency of a wide range of anesthetics, both volatile and nonvolatile, and the concentration required to augment the GABA response of cultured rat hippocampal neurons[77,78] (Figure 10.7), has been extended recently to a group of novel halogenated agents that do not act as anesthetics (despite their predicted potency based on their lipid solubility), using receptors expressed in *Xenopus* oocytes. It was found that these "nonanesthetics" did not potentiate responses to 5 μM GABA, but that enflurane and the anesthetic compound 1-chloro-1,2,2-trifluorocyclobutane did so, by shifting the GABA dose–response curve to the left.[62] Taken together, these findings strongly support a role for the $GABA_A$ receptor in producing general anesthesia.

10.2.3 Direct Activation Versus Modulation

Most of the discussion presented above has focused on the effects of volatile agents as modulators of the agonist response, whether agonist was applied exogenously or naturally (i.e., via synaptic release). However, direct activation of the $GABA_A$ receptor by a number of intravenous anesthetics has been well documented,[3] and although they have been studied less extensively, and results are less consistent, it is clear

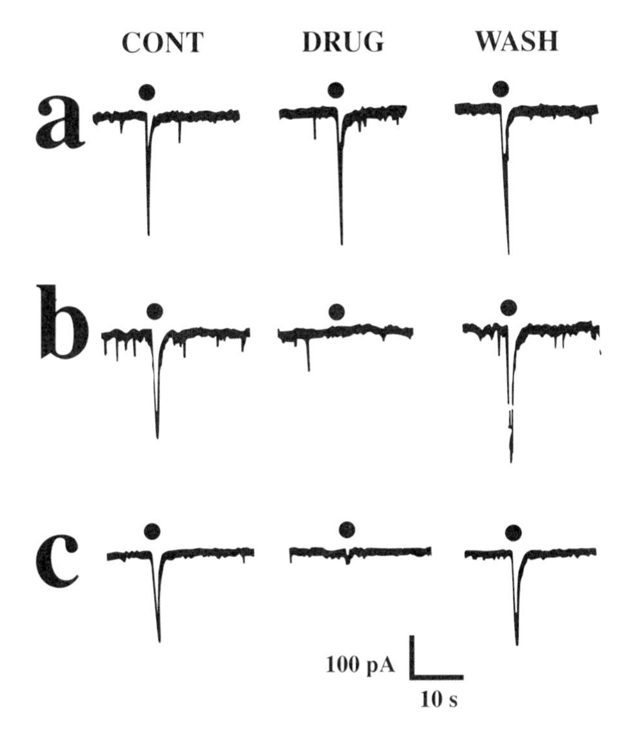

FIGURE 10.8 The isoflurane-gated current in cultured postnatal rat hippocampal neurons shows the pharmacologic properties of a $GABA_A$-gated current. Voltage clamp traces from cells held at –50 mV with CSCl internal solution. The bath solution contained 1 μ*M* TTX and 4 m*M* $MgCl_2$. Traces depict 100 ms × 20 psi pressure applications of isoflurane without (left), with (center), and after (right) coapplication of: (a) strychnine (10 μ*M*); (b) bicuculline (10 μ*M*); and (c) picrotoxinin (1 m*M*) with isoflurane (1.5 m*M*). (From Yang, J., Isenberg, K. E. & Zorumski, C. F. *FASEB J.* 6:914–918, 1992. With permission.)

that volatile agents also are able to activate $GABA_A$ receptors directly. In cultured postnatal rat hippocampal neurons, isoflurane (0.2 to 10 m*M*), enflurane (0.75 to 1.5 m*M*), and halothane (0.34 to 1.7 m*M*) all activated a bicuculline-sensitive anion current, the electrophysiologic properties and pharmacologic sensitivity of which indicated that the current was carried by $GABA_A$ receptors[38,79] (Figure 10.8). The effect was dose dependent and was half maximal at a concentration of 0.8 m*M* isoflurane (which is approximately 2.5 times the aqueous EC_{50} for this drug). Similarly, in recordings from hippocampal pyramidal neurons in brain slices, these three agents produced increases in baseline current noise that was blocked by bicuculline, consistent with direct channel activation. Isoflurane was most effective, enflurane produced a smaller increase, and halothane the least.[59] Sevoflurane (0.3 to 5 m*M*) also was found to directly activate GABA receptors on acutely dissociated CA1 hippocampal neurons, and this action was reversibly blocked by bicuculline, with an IC_{50} of 0.71 μ*M*.[67] Also, in neurons dissociated from the nucleus of the tractus solitarius, a small current was induced by halothane and enflurane, but not the volatile convulsant agent hexafluorodiethylether. Again, this current was not investigated in detail, but its properties are consistent with direct activation of $GABA_A$ or glycine receptors.[68] Thus, there is good evidence that volatile agents are able to activate $GABA_A$ receptors directly. Although the concentrations that do so are somewhat higher than those that produce anesthesia and augment responses to GABA, the relative importance of direct activation versus indirect modulation of synaptic responses has not been addressed experimentally.

Unlike the lack of subunit dependence for modulation of GABA responses, it has been reported that the direct activation of $GABA_A$ receptors by volatile agents does depend on subunit composition of receptors. Using an HEK 293 cell expression system, it was found that $\alpha_6\beta_2\gamma_{2s}$ but not $\alpha_1\beta_2\gamma_{2s}$ or $\alpha_1\beta_2$ receptors were directly activated by halothane.[74,75] Peak current amplitudes in response to halothane alone were much smaller than those activated by GABA alone, with an EC_{50} of 1.0 m*M* and Hill coefficient of 1.5 for halothane, similar to the Hill coefficient of 1.4 for activation of these receptors by GABA. This halothane-activated current was inhibited competitively by bicuculline, with an IC_{50} of 0.69 μ*M*, again similar to the IC_{50} for inhibition of GABA-activated currents (0.71 μ*M*). Because anesthetics are clearly able to potentiate GABA-induced responses of $\alpha_1\beta_2\gamma_{2s}$ receptors,[62] this result also suggests a difference in the mechanisms of direct activation and potentiation. However, a direct comparison of concentration-dependent responses using several anesthetics and the same expression system has not been performed using these different protocols. A subunit specificity in direct activation may account also for the finding that Purkinje neurons in cerebellar slices are not directly activated by enflurane (0.6 to 1.2 m*M*),[60] because granule cells but not Purkinje cells express α_6-containing receptors.[80]

10.2.4 Channel Block by Anesthetics

As noted above, there are a number of instances in which volatile anesthetics have been found to depress rather than enhance $GABA_A$-receptor-mediated responses. These include depression of fast evoked, spontaneous, and miniature IPSCs by enflurane in hippocampal CA1 neurons;[40,81] depression of mIPSCs by enflurane in cerebellar Purkinje neurons;[60] reduction by all volatile agents of the steady-state (desensitized) current elicited by high concentrations of GABA in cultured DRG neurons[65] and acutely dissociated neurons of the nucleus of the tractus solitarius and hippocampus;[67] depression of peak currents evoked by even low concentrations of GABA by high concentrations of sevoflurane and of currents evoked by high concentration of GABA by low concentrations of sevoflurane;[67] and depression by enflurane of peak responses to high concentrations of GABA in mouse cortical receptors expressed in *Xenopus* oocytes.[61] Quinlan et al.[82] found also that at high GABA concentrations, halothane decreased maximal chloride flux in purified brain microvesicles.[82] In addition, anesthetic block of channels was postulated to underly the "hump current" that is observed after coapplication of anesthetics and agonist[68,83,84] or anesthetic alone.[67]

It is not clear what the physiologic significance of anesthetic-induced depression of responses might be, or to what degree this action offsets the ability of volatile anesthetics to enhance synaptic inhibition. A common theme in the depression of responses is a high concentration of agonist, which is thought to be present during synaptic transmission at central GABAergic synapses.[85] Other drugs that block the $GABA_A$ receptor via competitive or noncompetitive mechanisms, including the volatile convulsants flurothyl and hexafluorodiethylether,[86,87] lead to seizure activity. It has been proposed that the ability of enflurane to reduce fast somatic $GABA_A$ synaptic currents underlies its proepileptic property.[40] In this particular case, the peak current amplitude may be crucial in curtailing the effect of excitatory synaptic inputs and preventing the generation of high-frequency bursts of action potentials. However, at other synapses, and with other anesthetics, more modest reductions in peak responses may be more than offset by a prolongation of inhibitory synaptic responses, leading to a greater overall charge transfer and a net increase in inhibition.

The mechanism of depression of GABA responses by volatile agents is also unclear. One hypothesis is that anesthetics produce an acceleration, or enhancement, of the normal process of desensitization, as proposed for the effects of neurosteroids on the $GABA_A$ receptor.[88] However, it is not clear that synaptic receptors desensitize to a significant degree during individual synaptic responses, or are in desensitized states prior to synaptic activation. An additional possibility is that the anesthetics directly block open channels, or bind to closed channels and reduce opening.[67] Direct block of other channels by volatile agents has been noted, including the related nAChR.[89] For this receptor, residues of the M2 region lining the pore of the channel have been implicated in channel block by long-chain alcohols and isoflurane,[90] and rapid rates of block have been measured.[91,92] It is possible that the mechanism by which nAChR channels are blocked applies also to $GABA_A$ receptors, and that results from this preparation will prove instructive.

10.3 MECHANISM OF ACTION

Considerable debate has taken place over the past decade concerning the biophysical mechanisms by which volatile agents alter ion channel function. Prior to this time, there was a general acceptance of nonspecific lipid-based theories, based on the correlation between anesthetic potency and lipid solubility. More recently, it has been suggested that specific anesthetic–protein interactions produce the observed alterations in channel function.[2] This suggestion is supported by observations that anesthetics interact with soluble proteins such as luciferase[93,94] and albumin;[95] that there is receptor specificity of actions;[2] and that volatile anesthetic actions are stereospecific (albeit with relatively small differences between stereoisomers), as observed for the $GABA_A$ receptor in electrophysiologic, binding, and chloride flux studies.[38,96-100]

Efforts to identify anesthetic binding sites on the $GABA_A$ receptor have been initiated by several investigators using the strategy of chimeric receptor subunits and site-directed mutagenesis, assuming that these agents exert their effects directly on the receptor/channel complex itself. Using chimeras composed of glycine α_1 and GABA ρ_1 subunits expressed as monomers, researchers have identified a critical region of 45 amino acids within the TM2 and TM3 domains as necessary and sufficient to confer enhancement by enflurane and ethanol of low-concentration GABA (EC_{10}) responses in these chimeric receptors.[101] Site-directed mutagenesis was then used to identify critical residues within this region that were necessary for enhancement by enflurane and ethanol of agonist responses, and analogous mutants were created in the $GABA_A$ receptor α and β subunits. Coexpression of the mutant receptor subunits in *Xenopus* oocytes produced receptors with normal gating by GABA, and modulation by propofol, but they were insensitive to enhancement by enflurane or isoflurane.[101,102] The finding that the volume of an amino acid residue at this critical site was inversely correlated with GABA EC_{50} and with isoflurane potentiation led to the suggestion that these residues may line a cavity that may be filled with either an anesthetic molecule or an amino acid side chain.[103] Further evidence that this site forms an anesthetic binding site has come from experiments in which an irreversible enhancement of receptor function was produced by covalently linking propanethiol (an unconventional anesthetic) to a cysteine residue inserted at the site.[104] Similar studies on related chimeric receptors[105] have suggested that halothane and isoflurane may bind to distinct sites to exert different effects, providing evidence against a unitary mechanism for all volatile agents on the $GABA_A$ receptor.

Alternative indirect mechanisms by which volatile agents might modulate $GABA_A$ receptor responses have also been considered. The finding that the action of halothane to prolong IPSCs is attenuated by reducing intracellular calcium via chelation with BAPTA or by preventing release of intracellular calcium stores using dantrolene[37] led to the suggestion that volatile agents act via an increase in intracellular calcium concentration. However, this hypothesis has not been supported by studies of other preparations.[61] Another indirect mechanism that has been proposed is alteration of channel activity via phosphorylation. Volatile agents have been found to alter the activity of protein kinase C,[106-108] which can phosphorylate the $GABA_A$ receptor and alter its activity.[109,110] However, the finding that enhancement of receptor function and channel block by halothane occur on the millisecond time scale, and in excised patches as in whole-cell recordings, suggests that they are produced via direct modulation of the channel.[84]

The kinetic mechanism by which volatile agents slow receptor deactivation has been investigated using recombinant $GABA_A$ receptors expressed in HEK293 cells. From a comparison of macroscopic currents obtained using rapid agonist application techniques and predicted effects of changes in individual kinetic steps derived from computer modeling, it was concluded that halothane slows the microscopic agonist unbinding rate but does not directly affect opening, closing, or desensitization.[111] A slowing of agonist dissociation rate was also found to be a component of the action of propofol[112] and benzodiazepines,[113] sugggesting that this may be a relatively general mechanism of action for intravenous as well as volatile agents.

10.3.1 Net Effects on the Nervous System

Steady progress has been made over the past decade in identifying the $GABA_A$ receptor as an important anesthetic target, and characterizing the actions of volatile and nonvolatile agents on these receptors. Just how the changes in receptor function that are produced by these drugs lead to the clinical state of "general anesthesia" is less easily answered, but efforts have been made toward answering these important questions also.

10.3.1.1 Anatomical Locus of Action

Although the $GABA_A$ receptor is associated primarily with higher CNS centers, and glycine receptors with the brainstem and spinal cord, there is a significant distribution of $GABA_A$ receptors in lower centers as well, including the spinal cord. Recent reports have demonstrated that the spinal cord is an important site at which volatile agents act to prevent movement in response to a noxious stimulus.[114-116] The findings that the ability of halothane to prevent such movement is antagonized by drugs that block the $GABA_A$ receptor (bicuculline and picrotoxin), but not by antagonists of $GABA_B$ receptors or glycine receptors,[117] and that $GABA_A$ antagonists did not by themselves alter the response latency, together suggest that the enhancement of GABA action in the spinal cord contributes to the antinocifensive action of halothane, and probably other volatile agents as well.

In addition to preventing movement in response to noxious stimuli, general anesthetics produce other changes in the nervous system, such as loss of consciousness and amnesia. These effects, which are likely to be produced by anesthetic actions on higher centers, generally require lower anesthetic concentrations (by a factor of 2 to 3) than those required to prevent movement.[118-120] Although effects of volatile agents on $GABA_A$ receptors of the cortex are certainly well documented, it remains unknown to what extent these actions can account for the observed changes in CNS function, in comparison with effects on other transmitter receptors or voltage-activated currents, or with effects on lower centers that indirectly would alter cortical processes, such as arousal pathways or thalamic relay nuclei. Nevertheless, it seems likely that at least some of the effects of volatile agents on higher order CNS functions are related to effects on cortical GABAergic circuitry, which is the predominant inhibitory system in the cortex.

10.3.1.2 Differing Effects on Functionally Distinct Circuits

Recent work on understanding the role of inhibition, and its modulation by volatile anesthetics, has been carried out in the hippocampus. This model cortical system has been studied extensively, due to both its relatively simple cytoarchitectural characteristics and the important role it plays in understanding learning and memory and the pathological generation of seizures. It has been found that many varieties of inhibitory interneurons exist, each with a characteristic dendritic architecture, axonal projection pattern, and complement of calcium binding proteins. In the CA1 region of the hippocampus, two physiologically distinct types of inhibitory currents in pyramidal neurons have been distinguished that are generated by these interneurons.[30] These currents differ in kinetic characteristics, anatomic properties, and pharmacologic sensitivity to a variety of drugs including volatile agents. One current, which has been termed $GABA_{A,slow}$, decays with a time constant of approximately 70 ms at 35°C, and corresponds to the $GABA_A$-mediated "early IPSP" that has been observed in hippocampal and other cortical pyramidal neurons.[30,121] This current underlies paired-pulse depression of the population spike,[40] which was found in previous studies to be prolonged by a wide variety of intravenous and volatile anesthetics.[44-46,122] By virtue of its dendritic localization and long-lasting conductance, $GABA_{A,slow}$ is able to regulate local synaptic plasticity by controlling the level of dendritic depolarization and thereby the voltage-dependent block of NMDA receptors[123,124] and back-propagation of dendritic calcium spikes.[125] Using glutamate receptor antagonists and local stimulation to study this current in isolation, researchers found that both halothane and enflurane prolonged the decay of the current without altering its amplitude.[40,58] Its prolongation by anesthetics was proposed to play a role in the amnesic effects of volatile agents.

A second inhibitory pathway that can be activated independently of $GABA_{A,slow}$ in CA1 pyramidal neurons is affected differently by volatile agents. $GABA_{A,fast}$, a rapidly decaying inhibitory current that is localized to the somatic region,[30] is prolonged by halothane without a change in its amplitude, but is blocked to a substantial degree by enflurane.[40] This fast somatic current is positioned to control the output of the neuron by preventing action potential generation, presumably by holding the cell body and adjacent axon hillock below the action potential threshold. The ability of enflurane to block this current was proposed to underlie its epileptogenic property, because reduction of inhibition by other drugs commonly leads to seizure activity. Indeed, it was found that selective blockade of $GABA_{A,fast}$ by furosemide leads to generation of multiple action potentials in intracellular recordings, and multiple population spikes in extracellular recordings, a hallmark of epileptic tissue.[30]

Anatomically segregated and physiologically distinct inhibitory synaptic inputs to pyramidal neurons have been described recently for the piriform cortex and neocortex as well.[126,127] This circuitry also provides the substrate for selective modulation by inhibition of synaptic plasticity in the dendritic regions and control of excitability via somatic inputs. The responses of these different inhibitory currents to anesthetics have not yet been tested.

10.3.1.3 Oscillations in the Brain

Relating the changes in receptor function that are produced by anesthetics, such as changes in IPSC time course and amplitude, to alterations in consciousness and memory will require an improved understanding of the role of inhibition in normal functioning of the cortex. One promising direction that is currently the topic of much research is the role of oscillations in the integrative functioning of the cortex and other brain regions.

Localized and transient gamma frequency oscillations (around 40 Hz) have been associated with information processing in various neocortical regions,[128,129] and theta oscillations (3 to 8 Hz) produced in the hippocampus during exploratory behaviors are believed to be integral to synaptic plasticity, which may underlie the formation of new memories.[130-132] There is evidence that these oscillations are produced or coordinated by GABAergic inhibitory circuits,[133-136] and that the time course and amplitude of inhibitory synaptic currents influence the frequency of oscillations.[137] Anesthetics produce characteristic alterations in the EEG pattern that include reductions in high-frequency and increases in low-frequency components, and it has been found recently that general anesthetics depress or slow the frequency of gamma oscillations.[138-142] Thus, it is tempting to speculate that alteration of coordinated oscillations by changes in the time course of $GABA_A$-mediated inhibition may underlie some of the changes in cognition and memory that are produced by general anesthetics.[142,143]

It must also be recognized, however, that factors other than alteration of GABA receptors themselves will also undoubtedly influence the activity of inhibitory circuits, such as alteration of intrinsic properties and synaptic excitation of interneurons.[56] At the present time, the relationship between anesthetic actions on $GABA_A$ receptors, changes in oscillations, and alterations in consciousness is intriguing, but purely speculative, and a much better understanding of the source and roles of oscillations will be required before a causal relationship can be established.

10.4 SUMMARY

Although the precise mechanisms by which volatile and other general anesthetics alter nervous system function to produce "anesthesia" remain unknown, it is likely that effects the $GABA_A$ receptor play an important role in the process. Research continues into the biophysical mechanisms by which alterations in receptor and channel function take place, the impact of actions on different receptor isoforms in different brain structures, and the relative contributions of $GABA_A$ receptors and other ion channels to integrative properties of neurons and circuits. The results will lead to advances in our understanding of the mechanisms of anesthesia, and also the workings of the brain itself.

REFERENCES

1. Tanelian, D. L., Kosek, P., Mody, I. & MacIver, M. B. The role of the $GABA_A$ receptor/chloride channel complex in anesthesia. *Anesthesiology* 78:757–776, 1993.
2. Franks, N. P. & Lieb, W. R. Molecular and cellular mechanisms of general anaesthesia. *Nature* 367:607–614, 1994.
3. Macdonald, R. L. & Olsen, R. W. $GABA_A$ receptor channels. *Annu. Rev. Neurosci.* 17:569–602, 1994.
4. Sieghart, W. Structure and pharmacology of $GABA_A$ receptor subtypes. *Pharmacol. Rev.* 47:181–234, 1995.
5. Guy, H. R. & Hucho, F. The ion channel of the nicotinic acetylcholine receptor. *Trends Neurosci.* 10:318–321, 1987.
6. Nayeem, N., Green, T. P., Martin, I. L. & Barnard, E. A. Quaternary structure of the native $GABA_A$ receptor determined by electron microscopic image analysis. *J. Neurochem.* 62:815–818, 1994.
7. Davies, P. A., Hanna, M. C., Hales, T. G. & Kirkness, E. F. Insensitivity to anaesthetic agents conferred by a class of $GABA_A$ receptor subunit. *Nature* 385:820–823, 1997.
8. Neelands, T. R. & Macdonald, R. L. Incorporation of the pi subunit into functional gamma-aminobutyric acid(A) receptors. *Mol. Pharmacol.* 56:598–610, 1999.
9. Draguhn, A., Verdorn, T. A., Ewert, M., Seeburg, P. H. & Sakmann, B. Functional and molecular distinction between recombinant rat $GABA_A$ receptor subtypes by Zn^{2+}. *Neuron* 5:781–788, 1990.
10. Tia, S., Wang, J. F., Kotchabhakdi, N. & Vicini, S. Distinct deactivation and desensitization kinetics of recombinant $GABA_A$ receptors. *Neuropharmacology* 35:1375–1382, 1996.

11. Haas, K. F. & Macdonald, R. L. $GABA_A$ receptor subunit gamma2 and delta subtypes confer unique kinetic properties on recombinant $GABA_A$ receptor currents in mouse fibroblasts. *J. Physiol.* 514:27–45, 1999.
12. Yeh, H. H. & Grigorenko, E. V. Mini review: Deciphering the native $GABA_A$ receptor: Is there hope? *J. Neurosci. Res.* 41:567–571, 1995.
13. McKernan, R. M. & Whiting, P. J. Which $GABA_A$-receptor subtypes really occur in the brain? *Trends Neurosci.* 19:139–143, 1996.
14. Chang, Y. C., Wang, R. P., Barot, S. & Weiss, D. S. Stoichiometry of a recombinant $GABA_A$ receptor. *J. Neurosci.* 16:5415–5424, 1996.
15. Verdoorn, T. A. Formation of heteromeric γ-aminobutyric acid type A receptors containing two different alpha subunits. *Mol. Pharmacol.* 45:475–480, 1994.
16. Maconochie, D. J., Zempel, J. M. & Steinbach, J. H. How quickly can $GABA_A$ receptors open? *Neuron* 12:61–71, 1994.
17. Jones, M. V. & Westbrook, G. L. Desensitized states prolong $GABA_A$ channel responses to brief agonist pulses. *Neuron* 15:181–191, 1995.
18. Puia, G., Costa, E. & Vicini, S. Functional diversity of GABA-activated Cl^- currents in Purkinje versus granule neurons in rat cerebellar slices. *Neuron* 12:117–126, 1994.
19. Oh, D. J. & Dichter, M. A. Desensitization of GABA-induced currents in cultured rat hippocampal neurons. *Neuroscience* 49:571–576, 1992.
20. Celentano, J. J. & Wong, R. K. Multiphasic desensitization of the $GABA_A$ receptor in outside-out patches. *Biophys. J.* 66:1039–1050, 1994.
21. Busch, C. & Sakmann, B. Synaptic transmission in hippocampal neurons: numerical reconstruction of quantal IPSCs. *Cold Spring Harb. Symp. Quant. Biol.* 55:69–80, 1990.
22. Macdonald, R. L., Rogers, C. J. & Twyman, R. E. Kinetic properties of the $GABA_A$ receptor main conductance state of mouse spinal cord neurones in culture. *J. Physiol. (Lond.)* 410:479–499, 1989.
23. Macdonald, R. L., Rogers, C. J. & Twyman, R. E. Barbiturate regulation of kinetic properties of the $GABA_A$ receptor channel of mouse spinal neurones in culture. *J. Physiol. (Lond.)* 417:483–500, 1989.
24. Weiss, D. S. & Magleby, K. L. Gating scheme for single GABA-activated Cl^- channels determined from stability plots, dwell-time distributions, and adjacent-interval durations. *J. Neurosci.* 9:1314–1324, 1989.
25. Twyman, R. E., Rogers, C. J. & Macdonald, R. L. Differential regulation of γ-aminobutyric acid receptor channels by diazepam and phenobarbital. *Ann. Neurol.* 25:213–220, 1989.
26. Eghbali, M., Curmi, J. P., Birnir, B. & Gage, P. W. Hippocampal $GABA_A$ channel conductance is increased by diazepam. *Nature* 388:71–75, 1997.
27. Birnir, B., Everitt, A. B. & Gage, P. W. Characteristics of $GABA_A$ channels in rat dentate gyrus. *J. Mem. Biol.* 142:93–102, 1994.
28. Staley, K. J., Soldo, B. L. & Proctor, W. R. Ionic mechanisms of neuronal excitation by inhibitory $GABA_A$ receptors. *Science* 269:977–981, 1995.
29. Edwards, F. A., Konnerth, A. & Sakmann, B. Quantal analysis of inhibitory synaptic transmission in the dentate gyrus of rat hippocampal slices: A patch-clamp study. *J. Physiol. (Lond.)* 430:213–249, 1990.
30. Pearce, R. A. Physiological evidence for two distinct $GABA_A$ responses in rat hippocampus. *Neuron* 10:189–200, 1993.
31. Tia, S., Wang, J. F., Kotchabhakdi, N. & Vicini, S. Developmental changes of inhibitory synaptic currents in cerebellar granule neurons — role of $GABA_A$ receptor alpha-6 subunit. *J. Neurosci.* 16:3630–3640, 1996.

32. Verdoorn, T. A., Draguhn, A., Ymer, S., Seeburg, P. H. & Sakmann, B. Functional properties of recombinant rat $GABA_A$ receptors depend upon subunit composition. *Neuron* 4:919–928, 1990.
33. Gingrich, K. J., Roberts, W. A. & Kass, R. S. Dependence of the $GABA_A$ receptor gating kinetics on the alpha-subunit isoform — implications for structure-function relations and synaptic transmission. *J. Physiol. (Lond.)* 489:529–543, 1995.
34. Nicoll, R. A., Eccles, J. C., Oshima, T. & Rubia, F. Prolongation of hippocampal inhibitory postsynaptic potentials by barbiturates. *Nature* 258:625–627, 1975.
35. Collingridge, G. L., Gage, P. W. & Robertson, B. Inhibitory post-synaptic currents in rat hippocampal CA1 neurones. *J. Physiol. (Lond.)* 356:551–564, 1984.
36. Gage, P. W. & Robertson, B. Prolongation of inhibitory postsynaptic currents by pentobarbitone, halothane and ketamine in CA1 pyramidal cells in rat hippocampus. *Br. J. Pharmacol.* 85:675–681, 1985.
37. Mody, I., Tanelian, D. L. & MacIver, M. B. Halothane enhances tonic neuronal inhibition by elevating intracellular calcium. *Brain Res.* 538:319–323, 1991.
38. Jones, M. V. & Harrison, N. L. Effects of volatile anesthetics on the kinetics of inhibitory postsynaptic currents in cultured rat hippocampal neurons. *J. Neurophysiol.* 70:1339–1349, 1993.
39. Orser, B. A., Wang, L. Y., Pennefather, P. S., and MacDonald, J. F. Propofol modulates activation and desensitization of $GABA_A$ receptors in cultured murine hippocampal neurons. *J. Neurosci.* 14:7747–7760, 1994.
40. Pearce, R. A. Volatile anesthetic enhancement of paired-pulse depression investigated in the rat hippocampus in vitro. *J. Physiol. (Lond.)* 492.3:823–840, 1996.
41. Galindo, A. Effects of procaine, pentobarbital and halothane on synaptic transmission in the central nervous system. *J. Pharmacol. Exp. Ther.* 169:185–195, 1969.
42. Rock, D. M. & Taylor, C. P. Effects of diazepam, pentobarbital, phenytoin and pentylenetetrazol on hippocampal paired-pulse inhibition *in vivo. Neurosci. Lett.* 65:265–270, 1986.
43. Wolf, P. & Haas, H. L. Effects of diazepines and barbiturates on hippocampal recurrent inhibition. *Naunyn-Schmiedebergs Arch. Pharmacol.* 299:211–218, 1977.
44. Dunwiddie, T. V., Worth, T. S. & Olsen, R. W. Facilitation of recurrent inhibition in rat hippocampus by barbiturate and related nonbarbiturate depressant drugs. *J. Pharmacol. Exp. Ther.* 238:564–575, 1986.
45. Proctor, W. R., Mynlieff, M. & Dunwiddie, T. V. Facilitatory action of etomidate and pentobarbital on recurrent inhibition in rat hippocampal pyramidal neurons. *J. Neurosci.* 6:3161–3168, 1986.
46. Pearce, R. A., Stringer, J. L. & Lothman, E. W. Effect of volatile anesthetics on synaptic transmission in the rat hippocampus. *Anesthesiology* 71:591–598, 1989.
47. Eccles, J. C., Schmidt, R. & Willis, W. D. Pharmacological studies on presynaptic inhibition. *J. Physiol. (Lond.)* 168:500–530, 1963.
48. Miyahara, J. T., Esplin, D. W. & Zablocka, B. Differential effects of depressant drugs on presynaptic inhibition. *J. Pharmacol. Exp. Ther.* 154:119–127, 1966.
49. Nicoll, R. A. The effects of anaesthetics on synaptic excitation and inhibition in the olfactory bulb. *J. Physiol. (Lond.)* 223:803–814, 1972.
50. Scholfield, C. N. Potentiation of inhibition by general anaesthetics in neurones of the olfactory cortex *in vitro. Pflugers. Arch.* 383:249–255, 1980.
51. Scholfield, C. N. A barbiturate induced intensification of the inhibitory potential in slices of guinea-pig olfactory cortex. *J. Physiol. (Lond.)* 275:559–566, 1978.
52. Yoshimura, M., Higashi, H., Fujita, S. & Shimoji, K. Selective depression of hippocampal inhibitory postsynaptic potentials and spontaneous firing by volatile anesthetics. *Brain Res.* 340:363–368, 1985.

53. Puil, E. & el Beheiry, H. Anaesthetic suppression of transmitter actions in neocortex. *Br. J. Pharmacol.* 101:61–66, 1990.
54. Sugiyama, K., Muteki, T. & Shimoji, K. Halothane-induced hyperpolarization and depression of postsynaptic potentials of guinea pig thalamic neurons *in vitro*. *Brain Res.* 576:97–103, 1992.
55. Miu, P. & Puil, E. Isoflurane-induced impairment of synaptic transmission in hippocampal neurons. *Exp. Brain Res.* 75:354–360, 1989.
56. Perouansky, M., Kirson, E. D. & Yaari, Y. Halothane blocks synaptic excitation of inhibitory interneurons. *Anesthesiology* 85:1431–1438, 1996.
57. MacIver, M. B., Mikulec, A. A., Amagasu, S. M. & Monroe, F. A. Volatile anesthetics depress glutamate transmission via presynaptic actions. *Anesthesiology* 85:823–834, 1996.
58. Lukatch, H. S. & MacIver, M. B. Voltage-clamp analysis of halothane effects on $GABA_{A,fast}$ and $GABA_{A,slow}$ inhibitory currents. *Brain Res.* 765:108–112, 1997.
59. Banks, M. I. & Pearce, R. A. Dual actions of volatile anesthetics on GABA(A) IPSCs: Dissociation of blocking and prolonging effects. *Anesthesiology* 90:120–134, 1999.
60. Antkowiak, B. & Heck, D. Effects of the volatile anesthetic enflurane on spontaneous discharge rate and $GABA_A$-mediated inhibition of Purkinje cells in rat cerebellar slices. *J. Neurophysiol.* 77:2525–2538, 1997.
61. Lin, L. H., Chen, L. L., Zirrolli, J. A. & Harris, R. A. General anesthetics potentiate gamma-aminobutyric acid actions on $GABA_A$ receptors expressed by *Xenopus* oocytes: lack of involvement of intracellular calcium. *J. Pharmacol. Exp. Ther.* 263:569–578, 1992.
62. Mihic, S. J., McQuilkin, S. J., Eger, E. I., II, Ionescu, P. & Harris, R. A. Potentiation of gamma-aminobutyric acid type a receptor-mediated chloride currents by novel halogenated compounds correlates with their abilities to induce general anesthesia. *Mol. Pharmacol.* 46:851–857, 1994.
63. MacIver, M. B. & Roth, S. H. Inhalation anaesthetics exhibit pathway-specific and differential actions on hippocampal synaptic responses *in vitro*. *Br. J. Anesth.* 60:680–691, 1988.
64. Harris, R. A., Mihic, S. J., Dildy-Mayfield, J. E. & Machu, T. K. Actions of anesthetics on ligand-gated ion channels: Role of receptor subunit composition. *FASEB J.* 9:1454–1462, 1995.
65. Nakahiro, M., Yeh, J. Z., Brunner, E. & Narahashi, T. General anesthetics modulate GABA receptor channel complex in rat dorsal root ganglion neurons. *FASEB J.* 3:1850–1854, 1989.
66. Yeh, J. Z., Quandt, F. N., Tanguy, J., Nakahiro, M., Narahashi, T. & Brunner, E. A. General anesthetic action on γ–aminobutyric acid-activated channels. *Ann. NY Acad. Sci.* 625:155–173, 1991.
67. Wu, J., Harata, N. & Akaike, N. Potentiation by sevoflurane of the γ-aminobutyric acid induced chloride current in acutely dissociated CA1 pyramidal neurons from rat hippocampus. *Br. J. Pharmacol.* 119:1013–1021, 1996.
68. Wakamori, M., Ikemoto, Y. & Akaike, N. Effects of two volatile anesthetics and a volatile convulsant on the excitatory and inhibitory amino acid responses in dissociated CNS neurons of the rat. *J. Neurophysiol.* 66:2014–2021, 1991.
69. Lambert, J. J., Belelli, D., Hill-Venning, C. & Peters, J. A. Neurosteroids and $GABA_A$ receptor function. *Trends Pharmacol. Sci.* 16:295–303, 1995.
70. Sanna, E., Mascia, M. P., Klein, R. L., Whiting, P. J., Biggio, G. & Harris, R. A. Actions of the general anesthetic propofol on recombinant human $GABA_A$ receptors: Influence of receptor subunits. *J. Pharmacol. Exp. Ther.* 274:353–360, 1995.

71. Hill-Venning, C., Belelli, D., Peters, J. A. & Lambert, J. J. Subunit-dependent interaction of the general anaesthetic etomidate with the $GABA_A$ receptor. *Br. J. Pharmacol.* 120:749–756, 1997.
72. Harrison, N. L., Kugler, J. L., Jones, M. V., Greenblatt, E. P. & Pritchett, D. B. Positive modulation of human $GABA_A$ and glycine receptors by the inhalation anesthetic isoflurane. *Mol. Pharmacol.* 44:628–632, 1993.
73. Lin, L. H., Whiting, P. & Harris, R. A. Molecular determinants of general anesthetic action: Role of $GABA_A$ receptor structure. *J. Neurochem.* 60:1548–1553, 1993.
74. Yeh, J. Z., Tanguy, J., Hamilton, B. J., Carter, D. B. & Brunner, E. A. Direct action of halothane on cloned rat $GABA_A$ receptors is subunit-dependent. *Anesthesiology* 81, A800, 1994.
75. Sincoff, R., Tanguy, J., Hamilton, B., Carter, D., Brunner, E. A. & Yeh, J. Z. Halothane acts as a partial agonist of the $\alpha_6\beta_2\gamma_{2s}$ $GABA_A$ receptor. *FASEB J.* 10:1539–1545, 1996.
76. Mihic, S. J. & Harris, R. A. Inhibition of rho(1) receptor GABAergic currents by alcohols and volatile anesthetics. *J. Pharmacol. Exp. Ther.* 277:411–416, 1996.
77. Jones, M. V., Brooks, P. A. & Harrison, N. L. Enhancement of GABA-activated Cl– currents in cultured rat hippocampal neurones by three volatile anaesthetics. *J. Physiol. (Lond.)* 449:279–293, 1992.
78. Zimmerman, S. A., Jones, M. V. & Harrison, N. L. Potentiation of $GABA_A$ receptor Cl^- current correlates with in vivo anesthetic potency. *J. Pharmacol. Exp. Ther.* 270:987–991, 1994.
79. Yang, J., Isenberg, K. E. & Zorumski, C. F. Volatile anesthetics gate a chloride current in postnatal rat hippocampal neurons. *FASEB J.* 6:914–918, 1992.
80. Santi, M. R., Vicini, S., Eldadah, B. & Neale, J. H. Analysis by polymerase chain reaction of α_1 and α_6 $GABA_A$ receptor subunit mRNA in individual cerebellar neurons after whole-cell recordings. *J. Neurochem.* 63:2357–2360, 1994.
81. Pearce, R. A. & Banks, M. B. Enflurane reduces IPSC amplitude by a postsynaptic mechanism. *Anesthesiology* 87, A628, 1997.
82. Quinlan, J. J., Gallaher, E. J. & Firestone, L. L. Halothane's effects on GABA-gated chloride flux in mice selectively bred for sensitivity or resistance to diazepam. *Brain Res.* 610:224–228, 1993.
83. Neumahr, S., Hapfelmeier, G., Scheller, M., Schneck, H., Franke, C. & Kochs, E. Dual action of isoflurane on the γ-aminobutyric acid (GABA)-mediated currents through recombinant $\alpha\beta_2\gamma_{2L}$-$GABA_A$-receptor channels. *Anesth. Analg.* 90:1184–1190, 2000.
84. Li, X., Czajkowski, C. & Pearce, R. A. Rapid and direct modulation of $GABA_A$ receptors by halothane. *Anesthesiology* 92:1366–1375, 2000.
85. Mody, I., De Koninck, Y., Otis, T. S. & Soltesz, I. Bridging the cleft at GABA synapses in the brain. *Trends Neurosci.* 17:517–525, 1994.
86. Koblin, D. D., Eger, E. I., II, Johnson, B. H., Collins, P., Terrell, R. C. & Speers, L. Are convulsant gases also anesthetics? *Anesth. Analg.* 60:464–470, 1981.
87. Yamashita, M., Ikemoto, Y., Nielsen, M. & Yano, T. Effects of isoflurane and hexafluorodiethyl ether on human recombinant $GABA_A$ receptors expressed in *Sf9* cells. *Eur. J. Pharmacol.* 378:223–231, 1999.
88. Zhu, W. J. & Vicini, S. Neurosteroid prolongs $GABA_A$ channel deactivation by altering kinetics of desensitized states. *J. Neurosci.* 17:4022–4031, 1997.
89. Dilger, J. P., Brett, R. S. & Lesko, L. A. Effects of isoflurane on acetylcholine receptor channels. 1. Single-channel currents. *Mol. Pharmacol.* 41:127–133, 1992.

90. Forman, S. A., Miller, K. W. & Yellen, G. A discrete site for general anesthetics on a postsynaptic receptor. *Mol. Pharmacol.* 48:574–581, 1995.
91. Dilger, J. P., Brett, R. S. & Mody, H. I. The effects of isoflurane on acetylcholine receptor channels. 2. Currents elicited by rapid perfusion of acetylcholine. *Mol. Pharmacol.* 44:1056–1063, 1993.
92. Dilger, J. P., Vidal, A. M., Mody, H. I. & Liu, Y. Evidence for direct actions of general anesthetics on an ion channel protein: A new look at a unified mechanism of action. *Anesthesiology* 81:431–442, 1994.
93. Franks, N. P. & Lieb, W. R. Do general anaesthetics act by competitive binding to specific receptors? *Nature* 310:599–601, 1984.
94. Franks, N. P. & Lieb, W. R. Mapping of general anaesthetic target sites provides a molecular basis for cutoff effects. *Nature* 316:349–351, 1985.
95. Eckenhoff, R. G. Amino acid resolution of halothane binding sites in serum albumin. *J. Biol. Chem.* 271:15521–15526, 1996.
96. Quinlan, J. J., Firestone, S. & Firestone, L. L. Isoflurane's enhancement of chloride flux through rat brain $GABA_A$ receptors is stereoselective. *Anesthesiology* 83:611–615, 1995.
97. Hall, A. C., Lieb, W. R. & Franks, N. P. Stereoselective and non-stereoselective actions of isoflurane on the $GABA_A$ receptor. *Br. J. Pharmacol.* 112 :906–910, 1994.
98. Moody, E. J., Harris, B. D. & Skolnick, P. Stereospecific actions of the inhalation anesthetic isoflurane at the $GABA_A$ receptor complex. *Brain Res.* 615:101–106, 1993.
99. Harris, B. D., Moody, E. J., Basile, A. S. & Skolnick, P. Volatile anesthetics bidirectionally and stereospecifically modulate ligand binding to GABA receptors. *Eur. J. Pharmacol.* 267:269–274, 1994.
100. Harris, B. D., Moody, E. J. & Skolnick, P. Stereoselective actions of halothane at GABA(A) receptors. *Eur. J. Pharmacol.* 341:349–352, 1998.
101. Mihic, S. J., Ye, Q., Wick, M. J., Koltchine, V. V., Krasowski, M. D., Finn, S. E., Mascia, M. P., Valenzuela, C. F., Hanson, K. K., Greenblatt, E. P., Harris, R. A. & Harrison, N. L. Sites of alcohol and volatile anaesthetic action on $GABA_A$ and glycine receptors. *Nature* 389:385–389, 1997.
102. Krasowski, M. D., Koltchine, V. V., Rick, C. E., Ye, Q., Finn, S. E. & Harrison, N. L. Propofol and other intravenous anesthetics have sites of action on the gamma-aminobutyric acid type A receptor distinct from that for isoflurane. *Mol. Pharmacol.* 53:530–538, 1998.
103. Koltchine, V. V., Finn, S. E., Jenkins, A., Nikolaeva, N., Lin, A. & Harrison, N. L. Agonist gating and isoflurane potentiation in the human gamma-aminobutyric acid type A receptor determined by the volume of a second transmembrane domain residue. *Mol. Pharmacol.* 56:1087–1093, 1999.
104. Mascia, M. P., Trudell, J. R. & Harris, A. Specific binding sites for alcohols and anesthetics on ligand-gated ion channels. *Proc. Natl. Acad. Sci. USA* 97:9305–9310, 2000.
105. Greenblatt, E. P. & Meng, X. Differential modulation of chimeric inhibitory receptors by halothane versus isoflurane. *Anesthesiology* 87, A704, 1997.
106. Slater, S. J., Cox, K. J., Lombardi, C. H., Kelly, M. B., Rubin, E. & Stubbs, C. D. Inhibition of protein kinase C by alcohols and anesthetics. *Nature* 364:82–84, 1993.
107. Hemmings, H. C., Jr. & Adamo, A. I. Effects of halothane and propofol on purified brain protein kinase C activation. *Anesthesiology* 81:147–155, 1994.
108. Hemmings, H. C., Jr. & Adamo, A. I. Activation of endogenous protein kinase C by halothane in synaptosomes. *Anesthesiology* 84:652–662, 1996.

109. Chen, Q. X., Stelzer, A., Kay, A. R. & Wong, R. K. $GABA_A$ receptor function is regulated by phosphorylation in acutely dissociated guinea-pig hippocampal neurones. *J. Physiol. (Lond.)* 420:207–221, 1990.
110. Lin, Y. F., Browning, M. D., Dudek, E. M. & Macdonald, R. L. Protein kinase C enhances recombinant bovine $\alpha_1\beta_1\gamma_{2L}$ $GABA_A$ receptor whole-cell currents expressed in L929 fibroblasts. *Neuron* 13:1421–1431, 1994.
111. Li, X. & Pearce, R. A. Effects of halothane on $GABA_A$ receptor kinetics: Evidence for slowed agonist unbinding. *J. Neurosci.* 20:899–907, 2000.
112. Bai, D., Pennefather, P. S., MacDonald, J. F. & Orser, B. A. The general anesthetic propofol slows deactivation and desensitization of $GABA_A$ receptors. *J. Neurosci.* 19:10635–10646, 1999.
113. Mellor, J. R. & Randall, A. D. Frequency-dependent actions of benzodiazepines on $GABA_A$ receptors in cultured murine cerebellar granule cells. *J. Physiol.* 503:353–369, 1997.
114. Rampil, I. J., Mason, P. & Singh, H. Anesthetic potency (MAC) is independent of forebrain structures in the rat. *Anesthesiology* 78:707–712, 1993.
115. Borges, M. & Antognini, J. F. Does the brain influence somatic responses to noxious stimuli during isoflurane anesthesia? *Anesthesiology* 81:1511–1515, 1994.
116. Antognini, J. F. & Schwartz, K. Exaggerated anesthetic requirements in the preferentially anesthetized brain. *Anesthesiology* 79:1244–1249, 1993.
117. Mason, P., Owens, C. A. & Hammond, D. L. Antagonism of the antinocifensive action of halothane by intrathecal administration of $GABA_A$ receptor antagonists. *Anesthesiology* 84:1205–1214, 1996.
118. Dwyer, R., Bennett, H. L., Eger, E. I. & Heilbron, D. Effects of isoflurane and nitrous oxide in subanesthetic concentrations on memory and responsiveness in volunteers. *Anesthesiology* 77:888–898, 1992.
119. Chortkoff, B. S., Bennett, H. L. & Eger, E. I., II. Subanesthetic concentrations of isoflurane suppress learning as defined by the category-example task. *Anesthesiology* 79:16–22, 1993.
120. Gonsowski, C. T., Chortkoff, B. S., Eger, E. I., II, Bennett, H. L. & Weiskopf, R. B. Subanesthetic concentrations of desflurane and isoflurane suppress explicit and implicit learning. *Anesth. Analg.* 80:568–572, 1995.
121. Nicoll, R. A., Malenka, R. C. & Kauer, J. A. Functional comparison of neurotransmitter receptor subtypes in mammalian central nervous system. *Physiol. Rev.* 70:513–565, 1990.
122. Lee, H. K., Dunwiddie, T. V. & Hoffer, B. J. Interaction of diazepam with synaptic transmission in the *in vitro* rat hippocampus. *Naunyn-Schmiedebergs Arch. Pharmacol.* 309:131–136, 1979.
123. Kanter, E. D., Kapur, A. & Haberly, L. B. A dendritic $GABA_A$ mediated IPSP regulates facilitation of NMDA mediated responses to burst stimulation of afferent fibers in piriform cortex. *J. Neurosci.* 16:307–312, 1996.
124. Kapur, A., Lytton, W. W., Ketchum, K. L. & Haberly, L. B. Regulation of the NMDA component of EPSPs by different components of postsynaptic GABAergic inhibition: Computer simulation analysis in piriform cortex. *J. Neurophysiol.* 78:2546–2559, 1997.
125. Tsubokawa, H. & Ross, W. N. IPSPs modulate spike backpropagation and associated (Ca^{2+})(I) changes in the dendrites of hippocampal CA1 pyramidal neurons. *J. Neurophysiol.* 76:2896–2906, 1996.
126. Kapur, A., Pearce, R. A., Lytton, W. W. & Haberly, L. B. $GABA_A$-mediated IPSCs in piriform cortex have fast and slow components with different properties and locations on pyramidal cells. *J. Neurophysiol.* 78:2531–2545, 1997.

127. Thomson, A. M., West, D. C., Hahn, J. & Deuchars, J. Single axon IPSPs elicited in pyramidal cells by three classes of interneurones in slices of rat neocortex. *J. Physiol. (Lond.)* 496:81–102, 1996.
128. Gray, C. M. Synchronous oscillations in neuronal systems: Mechanisms and functions. *J. Comp. Neurosci.* 1:11–38, 1994.
129. Joliot, M., Ribary, U. & Llinas, R. Human oscillatory brain activity near 40 Hz coexists with cognitive temporal binding. *Proc. Natl. Acad. Sci. USA* 91:11748–11751, 1994.
130. Huerta, P. T. & Lisman, J. E. Bidirectional synaptic plasticity induced by a single burst during cholinergic theta oscillation in CA1 *in vitro*. *Neuron* 15:1053–1063, 1995.
131. Bliss, T. V. & Collingridge, G. L. A synaptic model of memory: Long-term potentiation in the hippocampus. *Nature* 361:31–39, 1993.
132. Holscher, C., Anwyl, R. & Rowan, M. J. Stimulation on the positive phase of hippocampal theta rhythm induces long-term potentiation that can be depotentiated by stimulation on the negative phase in area CA1 *in vivo*. *J. Neurosci.* 17:6470–6477, 1997.
133. Traub, R. D., Whittington, M. A., Stanford, I. M. & Jefferys, J. G. R. A mechanism for generation of long-range synchronous fast oscillations in the cortex. *Nature* 383:621–624, 1996.
134. Whittington, M. A., Traub, R. D. & Jefferys, J. G. Synchronized oscillations in interneuron networks driven by metabotropic glutamate receptor activation. *Nature* 373:612–615, 1995.
135. Wang, X. J. & Buzsaki, G. Gamma oscillation by synaptic inhibition in a hippocampal interneuronal network model. *J. Neurosci.* 16:6402–6413, 1996.
136. Buzsaki, G. & Chrobak, J. J. Temporal structure in spatially organized neuronal ensembles: A role for interneuronal networks. *Curr. Opin. Neurobiol.* 5:504–510, 1995.
137. Traub, R. D., Whittington, M. A., Colling, S. B., Buzsaki, G. & Jefferys, J. G. R. Analysis of gamma rhythms in the rat hippocampus *in vitro* and *in vivo*. *J. Physiol. (Lond.)* 493:471–484, 1996.
138. Plourde, G. The effects of propofol on the 40-Hz auditory steady-state response and on the electroencephalogram in humans. *Anesth. Analg.* 82:1015–1022, 1996.
139. Plourde, G. & Villemure, C. Comparison of the effects of enflurane/N_2O on the 40-Hz auditory steady-state response versus the auditory middle-latency response. *Anesth. Analg.* 82:75–83, 1996.
140. Andrade, J., Sapsford, D. J., Jeevaratnum, D., Pickworth, A. J. & Jones, J. G. The coherent frequency in the electroencephalogram as an objective measure of cognitive function during propofol sedation. *Anesth. Analg.* 83:1279–1284, 1996.
141. Antkowiak, B. & Hentschke, H. Cellular mechanisms of gamma rhythms in rat neocortical brain slices probed by the volatile anaesthetic isoflurane. *Neurosci. Lett.* 231:87–90, 1997.
142. Dutton, R. C., Smith, W. D., Rampil, I. J., Chortkoff, B. S., & Eger, E. I., II. Forty-hertz midlatency auditory evoked potential activity predicts wakeful response during desflurane and propofol anesthesia in volunteers. *Anesthesiology* 91:1209–1220, 1999.
143. Whittington, M. A., Jefferys, J. G. & Traub, R. D. Effects of intravenous anaesthetic agents on fast inhibitory oscillations in the rat hippocampus in vitro. *Br. J. Pharmacol.* 118:1977–1986, 1996.

11 Neurochemical Actions of Anesthetics at the $GABA_A$ Receptors

Eric Moody and Phil Skolnick

CONTENTS

The $GABA_A$ receptor complex is the primary inhibitory neurotransmitter system in the vertebrate central nervous system. At the vast majority of GABAergic synapses, binding of GABA to its receptor on the macromolecular complex results in opening of an associated anion channel and chloride entry into the cell. This sequence of events hyperpolarizes the neuron, rendering further depolarization or transmission of impulses less likely. The augmentation of inhibitory pathways represents an attractive model for anesthetic mechanisms (see Chapter 1). Thus, it is not surprising that considerable attention has recently focused on anesthetic actions at the family of $GABA_A$ receptors. The $GABA_A$ receptors are members of a receptor superfamily that include other ligand-gated ion channels such as glycine and nicotonic acetylcholine receptors. Members of this receptor superfamily appear to share both a common transmembrane topology and significant sequence homology.[40]

$GABA_A$ receptors are comprised of distinct but related subunits arranged so that the four transmembrane-spanning domains form a central anion channel (Figure 11.1). The stoichiometry, composition, and arrangement of these subunits remain controversial. However, several lines of evidence are consistent with a pentameric arrangement.[19] This pentameric constitution, together with the large number subunits

0-8493-8555-5/01/$0.00+$.50

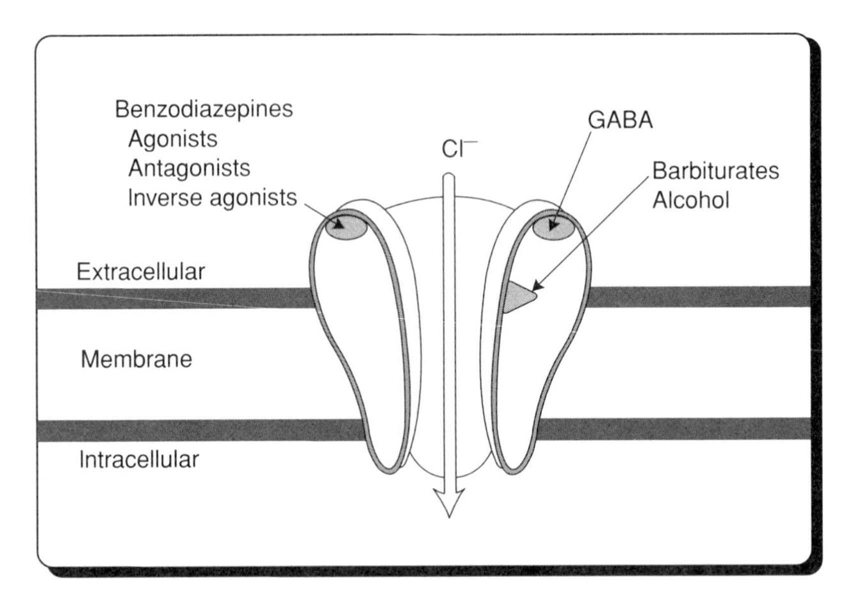

FIGURE 11.1 Structure of a $GABA_A$ receptor in the plasma membrane. Notice that the Cl^- channel is produced by membrane-spanning subunits, and the receptor has the same general structure as other ligand-gated channels such as ACh and glycine receptors.

and splice variants described to date, indicate there is considerable heterogeneity of $GABA_A$ receptor type, function, and number[26,45] in the central nervous system. This heterogeneity can have profound effects on both receptor physiology and pharmacology, and is likely to explain the differences in drug responses (including anesthetic agents) at $GABA_A$ receptors across brain regions.[12]

The major subunit classes described to date include α, β, γ, ρ, δ, and ε. There are subtypes of most of these receptor subunits arising from distinct gene products (e.g., $\beta_{1\text{-}4}$ and $\alpha_{1\text{-}6}$) as well as splice variants (e.g., γ_2 long and γ_2 short). Some of these subunits are very widely distributed in the brain, whereas others are expressed only in well-defined areas. For example, ρ subunits are expressed only in the retina, and α_6 subunits are primarily found in the cerebellum.[26] Thus, despite the possibility for the number of distinct $GABA_A$ receptors to range in the thousands based on the combinatorial principle, immunoprecipitation and immunohistocytochemistry studies indicate the number of receptors is likely less than 20, with receptors containing $\alpha_1\beta_{2/3}\gamma_2$ subunits representing ~50% of the total $GABA_A$ receptor pool.[26] Although the majority of $GABA_A$ receptors are ternary complexes containing α, β, and γ subunits, the stoichiometry of these receptors remains uncertain. Thus there are a number of possible combinations for ternary complexes containing five subunits: for example, two α, two β, and one γ; two α, one β, and two γ; and one α, two β, and two γ. Moreover, the ordering (i.e., positioning) of subunits within the receptor is unknown, and is undoubtedly critical because the extracellular domains of adjacent subunits likely combine to produce the binding domains for ligands such as GABA and benzodiazepines.

Although a number of studies have addressed these issues, there has been no definitive resolution.[56] These issues are far from academic because studies in recombinant systems have unequivocally demonstrated the critical role of subunit composition in defining receptor responses to drugs, including anesthetics. Exemplary of this principle, the type of α subunit appears to be the primary determinant of affinity for benzodiazepine-like molecules, whereas the γ subunit appears more important for determining the efficacy of this class of compounds.[20,49] Moreover, subunit composition also affects a variety of biophysical and electrophysiologic properties of the receptor (e.g., desensitization rate and rectifying properties).[10,14]

11.1 NEUROCHEMICAL STUDIES

To some extent, knowledge of volatile anesthetic actions at the $GABA_A$ receptors has followed insights obtained with nonvolatile agents. Intravenous agents are generally easier to utilize in assay systems because volatile agents have low aqueous solubility and low potency, and determining concentrations of these agents can be technically challenging. A number of initial neurochemical studies utilized subcellular preparations, such as synaptoneurosomes, to study anesthetic actions at these receptors. Synaptoneurosomes include both pre- and postsynaptic elements[15] and have been very useful in the neurochemical study of anesthetic action at the $GABA_A$ receptors. GABA, as well as agonists such as muscimol, activate the $GABA_A$ receptors in synaptoneurosomes, producing chloride movement through receptor-gated channels. This flux of chloride can be measured using $^{36}Cl^-$. In synaptoneurosomes loaded with radioactive chloride, vesicular efflux can be determined. Studies of this nature have been used to complement electrophysiologic studies demonstrating that barbiturates open GABA-gated chloride channels.[41] The use of synaptoneurosomes to measure $^{36}Cl^-$ influx resulted in less variability and a more robust signal. This preparation was used to demonstrate that anesthetic barbiturates increase chloride uptake at therapeutically relevant concentrations that correlate with anesthetic potencies.[42,43] Ethanol and other anesthetic alcohols also enhance $^{36}Cl^-$ uptake, and this action closely parallels their anesthetic potency.[42,43]

With similar techniques, volatile anesthetics have been observed to enhance $^{36}Cl^-$ uptake. Agents including diethylether, halothane, and isoflurane all increase $^{36}Cl^-$ uptake in the absence of GABA.[31] These increases are picrotoxin sensitive, indicating that the observed uptake of $^{36}Cl^-$ is due to activation of GABA-gated chloride channels. Other studies have determined that volatile anesthetics produce similar actions in the presence of GABA.[18] These data were from some of the first studies demonstrating that volatile anesthetics could perturb the $GABA_A$ receptors in a fashion similar to nonvolatile anesthetics. More recently the stereoisomers of isoflurane have been shown to increase $^{36}Cl^-$ uptake in a stereoselective fashion reinforcing the notion that these actions are receptor-mediated events (see Chapter 12).[36]

Limitations of the synaptoneurosome technique include the rapid desensitization of the GABA receptors[4] and the significant interassay variability resulting in relatively large standard error measurements. The procedure is also time consuming compared with radioligand binding techniques. Nonetheless, the data obtained from these synaptoneurosome studies acted as a catalyst to study anesthetic actions at $GABA_A$ receptors with other (e.g., electrophysiologic) techniques. Overall, the results of these early, neurochemical studies correlate very well with the ability of anesthetics to open GABA-gated chloride channels, as determined by electrophysiologic studies (see Chapter 10). The use of chloride-sensitive dyes, which change their emission characteristics with altered anion concentration, has provided an opportunity to determine intracellular chloride concentration using fluorescence techniques.[44] Fluorescent probes have been used to measure ethanol-induced Cl^- flux in cultured cells.[9] This technique also has been used to determine the physiologic responses of cells expressing recombinant receptors.[59] Volatile anesthetics and other anesthetic drugs have not been studied with this technique but it may be a promising method to determine action in cells containing mutated and recombinant receptors.

11.2 RADIOLIGAND BINDING STUDIES

Volatile anesthetics, as well as most other general anesthetics, are not useful as radioligands because of their low potency. Effective radioligands generally possess affinities in the nanomolar range to be used in filtration assays, although less potent ligands can be utilized in centrifugation and scintillation proximity assays. In contrast, volatile anesthetics have potencies in the 200 to 500 μM range (i.e., higher orders of magnitude lower than ideal radioligands). However, drugs that are not effective as radioligands themselves can be utilized to determine their effect on other radioligands at the receptor, and general anesthetics have been widely used in this manner. There are three general binding sites on the $GABA_A$ receptors, each with a class of radioligands that bind to it: the benzodiazepine receptor, the GABA receptor, and the Cl^- channel; each of these can be modulated by anesthetics.

11.3 MODULATION OF RADIOLIGAND BINDING AT THE GABA SITE

The binding of $[^3H]$GABA in recombinant cells is optimal with both α and β subunits.[63] Muscimol is a GABA mimetic that binds to the same site as GABA, resulting in opening of GABA-gated chloride channels. SR95531 is an antagonist that binds to the GABA recognition site but does not result in opening of the chloride channel. "Positive modulators" of the $GABA_A$ receptors (i.e., substances that increase Cl^- flux) usually result in an augmentation of $[^3H]$muscimol binding,[1] and this characteristic can be used as a surrogate marker for receptor modulation.[50] For example, intravenous anesthetics such as pentobarbital augment the binding of $[^3H]$muscimol at pharmacologically relevant concentrations.[25] Propofol also augments binding at this loci, indicating that this is a common feature of intravenous anesthetic action.[63] In addition, the volatile anesthetics isoflurane, enflurane, and halothane enhance $[^3H]$muscimol binding, indicating a similar mechanism of action to the intravenous agents[11,24] (Figure 11.2). Moreover, volatile anesthetics were shown to *inhibit* the binding of $[^3H]$SR95531. This bidirectional modulation of the $GABA_A$

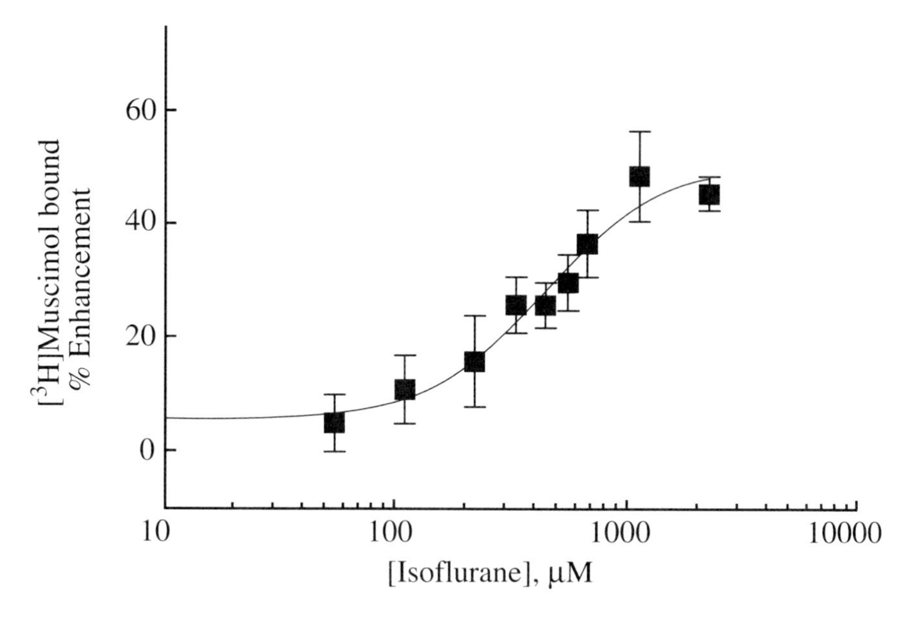

FIGURE 11.2 Augmentation of [^{3}H]muscimol binding by isoflurane in mouse cerebral cortex. Under equilibrium binding conditions, there were concentration-dependent increases in [^{3}H]muscimol binding. Methods are as described elsewhere.[11]

receptors by halothane indicates a degree of specificity for these interactions.[11] This ability of drugs to differentially affect the binding of agonists and antagonists has also been observed at other loci on $GABA_A$ receptors.[3,48,51,55] Saturation analysis revealed that halothane increased the B_{max} of [^{3}H]muscimol by approximately 70% without a significant action of the K_d.[11] These findings suggest that volatile anesthetics allosterically modulate $GABA_A$ receptors, possibly by the recruitment of low-affinity sites, in a manner similar to that observed with barbiturates.[33] This observed increase in B_{max} of [^{3}H]muscimol produced by halothane contrasts with the actions of volatile anesthetics on the increases in [^{3}H]flunitrazepam binding[4] and the inhibition of [^{35}S]TBPS binding,[12] which are effected via a change in K_d.

Isoflurane stereoisomers have also been reported to have differential actions in their ability to inhibit [^{3}H]SR95531 binding.[11] Although the significance of a stereoselective action is discussed in detail elsewhere (see Chapter 12), these findings reveal that the enantiomers of isoflurane can modulate the $GABA_A$ receptors with similar efficacy, albeit with different potencies. The anesthetic concentrations needed for volatile anesthetics to modulate [^{3}H]muscimol binding in these studies are relatively high, with EC_{50} values of approximately five to ten times the blood concentrations associated with clinical anesthesia. Nonetheless, these concentrations are similar to those necessary to produce direct stimulation of $^{36}Cl^-$ uptake in synaptoneurosomes.[31] In addition, although the EC_{50} values are above the pharmacologically relevant range in this model system, statistically significant modulation occurs at much lower concentrations.

11.4 MODULATION OF THE "BENZODIAZEPINE" SITE

Positive modulators at $GABA_A$ receptors enhance [^{3}H]benzodiazepine binding, which has been widely used as a marker for $GABA_A$ receptor activation. Anesthetic drugs, such as pentobarbital, exert dual actions at the benzodiazepine receptors: they augment the action of GABA, and in the absence of GABA, they directly increase channel opening.[47] The ability of agents to enhance benzodiazepine binding in the absence of GABA is also termed the direct action, and the enhancement in the absence of GABA is sometimes referred to as the indirect or augmentative action. A similar phenomenon is observed in electrophysiologic studies. Which of these actions is responsible for the pharmacologically relevant actions of anesthetics is unknown. However, augmentation of GABA actions generally is observed at lower concentrations than is direct gating of the channel. In addition, a number of agents, such as phenobarbital and benzodiazepine, which are not efficacious anesthetics, lack direct actions in electrophysiologic studies, suggesting direct channel opening may be necessary for anesthetic actions. Molecular biology studies, such as those detailed later in the chapter, may provide the opportunity to resolve this question.

These dual actions of anesthetics on [^{3}H]benzodiazepine binding were initially described with barbiturates. Thus, the anesthetic barbiturate, pentobarbital, was observed to augment binding of [^{3}H]benzodiazepines both in the presence and absence of GABA.[47] This property is shared by other intravenous anesthetics as well. For example, propofol, alphaxalone, and etomidate all augment [^{3}H]benzodiazepine binding, and do so via an increase in ligand affinity.[34,46] Volatile anesthetics also enhance [^{3}H]benzodiazepine binding in an analogous fashion. In mouse brain, enflurane, halothane, and isoflurane augment [^{3}H]flunitrazepam binding in a concentration-dependent manner.[12] This action was manifested in both cerebral cortex and cerebellum, although the magnitude of effect and anesthetic potency differed between these tissues, which likely represents a heterogeneity of $GABA_A$ receptors. In this study, volatile anesthetics did not augment the binding of the benzodiazepine antagonist [^{3}H]flumazenil and caused modest reductions in the binding of the benzodiazepine receptor inverse agonists. Other investigators, however, have reported that [^{3}H]flumazenil binding is augmented by volatile anesthetics.[35] There were differences in assay conditions (e.g., chloride concentration) that may account for these differences. Other investigators also have confirmed that volatile anesthetics modulate binding at the benzodiazepine loci.[32] Studies with the optical isomers of isoflurane have also demonstrated that both isomers augment [^{3}H]flunitrazepam binding and that there are modest differences in potency between the enantiomers.[28]

11.5 MODULATION OF CAGE CONVULSANT BINDING

A third group of radioligands available for study of the $GABA_A$ receptors is the cage convulsants, exemplified by picrotoxin and *t*-butylphosphorothionate (TBPS). These drugs bind in or near the channel and block it. TBPS is more potent than picrotoxin and can be readily used in filtration assays.[39] Agents such as GABA, which activate

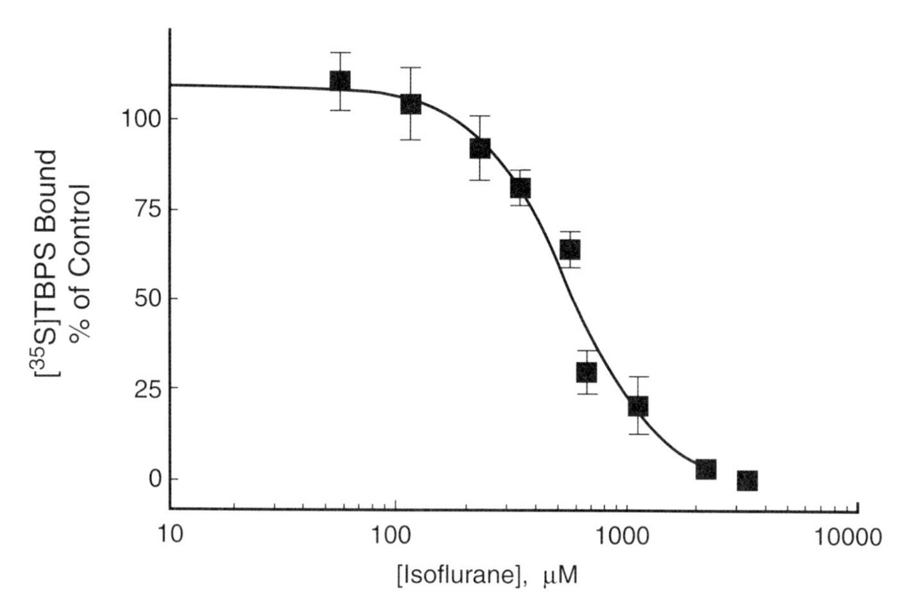

FIGURE 11.3 Inhibition of [^{35}S]TBPS binding by isoflurane in mouse cortical tissue. Under equilibrium conditions at 25°C isoflurane administration resulted in a concentration-dependent decrease in [^{35}S]TBPS binding. Methods are as previously described.[31]

GABA$_A$ receptors, inhibit [^{35}S]TBPS binding. Anesthetic barbiturates inhibit [^{35}S]TBPS binding in a manner that correlates with their anesthetic potency.[55] The enantiomers of some barbiturates have opposite effects, with one isomer having anesthetic properties and the other convulsant actions. These drugs have been reported to have different effects on [^{35}S]TBPS binding. Although both drugs inhibit [^{35}S]TBPS binding, the anesthetic isomers do so in a competitive fashion and the convulsant via a noncompetitive mechanism.[55] Other intravenous anesthetics including propofol, etomidate, chlormethiazole, and neurosteroids also inhibit [^{35}S]TBPS binding.[63] Inhalational anesthetics share this property as well and their potencies to competitively inhibit [^{35}S]TBPS binding correlate robustly with their anesthetic potencies[31] (Figure 11.3). Although both stereoisomers of isoflurane completely inhibited [^{35}S]TBPS binding to mouse cortical tissue, there was no significant potency difference between the isomers. This contrasts with the effects of these drugs at other loci on GABA$_A$ receptors.[28] The concentration of volatile agents needed to affect [^{35}S]TBPS is generally higher than that required to modulate binding at the other loci, although typically there is some inhibition within the clinical range. This suggests that this measure is somewhat less sensitive to perturbation than other sites.

11.6 NEUROCHEMICAL STUDIES USING RECOMBINANT RECEPTORS

As described above, the large number of GABA$_A$ subunits can potentially result in many receptors with distinct characteristics. Recombinant GABA$_A$ receptor subunits

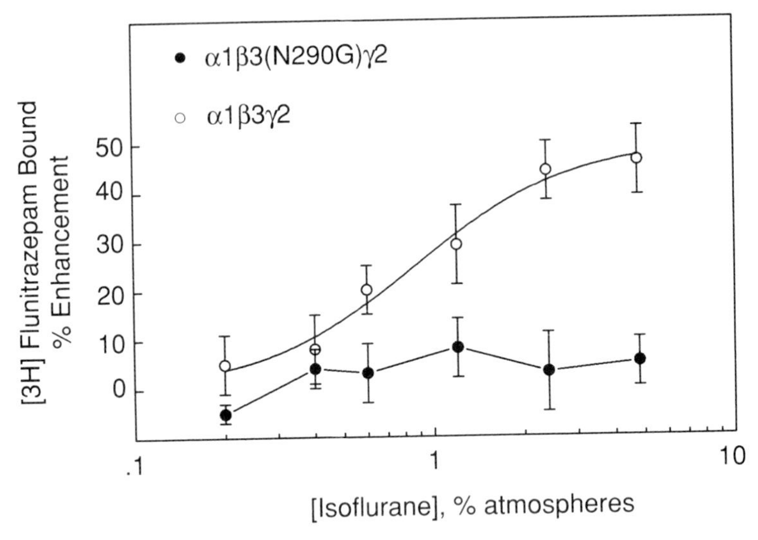

FIGURE 11.4 Isoflurane action is modulated by mutation of residue 290 in the rat β_3 subunit. Using site-directed mutagenesis, a glycine was substituted for asparagine. [^{3}H]Flunitrazepam binding was used as a marker for positive modulation of the receptor.[14] In HEK cells transfected with $\alpha_1\beta_3\gamma_2$ subunits, isoflurane administration resulted in a robust enhancement of binding with the wild-type β_3 subunit. However, in the presence of the glycine mutation there was no enhancement.

expressed in mammalian cell lines (e.g., HEK 293 cells) can be used in radioligand binding studies with the three classes of ligands discussed above (Figure 11.4). For [^{3}H]benzodiazepine binding, α and γ subunits must be coexpressed, whereas for [^{3}H]muscimol and [^{35}S]TBPS binding, α and β subunits are needed. Using [^{3}H]flunitrazepam binding as a marker for positive modulation of $GABA_A$ receptors, the volatile anesthetics were shown to augment binding in the presence of $\alpha_1\gamma_2$ and $\alpha_1\beta_3\gamma_2$ subunits.[14] There were similar requirements for halothane, isoflurane, and enflurane. In the presence of the β_3 subunit, action with halothane was enhanced over that with $\alpha\gamma$ subunits alone. The subunit requirements of the volatile anesthetics contrasted with those for the intravenous anesthetics pentobarbital and alphaxalone. The latter agents required a β_2 or β_3 subunit in order to augment [^{3}H]flunitrazepam binding. Thus, different classes of anesthetics have different subunit requirements for their pharmacologic actions, which may account for differences observed among general anesthetics.

11.7 MUTATIONAL STUDIES WITH $GABA_A$ RECEPTORS

The above discussion indicates that all $GABA_A$ receptors and their subunits are *not* equally sensitive to anesthetic perturbation. Specifically for the barbiturates, a β_2 or β_3 subunit seemed to be required for their actions. On the other hand, for volatile anesthetics, the β subunits were not required, but isoflurane actions were present in

$\alpha\gamma$ combinations. Yang reported that in homomeric receptors comprised of β_3 subunits, barbiturate-induced currents were present, whereas they were absent in β_1-expressing cells.[5,62] We hypothesized that the transmembrane portions of the subunit would be most likely to contain putative active loci, and that the TM2 region that lines the channel would be a likely target for the anesthetics. β_3 and β_1 subunits have extensive sequence homology and in the TM2/TM3 regions they vary only at a single amino acid. This residue was mutated in the β_3 subunit to convert it to the residue of the β_1 subunit to produce β_3 (A290S). When coexpressed with α_1 and γ_2, β_3(A290S)-containing cells retained normal responses to GABA, volatile anesthetics, as well as propofol, pentobarbital, and alphaxalone. However, the direct actions of etomidate were absent in both electrophysiologic and neurochemical measures.[29] This indicated that the asparagine residue at position 290 was crucial for the direct actions of etomidate. Others have made similar mutations,[2,6,27,62] although Belelli and colleagues[2] report a complete loss of direct and indirect action of etomidate with the serine substitution. The reasons for these discrepancies are not clear because similar preparations were used. Yang and Cestari[62] reported loss of barbiturate responses with their mutations, but that may represent a characteristic unique to the β homomeric receptors rather than the heteromeric receptors used by others.

Other investigators have used chimeric receptors to narrow down the active site for volatile anesthetic to a portion of the transmembrane regions of the α subunit.[27] Recombinant subunits were comprised of portions of the ρ subunit, which is not modulated by anesthetic and the α_2. With subsequent mutational analysis, it was determined that two distinct amino acids, α_2(270) and α_2(291), were crucial for the actions of enflurane and isoflurane.[27] These loci are located on the TM2 and TM3 portions of the α_2 subunit. Mutations at these loci did not alter the actions of GABA or propofol. Subsequent mutations have identified a homologous site on the β_1 subunit that is crucial for the action of isoflurane.[22] They also identified a locus that was distinct from that for isoflurane that was crucial for propofol. The observation that two amino acids separated by 21 residues were both crucial for the actions of the volatile anesthetics led the authors to hypothesize that these two loci, on TM2 and TM3, respectively, are adjacent in the tertiary structure of the receptors, and comprise a binding pocket for the volatile anesthetics. Of interest is that the mutations on the α_2 subunit for the volatile agents and those identified on the β_3 subunit for etomidate are at homologous locations: both are located on TM2 approximately five residues from the extracellular surface. This suggests that this general location may be a common site (a "hot spot") for modulation of the receptor by anesthetic drugs (Figure 11.5).

11.8 PHARMACOLOGIC STUDIES AT THE $GABA_A$ RECEPTORS

Long sleep (LS) and short sleep (SS) mouse lines were selectively bred based on their response to ethanol and their altered $GABA_A$ receptors. Anesthetic potencies have been determined in these mouse lines, and LS mice, which are more sensitive to the effects of ethanol, are also more sensitive to most anesthetics.[57] Although the precise differences in $GABA_A$ receptors between these mouse lines is unknown,

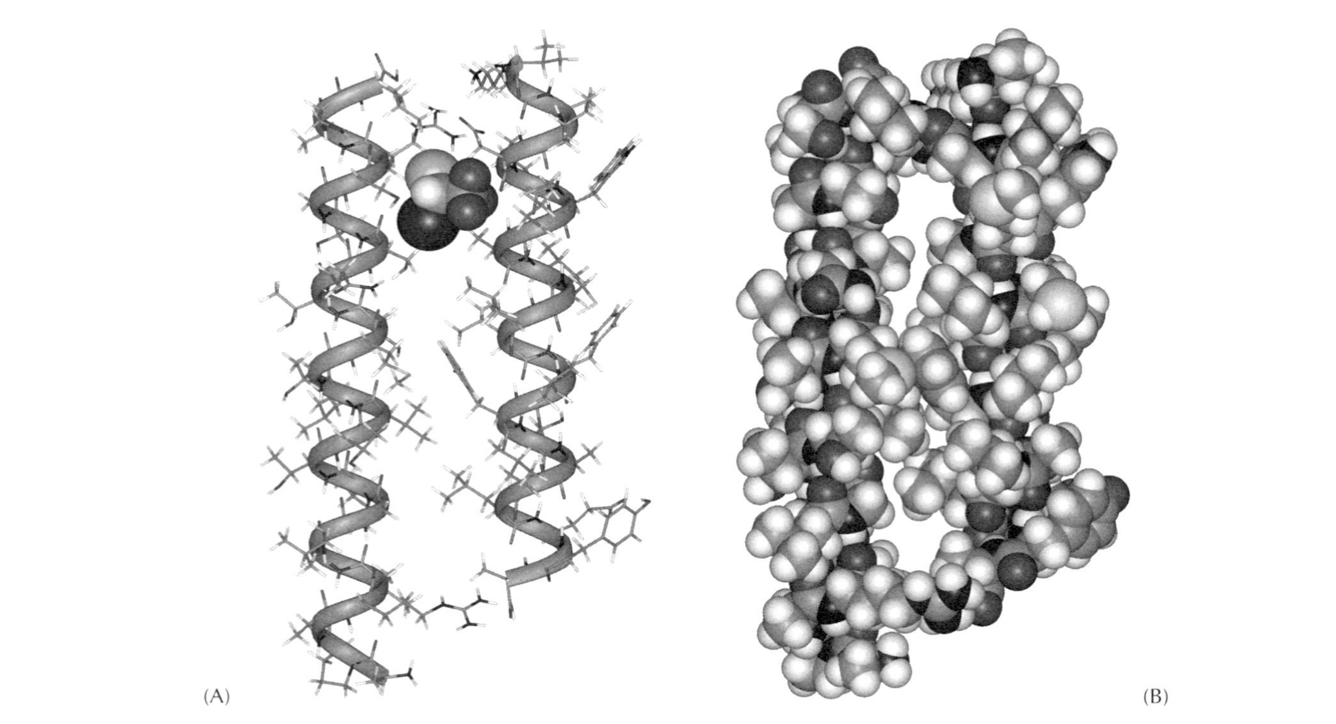

FIGURE 11.5 A molecular model of the crucial site for volatile anesthetics on the TM2/3 subunits of the GABA alpha 1 subunit. TM2 is on the left and TM3 on the right. The ion channel pore is located to the left of TM2. (A) The amino acids are rendered as stick structures overlaying the alpha helices. Halothane is depicted as a space-filling (van der Waals) surface molecule in the site of the discussed mutations. (B) The same structure depicted as a space filling molecule. The anesthetic is omitted to demonstrate the space that is occupied in panel A. (This figure was produced by Dr. James Trudell, Stanford University and is used with his permission.)

these data reinforce the hypothesis that alteration in the $GABA_A$ receptor function affects sensitivity to anesthetics.

The DS and DI (diazepam-sensitive and diazepam-insensitive) mouse lines have been selectively bred for differential sensitivity to diazepam. In these mouse lines halothane is more potent in diazepam-sensitive animals,[37] which is again consistent with the hypothesis that $GABA_A$ receptors are involved in anesthesia. The AT/ANT rat lines were bred to be alcohol tolerant and alcohol nontolerant. A genetic difference has been identified between these lines at a specific amino acid on the α_6 subunit, and it has been hypothesized that this mutation is responsible for the phenotypic differences in alcohol response.[21] It will be of interest to determine whether these strains have differential sensitivity to anesthetics, especially in view of the observation that a knock-out of the α_6 does not alter anesthetic potency.[17] Other investigators have found that rats tolerant to ethanol show relative resistance to barbiturates.[7] This cross-tolerance suggests a common mechanism of action. High-alcohol-sensitive (HAS) rats have also been found to be more sensitive to both intravenous and inhalational anesthetics.[8]

The benzodiazepine Ro 15-4513 is an inverse agonist that has been reported to have antialcohol properties both *in vivo* and *in vitro*.[13,23,52] Because of these properties, it was hypothesized that this drug might reverse the anesthetic effects of volatile anesthetics as well. In methoxyflurane-anesthetized mice, Ro 15-4513 shortened the return of the righting reflex in a dose-dependent fashion.[30] This reversal of volatile anesthetic effects by Ro 15-4513 was not complete, decreasing the sleep time by approximately 40%. The significance of these findings is unclear, however, because full inverse benzodiazepine agonists and antagonists did not alter the recovery time. Nonetheless, because the antagonist flumazenil blocked these actions of Ro 15-4513, the effect was mediated via benzodiazepine receptors. The identification of "diazepam-insensitive" benzodiazepine receptors probably containing α_6 subunits may account for some of these effects.[60,61]

11.9 KNOCK-OUT STUDIES

Several homologous recombination studies have produced mice lacking specific $GABA_A$ receptor subunits. A "knock-out" of the α_6 subunit was performed because a mutation in that subunit has been implicated in the AT/ANT rat lines bred for tolerance to alcohol. Mice who do not express the α_6 subunit did not have altered sensitivity to anesthetics, suggesting that the α_6 subunit is not crucial for anesthetic action.[17] It is possible, however, that this subunit influences anesthetic action in the normal state, but in the animal lacking the subunit its absence has been compensated for by either another subunit or some postreceptor phenomenon. This suggests the possibility that the altered anesthetic sensitivity of the AT/ANT rat strains is due not to the known mutation in the α_6 subunit but perhaps to some other difference(s) between the strains.

Knock-out mice lacking a β_3 subunit also have been produced.[16] This subunit has been implicated in the actions of general anesthetics in studies utilizing recombinant receptors. Moreover, the β_3 subunit is expressed in many brain areas and is likely an important constituent of native $GABA_A$ receptors.[14] Knock-out of the β_3

subunit was lethal in more than 90% of offspring. However, those surviving to adulthood were observed to be hyperactive and tolerant to etomidate.[38] Although the interpretation of these data are difficult in view of the significant neurologic impairment of these animals, the finding of relative resistance to etomidate is consistent with the mutational studies discussed above, which identified the β_3 subunit as being crucial for intravenous anesthetic action *in vivo*. Whether these animals exhibit altered sensitivity to volatile anesthetics is unknown. Such a determination, however, would be very useful, because it may clarify the relative contribution of the β_3 subunit to anesthesia. *In vitro* studies with recombinant receptors have shown that volatile anesthetics do not require the β subunit for action.[14] Nonetheless, the presence of a β subunit in α/γ combinations enhances the actions of volatile anesthetics. Therefore, altered volatile anesthetic sensitivity in these knock-out mice would suggest an additional loci for volatile anesthetics on the β subunit. Alternatively, the β subunit could result in the expression of different numbers of receptors. The production of mice strains that contain specific mutations in their receptors will be useful in resolving these issues.

These knock-out mice lines have a number of limitations. Deletion of an entire subunit may alter underlying receptor function throughout the brain, with unexpected consequences. In addition, other subunits may compensate for the lack of a single subunit, such as α_2 substituting for β_3, which may obscure expected alterations in drug sensitivity. The use of "knock-in" techniques, in which specific amino acids are altered in a receptor that is otherwise normal, avoids some of these difficulties. Now that specific loci implicated in anesthetic action have been identified *in vitro*, it is possible to engineer other mouse lines with altered receptors. This will provide an opportunity to determine the significance of these loci with regard to anesthetic action *in vivo*. This will allow evaluation of the role of the $GABA_A$ receptors in anesthesia as well as rationally guide future anesthetic drug development.

11.10 FUTURE DIRECTIONS

Identification of the specific sites of general anesthetic action on the $GABA_A$ receptors may have multiple benefits. It will allow determination of whether mutations at these loci are responsible for the variations in general anesthetic sensitivity that are clinically observed. If, for instance, amino acid heterogeneity at anesthetic binding sites correlates with anesthetic potency, then screening tests could be developed that would identify individuals at risk for altered anesthetic susceptibility under clinical situations. Moreover, identification of the requirements for anesthetic interaction at the molecular level will provide insight into the anesthetic actions at other related loci. As noted above, the $GABA_A$ receptors have extensive sequence homology with 5HT, glycine, and ACh receptors, and anesthetics are known to modulate those receptor systems. Comparison of the identified sites on the $GABA_A$ receptors to homologous sequences on these related receptors will be useful in defining anesthetic actions at these other sites.

General anesthetics have significant side effects, which are reflected in their low therapeutic indices.[58] A major side effect of general anesthetics is their cardiovascular depression. Knowledge of the molecular requirements of anesthetic actions will enable better prediction of anesthetic interactions at these side effect loci. For example, anesthetics are known to modulate L-type calcium channels, and it is possible that their sites of interaction in heart muscle are similar to those on CNS receptors. Insights of this nature could lead to more powerful modeling techniques as well as *in vitro* screening tests that would correlate with clinical side effects. In any event, once the relevant receptors responsible for general anesthesia are elucidated, the role of specific neural pathways and anatomic regions of the CNS can be addressed.

REFERENCES

1. Beaumont, K., W. Chilton, et al. (1978). "Muscimol binding in rat brain: Association with synaptic GABA receptors." *Brain Res.* 148: 153–162.
2. Belelli, D., J. J. Lambert, et al. (1997). "The interactions of the general anesthetic etomidate with the gamma-aminobutyric acid type A receptor is influenced by a single amino acid." *Proc. Natl. Acad. Sci. USA* 94: 11031–11036.
3. Braestrup, C., R. Schmiechen, et al. (1982). "Interaction of convulsive ligands with benzodiazepine receptors." *Science* 216: 1241–1243.
4. Cash, D. J. and K. Subbarao. (1987). "Desensitization of aminobutyric acid receptor from rat brain: Two distinguishable receptors on the same membrane." *Biochemistry* 26: 7556–7562.
5. Cestari, I. N., I. Uchida, et al. (1996). "The agonistic action of pentobarbital on $GABA_A$ beta-subunit homomeric receptors." *Neuroreport* 7: 943–947.
6. Cestari, I. N. and J. Yang. (1996). "Pentobarbital opening of a homomeric chloride ion channel requires a specific asparagine residue." *Abstr. Soc. Neurosci.* 22: #511.4.
7. Curran, M., M. Newman, et al. (1988). "Barbiturate anesthesia and alcohol tolerance in a rat model." *Anesth. Analg.* 67: 868–871.
8. Deitrich, R., L. Draski, et al. (1994). "Effect of pentobarbital and gaseous anesthetics on rats selectively bred for ethanol sensitivity." *Pharmacol. Biochem. Behav.* 47: 721–725.
9. Engblom, A. C., I. Halopainen, et al. (1991). "Ethanol-induced Cl^- flux in rat cerebellar granule cells as measured by a flourescent probe." *Brain Res.* 568: 55–60.
10. Granja, R. S. M., C.S. Knauer, P. Skolnick. (1998). "Anomalous rectifying properties of 'diazepam-insensitive' $GABA_A$ receptors." *Eur. J. Pharmacol.* 345(3): 315–321.
11. Harris, B., E. J. Moody, et al. (1994). "Volatile anesthetics bidirectionally and stereospecifically modulate ligand binding at $GABA_A$ receptors." *Eur. J. Pharmacol. Mol. Pharmacol.* Section 267: 269–274.
12. Harris, B., G. Wong, et al. (1993). "Neurochemical actions of inhalational anesthetics at the $GABA_A$ receptor complex." *J. Pharmacol. Exp. Ther.* 265: 1392–1398.
13. Harris, B. D., E. J. Moody, et al. (1995). "Contribution of "diazepam-insensitive" $GABA_A$ receptors to the alcohol antagonist properties of Ro 15–4513 and related imidazobenzodiazepines." *Pharmacol. Biochem. Behav.* 52: 113–118.
14. Harris, B. D., G. Wong, et al. (1995). "Different subunit requirements for volatile and nonvolatile anesthetics at gamma-aminobutyric acid type A receptors." *Mol. Pharmacol.* 47: 363–367.
15. Hollingsworth, E., E. McNeal, et al. (1985). "Biochemical characteristization of a filtered synaptoneurosome preperation from guinea pig cerebral cortex: Cyclic adenosine 3' 5' -monophosphate generating systems, receptors and enzymes." *J. Neurosci.* 5: 2240–2253.

16. Homanics, G. E., T. M. DeLorey, et al. (1997). "Mice devoid of gamma-aminobutyrate type A receptor beta$_3$ subunit have epilepsy, cleft palate, and hypersensitive behavior." *Proc. Natl. Acad. Sci. USA* 94(8): 4143–4148.
17. Homanics, G. E., C. Ferguson, et al. (1997). "Gene knockout of the alpha$_6$ subunit of the gamma-aminobutyric acid type A receptor: Lack of effect on responses to ethanol, pentobarbital, and general anesthetics." *Mol. Pharmacol.* 51: 588–596.
18. Huidobro-Toro, J. P., V. Bleck, et al. (1987). "Neurochemical actions of anesthetic drugs on the gamma-aminobutyric acid receptor-chloride channel complex." *J. Pharmacol. Exp. Ther.* 242: 963–969.
19. Knight, A. R., Hartnett, C., Marks, M., Brown, D., Gallager, J., Tallman, T., and Rambhadran, V. (1998). "Molecular size of recombinant alpha$_1$beta$_1$ and alpha$_1$beta$_1$gamma$_2$ GABA$_A$ receptors expressed in Sf9 cells." *Receptors Channels.* 6(1): 1–18.
20. Knoflach, F., K. H. Backus, et al. (1992). "Pharmacological and electrophysiological properties of recombinant GABA$_A$ receptor comprising the α_3, β_1, and γ_2 subunits." *Eur. J. Neurosci.* 4: 1–9.
21. Korpi, E. R., C. Kleingoor, et al. (1993). "Benzodiazepine-induced motor impairment linked to point mutation in cerebellar GABA$_A$ receptor." *Nature* 361: 356–359.
22. Krasowski, M., S. Finn, et al. (1997). "Distinct molecular sites of action of isoflurane and propofol in the human GABA$_A$ receptor." *Anesthesiology* 87(3A): A614.
23. Lister, R. (1988). "Antagonism of the behavioral effects of ethanol, sodium pentobarbital, and Ro 15-4513 by the imidazodiazepine Ro 15-3505." *Neurosci. Res. Comm.* 2: 85–92.
24. Longoni, B., G. C. Demontis, et al. (1993). "Enhancement of γ-aminobutyric acid A receptor function and binding by the volatile anesthetic halothane." *J. Pharmacol. Exp. Ther.* 266: 153–159.
25. Maksay, G., M. Nielsen, et al. (1986). "The enhancement of diazepam and muscimol binding by pentobarbital and (+)-etomidate: Size of the molecular arrangement estimated by electron irradation inactivation of rat cortex." *Neurosci. Lett.* 70: 116–120.
26. McKernan, R. M. and P. J. Whiting. (1996). "Which GABA-receptor subtypes really occur in the brain?" *Trends Neurosci.* 19: 139–143.
27. Mihic, S. J., Q. Ye, et al. (1997). "Sites of alcohol and volatile anaesthetic action on GABA$_A$ and glycine receptors." *Nature* 389: 385–389.
28. Moody, E. J., B. Harris, et al. (1993). "Stereospecific actions of the inhalation anesthetic isoflurane at the GABA$_A$ receptor complex." *Brain Res.* 615: 101–106.
29. Moody, E. J., C. Knauer, et al. (1997). "Distinct loci mediate the direct and indirect actions of the anesthetic etomidate at GABA$_A$ receptors." *J. Neurochem.* 69: 1310–1313.
30. Moody, E. J. and P. Skolnick. (1988). "The imidazodiazepine Ro 15-4513 antagonizes methoxyflurane anesthesia." *Life Sci.* 43: 1269–1276.
31. Moody, E. J., P. D. Suzdak, et al. (1988). "Modulation of the benzodiazepine/gamma-aminobutyric acid receptor complex by inhalational anesthetics." *J. Neurochem.* 51: 1386–1393.
32. Nakao, S., T. Arai, et al. (1991). "Halothane enhances the binding of diazepam to synaptic membranes from rat cerebral cortex." *Acta Anaesth. Scand.* 35: 205–207.
33. Olsen, R. W. and A. M. Snowman. (1982). "Chloride-dependent enhancement by barbiturates of γ-aminobutyric acid receptor binding." *J. Neurosci.* 2: 1812–1821.
34. Prince, R. J. and M. A. Simmonds. (1992). "Propofol potentiates the binding of [^{3}H]flunitrazepam to the GABA$_A$ receptor." *Brain Res.* 596: 238–242.
35. Quinlan, J. J. and L. L. Firestone. (1992). "Ligand-dependent effects of ethanol and diethylether at brain benzodiazepine receptors." *Pharmacol. Biochem. Behav.* 42: 787–790.

36. Quinlan, J. J., S. Firestone, et al. (1995). "Isoflurane's enhancement of chloride flux through rat brain gamma-aminobutyric acid type A receptors is stereoselective." *Anesthesiology* 83: 611–615.
37. Quinlan, J., G. Homanics, et al. (1997). "Sensitivity to intravenous anesthetics in mice lacking the $beta_3$ or $alpha_6$ subunit." *Fifth International Conference on Molecular and Celllular Mechanisms of Anaesthesia* 5(T-76).
38. Quinlan, J. J., K. Jin, et al. (1994). "Halothane sensitivity in replicate mouse lines selected for diazepam sensitivity or resistance." *Anesth. Analog.* 79(5): 927–932.
39. Ramanjaneyulu, R. and M. Ticku. (1984). "Binding characteristics and interactions of depressant drugs with [^{35}S] t-butylbicyclophosphorothionate, a ligand that binds to the picrotoxin site." *J. Neurochem.* 42: 221–229.
40. Schofield, P. R., M. G. Darlison, et al. (1987). "Sequence and functional expression of the $GABA_A$ receptor shows a ligand-gated receptor super-family." *Nature* 328: 221–227.
41. Schwartz, R., J. Jackson, et al. (1985). "Characterization of barbiturate-stimulated chloride efflux from rat brain synaptoneurosomes." *J. Neurosci.* 5: 2963–2970.
42. Schwartz, R. B., P. D. Suzdak, et al. (1986). "GABA and barbiturate mediated $^{36}Cl^-$ uptake in rat brain synaptoneurosomes: Evidence for rapid desensitization of the GABA receptor coupled Cl{+-} ion channel." *Mol. Pharmacol.* 30: 419–426.
43. Schwartz, R. D., P. D. Suzdak, et al. (1988). "Gamma-aminobutyric acid (GABA) and barbiturate-medicated [$^{36}Cl^-$] uptake in rat brain synaptoneurosomes: Evidence for rapid desensitization of the GABA receptor-coupled chloride ion channel." *Mol. Pharmacol.* 30: 419–426.
44. Schwartz, R. D. and Y. Xiao. (1995). "Optical imaging of intracellular chloride in living brain slices." *J. Neurosci. Meth.* 62: 185–192.
45. Sieghart, W. (1989). "Multiplicity of $GABA_A$-benzodiazepine receptors." *Trends Pharmacol. Sci.* 10: 407–411.
46. Simmonds, M. (1991). "Modulation of the $GABA_A$ receptor by steroids." *Semin. Neurosci.* 3: 231–239.
47. Skolnick, P., V. Moncada, et al. (1981). "Pentobarbital: Dual actions to increase brain benzodiazepine receptor affinity." *Science* 211: 1448–1450.
48. Skolnick, P., M. Schweri, et al. (1982). "An *in vitro* test which differentiates benzodiazepine "agonist" and "antagonist"." *Eur. J. Pharmacol. Mol. Pharmacol. Section* 78: 133–136.
49. Slany, A., J. Zezula, et al. (1995). "Allosteric modulation of [^3H]flunitrazepam binding to recombinant $GABA_A$ receptors." *Eur. J. Pharmacol. Mol. Pharmacol. Section* 291: 99–105.
50. Squires, R. (1983). "Benzodiazepine receptor multiplicity." *Neuropharmacology.* 22(12B): 1443–50.
51. Supavilai, P. and M. Karobath. (1984). [^{35}S]T-butylbicyclophosphorothionate binding sites are constituents of the γ-aminobutyric acid benzodiazepine receptor complex." *Mol. Pharmacol.* 4: 1193–1200.
52. Suzdak, P. D., J. R. Glowa, et al. (1986). "A selective imidazobenzodiazepine antagonist of ethanol in the rat." *Science* 234: 1243–1247.
53. Suzdak, P. D., R. Schwartz, et al. (1988). "Alcohols stimulate γ-aminobutyric acid receptor mediated chloride uptake in brain vesicles: Correlation with intoxication potency." *Brain Res.* 444: 340–345.
54. Suzdak, P. D., R. D. Schwartz, et al. (1986). "Ethanol stimulates γ-aminobutyric acid-mediated chloride transport in rat brain synaptoneurosomes." *Proc. Natl. Acad. Sci. USA* 83: 4971–4975.

55. Ticku, M. and R. Ramanjaneyulu. (1984). "Differential interactions of GABA agonists, depressant and convulsant drugs with [^{35}S]-t-butylbicyclophosphorothionate binding sites in cortex and cerebellum." *Pharmacol. Biochem. Behav.* 21: 151–158.
56. Tretter, V., E. Noosha, et al. (1997). "Stoichiometry and assembly of a recombinant $GABA_A$ receptor subtype." *J. Neurosci.* 17: 2728–2737.
57. Wehner, J., J. Pounder, et al. (1992). "A recombinant inbred strain analysis of sleep-time responses to several sedative-hypnotics." *Alcohol Clin. Exp. Res.* 16: 522–528.
58. Wolfson, B., W. D. Hetrick, et al. (1978). "Anesthetic indices — further data." *Anesthesiology* 48: 187–190.
59. Wong, G., Y. Sei, et al. (1992). "Stable expression of type 1 gamma-aminobutyric $acid_A$/benzodiazepine receptors in a transfected cell line." *Mol. Pharmacol.* 42: 996–1003.
60. Wong, G. and P. Skolnick. (1992). "High affinity ligands for "diazepam-insensitive" benzodiazepine receptors." *Eur. J. Pharmacol. Mol. Pharmacol. Section* 225: 63–68.
61. Wong, G. and P. Skolnick. (1992). "Ro 15-4513 binding to $GABA_A$ receptors: Subunit composition determines ligand efficacy." *Pharmacol. Biochem. Behav.* 42: 107–110.
62. Yang, J. and I. N. Cestari. (1996). "Pentobarbital opening of a homomeric chloride ion channel requires a specific asparagine residue in the M2-M3 extracellular linker region." *Abstr. Amer. Soc. Anesthesiol.* 65: A627.
63. Zezula, J., A. Slany, et al. (1996). "Interaction of allosteric ligand with $GABA_A$ receptors containing one, two, or three different subunits." *Eur. J. Pharmacol.* 310: 207–214.

12 Stereoselective Actions of Volatile Anesthetics

Phil Skolnick and Eric J. Moody

CONTENTS

12.1 INTRODUCTION

Despite a history of clinical use spanning two centuries, the molecular mechanisms by which inhalation agents produce anesthesia remain a subject of controversy. Several clinically useful inhalation agents possess centers of asymmetry, but the optically resolved isomers of halothane and isoflurane have only recently become available.[20,27] The demonstration of a significant difference in the anesthetic potencies of (+) and (–) isoflurane provides compelling evidence for the hypothesis that proteins, rather than lipids, are the primary sites of anesthetic action. Moreover, the optically active isomers of volatile anesthetics provide new tools to discriminate among putative molecular targets of anesthesia. A difference in the anesthetic potencies of (+) and (–) isoflurane, when taken together with an apparent lack of stereoselectivity in cardiac tissue, raises the possibility that an optically active volatile agent may have clinical advantages over currently available racemic mixtures.

0-8493-8555-5/01/$0.00+$.50

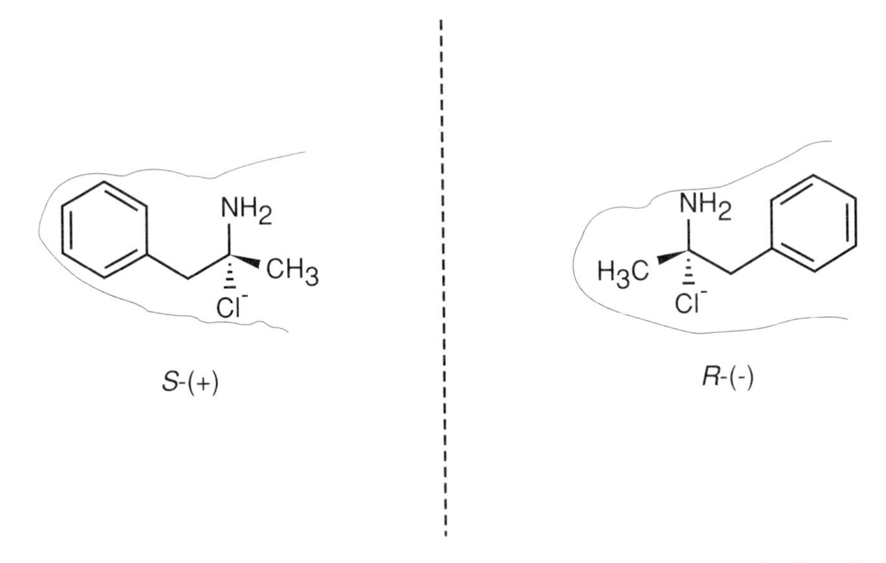

FIGURE 12.1 Principle of stereoselectivity. Here the two forms of a compound with a chiral center shown within a receptor pocket. This demonstrates that different stereoisomers can have different interactions at a receptor.

12.2 BACKGROUND

The stereoisomers of drugs that possess one or more chiral centers (Figure 12.1) often exhibit marked affinity differences at pharmacologically relevant loci, including enzymes and receptors. Such stereoselective actions are readily demonstrable both *in vivo* and *in vitro*,[14,22,28,31] and have proven useful in confirming a receptor-based action for drugs such as the opiates and benzodiazepines. Although the structures of commonly used inhalational anesthetics are simpler than most agents, several possess chiral centers (Figure 12.2). The availability of optically pure inhalation anesthetics provides both novel tools for defining the loci of action of these drugs and the potential for anesthetics with a greater safety margin than currently available agents. This chapter will review recent *in vivo* and *in vitro* evidence that inhalation agents exhibit stereoselective actions.

12.3 *IN VIVO*

12.3.1 STUDIES IN MAMMALS

Harris et al.[17] provided the first *in vivo* evidence indicating the anesthetic action of isoflurane was stereoselective. In this study, sleep times were measured in groups of mice following a single intraperitoneal injection (ranging from 5.4 to 10.8 mmol/kg) of either (+) or (–) isoflurane. Although both stereoisomers produced a dose-dependent increase in sleep time, the sleep times of mice receiving (+) isoflurane were longer (ranging between 21 and 56%) than in animals administered the (–) isomer. The study design was constrained by the limited quantities of optically

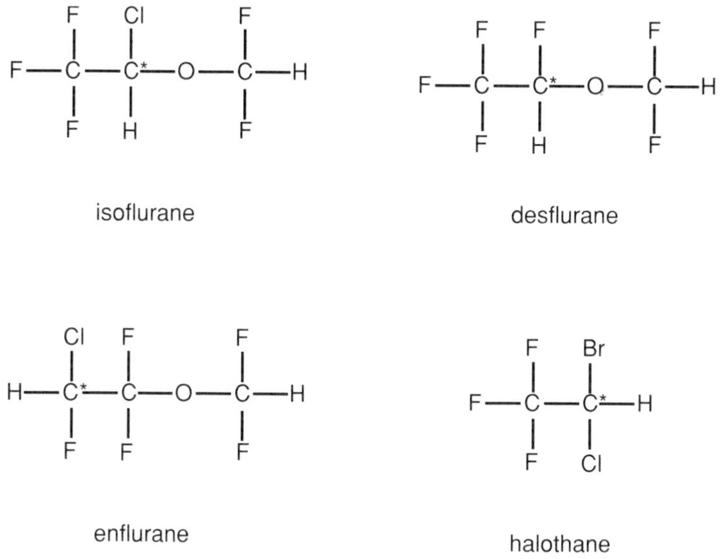

isoflurane

desflurane

enflurane

halothane

FIGURE 12.2 The structures of four clinically used volatile anesthetics that contain a chiral center (*).

pure isoflurane, and although provocative, it could not be considered definitive evidence for a stereoselective difference in *anesthesia*. Thus, neither sleep time nor return of the righting reflex (the end point used in this study) can be considered equivalent to more traditional measures using response to painful stimuli as an end point.[6] Moreover, these data were not obtained under equilibrium conditions, and the apparent stereoselectivity that was observed could be attributable to phamacokinetic [e.g., differences in either the distribution or metabolism of (+) and (–) isoflurane] rather than pharmacodynamic differences.[37]

Subsequently, Lysko et al.[26] determined the minimum alveolar concentration (MAC) of (+) or (–) isoflurane in rats under equilibrium conditions using the response to tail clamp as an end point. The MAC of (+) isoflurane was ~60% lower ($p < 0.05$) than that of the (–) isomer ($1.06 \pm 0.07\%$ and $1.62 \pm 0.02\%$, respectively); the MAC for racemic isoflurane ($1.32 \pm 0.03\%$) was midway between the values obtained for the optically pure isomer (Figure 12.3). Mean arterial blood pressure was also measured in this study. However, because the MAC values were not known in advance, these determinations were made at a fixed rather than at an equi-efficacious anesthetic concentration. The reductions in mean arterial blood pressure produced by (+), (–), and (±) isoflurane at an end-tidal concentration of 2% were statistically significantly different ($p < 0.05$) from baseline (i.e., when the animals were conscious) but not different from each other. This latter finding suggests that the cardiovascular actions of isoflurane may not be stereoselective, a hypothesis consistent with two independent lines of *in vitro* evidence.[13,29]

Two other studies examined the potency of volatile anesthetic stereoisomers. The first is a brief mention, a small ($n = 3$ rats/group), otherwise unpublished study comparing desflurane stereoisomers in mice and rats. In those experiments no

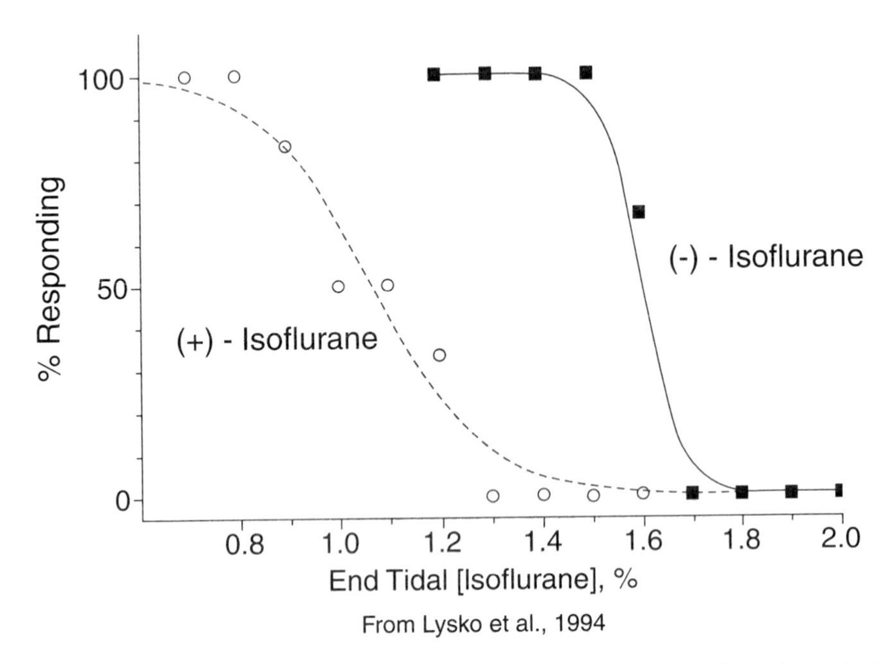

FIGURE 12.3 Stereoselectivity of isoflurane anesthesia. The concentration of anesthetic agent (abscissa) required to prevent movement in response to a noxious stimulus (tail clamp) was measured in groups of six rats. The MAC value for (+) isoflurane was significantly lower than for either the (–) isomer or racemic mixture ($p < 0.05$); the MAC value for (–) isoflurane was significantly greater than for the racemic mixture ($p < 0.05$). (Data are redrawn from Lysko, G.S., Robinson, J.L., Casto, R., and Ferrone, R.A., *Eur. J. Pharmacol.* 263, 25, 1994. With permission.)

statistically significant differences were observed between groups, although the size of the groups indicates that a type II error could have obscured a difference. A larger study also compared anesthetic potency with isoflurane isomers in a tail clamp assay. This investigation failed to demonstrate any difference between either the isomers or the racemic mixture. There is no easy explanation for the apparent discrepancy between these findings and those above, although there are minor procedural differences. Of note are the data discussed below, for which consistent differences are noted between stereoisomers *in vitro*. If in fact there is not an *in vivo* difference in anesthetic potency, the significance of the targets identified *in vitro* may be questioned. However, converging lines of evidence provide quite compelling confirmation about the importance of these ion channels. Further availability of the stereoisomers of these volatile anesthetics may clarify this issue by allowing larger studies.

12.3.2 Studies in Other Vertebrates and Invertebrates

In contrast to the stereoselectivity observed with (+) and (–) isoflurane in rats and mice, Firestone et al.[8] reported that these isomers were equipotent in producing a loss in righting reflex in tadpoles. Sedensky et al.[36] examined the anesthetic potencies of (+) and (–) halothane on wild-type and mutant variants of the nematode

Caenorhabditis elegans. The concentration of (±) halothane required to immobilize (the end point used in this study) the N2 wild-type strain of *C. elegans* (EC_{50} 3.52 ± 0.03%) is severalfold higher than the reported MAC values (~0.75 to 1% atm) in mammals. Nonetheless, (+) halothane was –12% more potent that the (–) isomer in this wild-type strain of *C. elegans*; a similar potency difference between isomers was also observed in other *C. elegans* strains exhibiting comparable (i.e., low) sensitivities to racemic halothane. A different picture emerged when the potency of halothane was measured in strains of *C. elegans* that are more sensitive to halothane (EC_{50} values for racemic halothane ranging between ~0.93 and 1.21%). Several of these strains exhibited no stereoselectivity, and in two strains the racemic mixture was more potent than either isomer. In strain *fc*34, (+) halothane was threefold more potent than the (–) isomer; the (–) isomer was also significantly less potent than the racemic mixture.

12.3.3 Can the Findings in Mammals and Other Phyla Be Reconciled?

The most obvious explanation for the inability to detect a stereoselective action of isoflurane in tadpoles may be related to the sensitivity of the end point used to determine "anesthesia." That is, it could be argued that loss of righting reflex in tadpoles may not be a sufficiently sensitive measure to detect the potency differences of 60% reported in rats using conventional end points.[26] It is difficult to make a similar case in the *C. elegans* study; despite the use of immobility (in some mutants, immobility is accompanied by a remarkable shrinking that can reduce body length by more than half) as an end point, statistically significant differences in the potency of halothane isomers (on the order of ~12%) were evinced.[36] The apparent strain dependence of stereoselectivity in *C. elegans* mutants has been interpreted as evidence that halothane and its isomers can act at multiple sites. Indeed, because pharmacologically relevant concentrations of inhalation agents can affect multiple targets in the central nervous system[10,11,25] (see Chapters 1, 6, 9, and 11), the *process* (i.e., the contribution of individual cellular and molecular pathways producing the end result) of anesthesia may encompass a distinct (albeit overlapping) set of targets in homeotherms and poikilotherms. Thus, differences in the *relative* contributions of individual cellular pathways (e.g., the importance of GABAergic pathways in rodents relative to their importance in wild-type and mutant strains of *C. elegans*) could plausibly reconcile these *in vivo* findings. Even if it were hypothesized that the process of anesthesia was strictly conserved across evolution, it is likely the molecular targets themselves (e.g., $GABA_A$ receptors) are sufficiently different to obscure (or in some cases magnify) a subtle effect between optical isomers, making extrapolation from invertebrates to mammals problematic. The principal argument against this latter hypothesis is that several of the putative molecular targets of anesthesia appear to be highly conserved across evolution. Using $GABA_A$ receptor subunits as an example of this remarkable evolutionary conservation, there is a 65% base pain match between DNA coding for the mouse β_2 subunit and a β subunit found in the pond snail *L. stagnalis* (which has been used to study stereospecificity of isoflurane — see below). For comparison, there is a 72 to 73% base pair match among the three identified β subunits in mouse. Despite the conservation, there are dramatic differences in the pharmacologic and biochemical properties of $GABA_A$ receptors between mammals and invertebrates.[35] Thus, although GABA acts as an inhibitory transmitter throughout the animal kingdom, there are prominent differences in the "details" of $GABA_A$ receptor pharmacology across phylogeny. For example, the affinities of benzodiazepines and neuroactive steroids at $GABA_A$ receptors of the cockroach are two to five orders of magnitude lower than in the mammalian central nervous system. Such striking differences should not be viewed as surprising based on the ability of a single amino acid substitution to reduce the affinities of benzodiazepines (such as diazepam) at $GABA_A$ receptors by more than three orders of magnitude.[39,40] Thus, even a single base substitution on the DNA encoding a putative molecular target of anesthesia could affect the interaction of this anesthetic with its target, most certainly to an extent that could collapse (or perhaps magnify) the modest stereoselective effect observed in rodents.

12.4 *IN VITRO* STUDIES

Optically active inhalation agents that exhibit stereoselectivity *in vivo* represent novel tools to identify potential targets of anesthesia. The utility of these agents goes beyond a potential resolution of the lipid versus protein controversy[9] to an examination of the contribution of a particular pathway, subunit, or domain to the process of anesthesia. Thus, the demonstration of a potency difference between the stereoisomers of an inhalation agent corresponding to the *in vivo* situation [e.g., (+) isoflurane is more potent than (–) isoflurane] is evidence of a potential target of anesthetic action. Nonetheless, if the process of anesthesia involves multiple cellular targets, then the failure to evince stereoselectivity cannot be considered prima facie evidence that a model system does not contribute to this process, particularly if the concentrations of inhalation agent are pharmacologically relevant.

12.4.1 Ion Channels in *L. stagnalis*

Converging lines of evidence suggest ion channels are molecular targets for inhalational agents (reviewed in Refs. 11 and 33; see Chapters 1, 6, and 8–11). Franks and Lieb (1991) first reported a stereoselective action of (+) and (–) isoflurane at two different ion channels in neurons of the pond snail *L. stagnalis*. At anesthetic concentrations that fall within the clinically useful range (0.5 to 2.4 MAC), (+) isoflurane evoked potassium currents that were twice as large as those produced by the (–) isomer. The (+) isomer was also ~50% more potent than the (–) isomer as an inhibitor of acetylcholine-evoked currents at nicotinic acetylcholine receptors. However, a corresponding stereoselectivity was not manifested for inhibition of the voltage-gated fast transient potassium currents in these neurons. Although the latter currents are far less sensitive to isoflurane than those exhibiting stereoselectivity, these observations demonstrate that the stereoselective effects of isoflurane are not "promiscuous." This point has been reinforced in studies with other ion channels, neurotransmitter transport proteins, and lipid bilayers (see below). This seminal study also demonstrated there was no difference in the ability of (+) and (–) isoflurane to depress the melting point of artificial lipid bilayers, which is a sensitive indicator of bilayer disruption. Although artificial lipid bilayers may not mimic the complex arrangement of heterogeneous lipids present in neural membranes, this lack of stereoselectivity is compelling support for the hypothesis that proteins rather than lipids are the primary target of inhalation anesthetics. The inability of (+) and (–) isoflurane to exhibit stereoselectivity in a lipid environment is consistent with a subsequent study by these authors,[2] who reported equal partitioning of these isomers between lipid bilayers (consisting of phosphatidylcholine, phosphatidic acid, and cholesterol) possessing chiral centers and water. The failure to demonstrate a stereoselective effect of an inhalation anesthetic in a lipid environment supports the early studies by Kendig et al.,[24] who reported no difference in the effects of (+) and (–) halothane on synthetic phospholipid bilayer membranes. However, the findings of Kendig et al.[24] had been considered questionable because these isomers were not optically pure [e.g., (+) halothane contained ~25% of the (–) isomer]; this adulteration might thus obscure a modest stereoselectivity.

12.4.2 $GABA_A$ Receptors

Neurochemical, electrophysiologic, and behavioral studies have shown that pharmacologically relevant concentrations of anesthetics,[38] including inhalational agents, act as positive modulators (i.e., increase or facilitate the movement of chloride currents) at $GABA_A$ receptors (see Chapters 10 and 11). This family of ligand-gated ion channels is, therefore, a logical candidate for examining the stereoselectivity of inhalation agents. Several electrophysiologic and neurochemical studies have now demonstrated significant differences in the potency and/or efficacy of (+) and (–) isoflurane at $GABA_A$ receptors. The large, fast inhibitory postsynaptic potentials (IPSPs) recorded from many brain regions are mediated by activation of $GABA_A$ receptors. Jones and Harrison[23] reported that although both (+) and (–) isoflurane prolonged the duration of IPSPs recorded from primary hippocampal neuron cultures, this increase was approximately twice as large with the (+) isomer. Hall et al.[15] reported that the (+) isomer of isoflurane was between 1.5- and 2-fold more effective than the (–) isomer in potentiating the chloride currents produced by a nondesensitizing concentration of GABA in acutely dissociated cerebellar Purkinje neurons (Figure 12.4). This stereoselective effect was optimum at isoflurane concentrations approximating MAC in humans (310 μM; and appeared to plateau at concentrations up to ~3 MAC), diminished at lower concentrations, and was obscured at subanesthetic concentrations (77 μM). After application of desensitizing concentrations of GABA, there is a sustained residual current that is *inhibited* by relatively high concentrations of anesthetics. Although not examined in as great detail as the augmentation of GABA currents, this latter action did not appear to exhibit stereoselectivity. The observation that different $GABA_A$-receptor-mediated responses are not uniformly affected by (+) and (–) isoflurane provides additional evidence that these stereoisomers act at a specific locus (or loci), a hypothesis consistent with neurochemical evidence demonstrating stereoselective actions of isoflurane on some, but not all, allosteric modulatory sites on $GABA_A$ receptors (see below). Consistent with the ability of inhalational agents to enhance GABA currents,

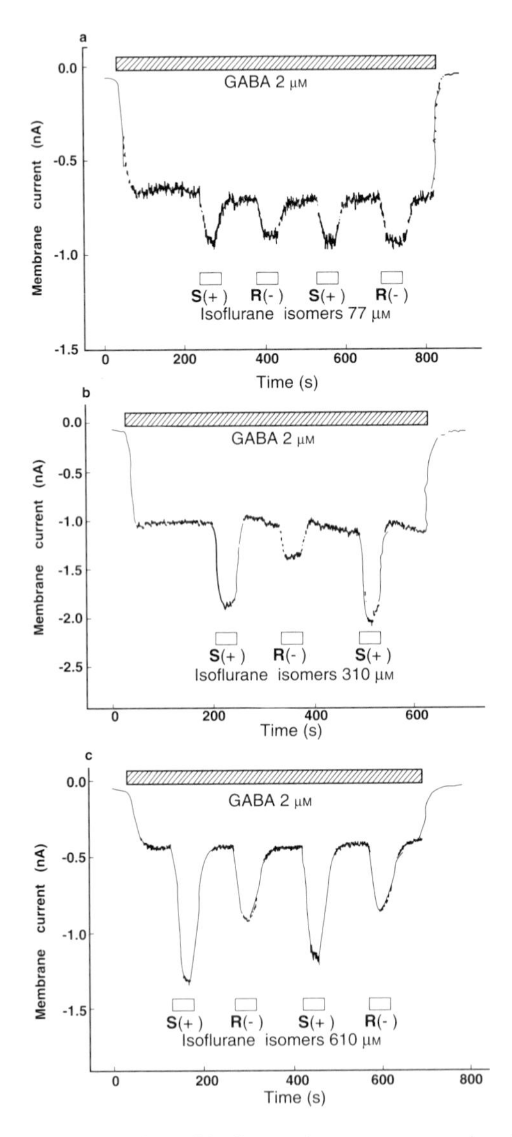

FIGURE 12.4 Stereoselective action of isoflurane isomers to potentiate GABA-induced currents in acutely dissociated rat cerebellar Purkinje neurons. Top panel: At a subanesthetic concentration (77 μ*M*), both (+) and (–) isoflurane enhance GABA currents to the same extent. At concentrations well within the anesthetic range (middle panel: 310 μ*M*; bottom panel: 610 μ*M*), (+) isoflurane is approximately twice as effective as (–) isoflurane in potentiating GABA-induced currents. The currents obtained in response to the higher concentrations of anesthetic appear to desensitize (compare ordinates of middle and bottom panels). (Data are from Hall, A.C., Lieb, W.R., and Franks, N.P., *Br. J. Pharmacol.* 112, 906, 1994. With permission.)

these anesthetics also increase $GABA_A$-receptor-mediated chloride uptake in synaptic microvesicles (synaptoneurosomes).[21,32] Using synaptoneurosomes prepared from (whole) rat brain, Quinlin et al.[34] demonstrated that at a fixed concentration (140 μ*M*), (+) isoflurane was significantly more effective than (–) isoflurane in reducing the EC_{50} (i.e., increasing the potency) of GABA to stimulate $^{36}Cl^-$ uptake. Thus, (+) isoflurane reduced the EC_{50} of GABA from 12.7 ± 1.0 μ*M*, compared with 9.6 ± 1.0 μ*M* with the (–) isomer. Using a fixed concentration of GABA (5 μ*M*) well below its EC_{50}, these investigators demonstrated that the (+) isomer was approximately 1.6-fold more potent than the (–) isomer to increase $^{36}Cl^-$ flux (79 ± 11 μ*M* vs. 130 ± 17 μ*M*, respectively; $p = 0.01$). Moreover, under these conditions the maximal $^{36}Cl^-$ flux (i.e., E_{max}) produced by (+) isoflurane was significantly greater (~35%; $p < 0.001$) than in the presence of the (–) isomer.

The stereoselective actions of isoflurane at $GABA_A$ receptors were first evinced using radioligand binding techniques.[30] The multiple, allosteric regulatory sites on $GABA_A$ receptors are readily amenable to study using these methodologies, and such studies have provided insight into the molecular mechanism responsible for the GABA-positive actions of inhalation agents. Like other GABA-positive substances (e.g., barbiturates such as pentobarbital; etomidate; propofol; and the GABA mimetic, muscimol), inhalational anesthetics increase [^{3}H]benzodiazepine binding to $GABA_A$ receptors; this effect is highly dependent on the presence of Cl^- (or other anions that are permeable through GABA-gated chloride channels). We initially compared the effects of (+) and (–) isoflurane on [^{3}H]flunitrazepam binding to well-washed mouse cortical membranes.[30] Although the potencies of (+) and (–) isoflurane to enhance [^{3}H]flunitrazepam binding were virtually identical in the absence of Cl^-, (+) isoflurane was ~35% more potent than the (–) isomer in the presence of Cl^- (50 to 200 m*M*). This enhancement in [^{3}H]flunitrazepam binding is effected through an increase in the apparent affinity (a decrease in K_d), with no change in the number of [^{3}H]flunitrazepam binding sites (B_{max}). In a variation of this procedure, both isomers (at concentrations of 440 and 880 μ*M*) were found to significantly increase the potency of Cl^- to enhance [^{3}H]flunitrazepam binding (an action common to other general anesthetics such as barbiturates). However, the increase in potency (i.e., decrease in the EC_{50}) of Cl^- was twice as great ($p < 0.05$) in the presence of the (+) compared to the (–) isomer (EC_{50} values for Cl^-: 58 ± 8 m*M*, 18 ± 2 m*M*, and 40 ± 5 m*M* in the absence and presence of 880 μ*M* (+) and (–) isoflurane, respectively). Isoflurane also produced a significant increase in the E_{max} of Cl^--enhanced [^{3}H]flunitrazepam binding (i.e., the amount of [^{3}H]flunitrazepam bound), and in this measure, the (+) isomer was more efficacious than the (–) isomer. Both volatile and nonvolatile anesthetics also increase the binding of [^{3}H]GABA and the GABA mimetic [^{3}H]muscimol to GABA receptors (see Chapter 11). The enhancement of [^{3}H]muscimol binding to mouse cortical membranes was stereoselective,[16] with (+) isoflurane approximately twice as potent as the (–) isomer ($p < 0.01$). The ability of drugs to bidirectionally modulate the binding agonist and antagonist ligands to $GABA_A$ receptors is well described. Both (+) and (–) isoflurane produced a concentration-dependent and stereoselective inhibition on the binding of a GABA antagonist ([^{3}H]SR 95531). (+) Isoflurane was ~50% more potent than (–) isoflurane in this measure ($p < 0.01$). In contrast to the stereospecific effects of isoflurane on radioligand binding to benzodiazepine and GABA binding sites, no significant differences were observed in the potencies of (+) and (–) isoflurane to inhibit the binding of [^{35}S]*t*-butylbicyclophosphorothionate, a "cage" convulsant presumed to act at or near GABA-gated chloride channels. The failure to demonstrate stereoselectivity at each of the allosteric modulatory sites examined suggests that the stereoselective effects of isoflurane at $GABA_A$ receptors result from an action at a specific locus (or loci) rather than a nonselective perturbation at, for example, a protein–lipid interface. Another means of testing this hypothesis is to remove $GABA_A$ receptors from their endogenous membrane milieu and determine whether the stereoselective action of isoflurane remains. This may be accomplished by techniques such as incorporating wild-type $GABA_A$ receptors in an artificial membrane and expressing recombinant $GABA_A$ receptors in a membrane environment unlike that of the central nervous system. We selected the latter approach, because studies using recombinant $GABA_A$ receptors of known composition already have provided valuable information about differences in the subunit requirements for volatile and nonvolatile anesthetics.[18] For these studies, we used the WSS-1 cell line [this cell line is derived from H(uman)E(mbryonic)K(idney) 293 cells], which contains cells that stably express $GABA_A$ receptors composed of rat α_1 and γ_2 subunits.[43] Although the quantity of optically resolved isoflurane that was available limited the scope of these studies, there was clear-cut evidence of stereoselectivity between (+) and (–) isoflurane to enhance [^{3}H]flunitrazepam binding in membranes prepared from these cells. Thus, in the absence of Cl^-, addition of 0.5% (+) isoflurane stimulated [^{3}H]flunitrazepam binding to a greater extent ($p < 0.05$) than did the (–) isomer [basal binding no drug, 21.2 ± 2 gmol/mg protein; with (+) isoflurane, 27.4 ± 1.7 fmol/mg protein; with (–) isoflurane, 19.6 ± 0.1 fmol/mg protein]. Moreover, in the presence of increasing concentrations of Cl^-, the maximum increase in [^{3}H]flunitrazepam binding was significantly higher ($p < 0.05$) in the presence of (+) compared to (–) isoflurane [E_{max}:no drug (Cl^- only), 31.4 ± 2.1 fmol/mg protein; (+) isoflurane, 70 ± 3.1 fmol/mg protein; (–) isoflurane, 51 ± 0.9 fmol/mg protein]. Although these (α_1, γ_2) $GABA_A$ receptors represent a very minor constituent of the total receptor pool,[12] these studies are consistent with the view that the stereoselective activation of $GABA_A$ receptors observed with the isomers of isoflurane is produced at distinct amino acid sequences rather than by perturbation of membrane lipids.

12.4.3 Voltage Operated Calcium Channels

The depression of myocardial contractility produced by inhalation agents has been linked to an action at L-type Ca^{2+} channels.[3,19] Studies in isolated guinea pig heart[13] have shown that the optical isomers of both isoflurane and desflurane (at concentrations equivalent to 1 to 2 MAC) decreased left ventricular pressure, heart rate, and percent oxygen extraction in a concentration-dependent fashion. There was no evidence of stereoselectivity between the isomers of either isoflurane or desflurane. Consistent with this study, the stereoisomers of isoflurane inhibit [^{3}H]isradipine binding to L-type Ca^{2+} channels from mouse heart in a concentration-dependent fashion, but no evidence of stereoselectivity [IC_{50} values of 0.48 ± 0.02% and 0.46 ± 0.01% for (+) and (–) isoflurane, respectively] was obtained. Similarly, no evidence of stereoselectivity was obtained for inhibition of [^{3}H]isradipine binding to L-type Ca^{2+} channels from mouse brain.[29] Based on these findings, the optical isomers of isoflurane should be equipotent in depressing myocardial function.

12.4.4 Other Systems

El-Maghrabi and Eckenhoff[7] have demonstrated a lack of stereoselectivity between the isomers of isoflurane to inhibit [^{3}H]dopamine uptake in rat brain synaptosomes. Eckenhoff and Shuman[5] have demonstrated that [^{14}C]halothane can be used as a photoaffinity label, binding to soluble proteins such as bovine serum albumin with a K_d between 0.3 and 0.5 m*M*. These investigators reported that, like other anesthetics, isoflurane can block this labeling, and that (+) isoflurane was approximately 50% more potent than (–) isoflurane (K_i 1.42 ± 0.15 m*M* vs. 2.12 ± 0.33 m*M*, respectively; $p < 0.035$). Eckenhoff[4] subsequently identified two amino acid sequences on bovine serum albumin that are photoaffinity labeled by [^{14}C]halothane; both of these sequences contained tryptophan residues.

12.5 WHY ARE THE POTENCY DIFFERENCES BETWEEN (+) AND (–) ISOFLURANE SO MODEST?

Despite the use of different systems, methods, and end points to study the stereoselectivity of inhalation agents, the conclusions are similar: when manifested, the potency (or efficacy) difference between the stereoisomers of inhalation agents is generally less than twofold. This very modest difference pales by comparison with the potency differences between enantiomeric pairs of complex heterocycles, which routinely exceed two orders of magnitude. For example the affinities of (+) and (–) SNC 80 (a highly substituted *N,N*-diethylbenzamide) at δ opiate receptors are ~1 and >2000 n*M*, respectively,[1] whereas the affinities of a pair of 1,4-benzodiazepines [designated B10 (+) and (–)] at $GABA_A$ receptors differ by >100-fold.[28] If potency differences between stereoisomers are the result of a unique interaction between a drug and its biological substrate (see Figure 12.1), then the relatively simple halogenated hydrocarbon structures of inhalation agents (Figure 12.2) would likely constrain the degree/extent of stereoselectivity. Thus, the affinity of a ligand for its recognition site is determined by factors such as steric (size) constraints, hydrogen bonding, and van der Waals forces. Stereoselectivity arises from the ability of one isomer to more closely satisfy the size and charge requirements imposed by the recognition site compared with a nonsuperimposable mirror image (Figure 12.1). If this assumption is valid then the synthesis of inhalation agents more complex than the typical halogenated ether should increase the potency difference between stereoisomers. As a corollary, it would be predicted that in a direct comparison of the potency difference between stereoisomers of isoflurane and halothane, the comparatively simple structure of halothane (a halogenated ethane derivative) would result in an even more modest stereoselectivity than observed with isoflurane (a halogenated methyl ethyl ether).

If multiple pathways contribute to the anesthesia induced by inhalation agents, then the modest potency difference between (+) and (–) isoflurane in rodents[17,26] may also reflect an averaging of effects, with the stereoisomers exhibiting a twofold (or greater) potency difference at some relevant targets and little or no potency difference at others. Although the failure to consistently demonstrate stereoselectivity across magnitude systems indicates some degree of specificity, the observation that (+) isoflurane is –50% more potent than (–) isoflurane to inhibit photoaffinity labeling of halothane to bovine serum albumin[5] merits comment. Based on this observation, it could be argued that the demonstration of stereoselectivity at a soluble protein unrelated to anesthesia suggests that even if not a totally random event, the potency difference between, for example, (+) and (–) isoflurane may not be useful in delineating relevant targets of anesthesia. However, resolution of the sites on human and bovine serum albumin that are photolabeled by halothane suggests the presence of specific and discrete sites containing aromatic residues such as tryptophan.[4] Given the lipophilic nature of inhalation anesthetics, Eckenhoff's findings demonstrate that aromatic residues may form part of a hydrophobic binding motif common to many, but not all proteins.

12.6 SAFER ANESTHETICS THROUGH STEREOCHEMISTRY?

Inhalation agents such as halothane and isoflurane are the mainstay of clinical anesthesia despite the low therapeutic indices common to this class of drugs.[42] This narrow safety margin may be attributed to a significant depression of cardiovascular function as well as deleterious effects on respiration, airway reflexes, and temperature. The management of these side effects represents a significant challenge in the patient undergoing anesthesia and may limit surgery in compromised patients. Based on preclinical data, we and others have postulated[26,29] that even a modest increase in anesthetic potency with no concomitant increase in toxicity could have a significant impact on the practice of anesthesiology and surgery. The most stringent test of this hypothesis requires a direct comparison between the more potent of a pair of optically pure inhalation agents [e.g., (+) isoflurane] and the racemic mixture rather than the less potent isomer. This comparison is the most clinically relevant because there must be a significant improvement in the therapeutic index over the chemical form that would most likely be produced by standard synthetic approaches (all currently available inhalation agents with a chiral center are racemic mixtures). At present, there are no commercially feasible means of synthesizing (or resolving) optically active volatile anesthetics in sufficient quantities for clinical testing. However, based on the ~30% difference in anesthetic potency between (+) and (–) isoflurane in rates[26] and the apparent lack of stereoselectivity in depressing cardiac function,[13] the therapeutic advantage of an optically pure isomer can be approximated (Figure 12.5). The isoflurane model may be considered a "typical" representation based on the assumption that structurally related inhalation agents act through common mechanisms, with some difference in anesthetic potency manifested by each pair of enantiomers. Both the relative simplicity of chiral inhalation agents (Figure 12.2) and the modest difference in anesthetic potency between (+) and (–) isoflurane suggest that the optical isomers of other currently used inhalation agents will not exhibit more remarkable potency differences. At clinically useful

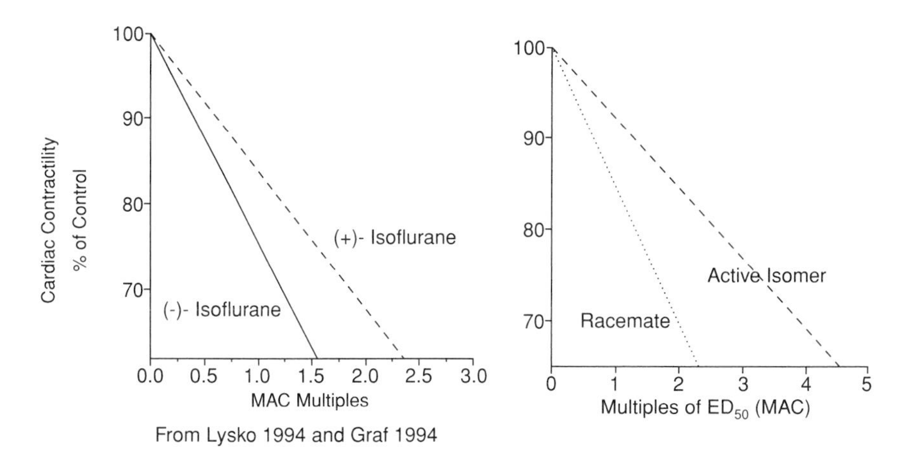

FIGURE 12.5 Decrease in cardiac contractility versus MAC multiples for (+), (–), and (±) isoflurane (left panel) and for a hypothetical anesthetic agent with enantiomers that differ in anesthetic potency by 1000-fold (right panel). The data for isoflurane were taken from Lysko et al.[26] (MAC multiples in rat) and Graf et al.[13] (decrease in cardiac contractility in isolated guinea pig hearts). The data for the hypothetical compound assume the same cardiac depressant profile as for isoflurane. Note that the most clinically relevant comparison is between the more potent isomer and the racemate (which would be the form produced by usual chemical means of synthesis) rather than the less potent isomer. In the right panel the racemate line is shown. In the left panel it would be halfway between the two lines. It is apparent from comparing equal anesthetic concentrations (i.e., same MAC multiple on the x axis) that even the more active isomer of an anesthetic with great stereoselectivity will result in only a modest decrease in cardiac suppression. Note that relevant comparisons should be made in the 1 to 1.5 MAC (or less) range. A similar graph can easily be envisaged. For example, an agent with great cardiac depression will have an even steeper slope, and the isomers would need to be very different from the racemate to have a significant therapeutic advantage.

concentrations (1 to 1.5 MAC), the decrease in cardiac contractility produced by (+) isoflurane is only ~3 to 4% less than that produced by racemate (Figure 12.5, left).

What would occur if a pair of optically active inhalation agents were synthesized with a 1000-fold difference in anesthetic potency? First, the difference in anesthetic potency between the active isomer and racemate would be only approximately twofold because the potency of the racemate approximates that obtained with equal parts of active isomer and inert material. Assuming that the MAC and degree of cardiovascular depression produced by the racemate are equivalent to isoflurane, at 1 MAC the active isomer would produce ~7 to 8% less cardiac depression than the racemate. The cardiac depression produced by the active isomer would be ~11% lower than the racemate at 1.5 MAC (Figure 12.5, right). A similar argument would pertain to a racemic anesthetic with greater cardiotoxicity. In this case the more potent isomer would result in less cardiac depression at an anesthetizing concentration than the racemate. However, because we are assuming a lack of stereoselectivity at the nonanesthetic targets (i.e., the heart), there would still be significant cardiotoxicity, making this compound unlikely to be superior (safer) than racemic isoflurane.

These estimates demonstrate the very modest potential therapeutic advantage that would be gained with an optically active inhalation agent such as (+) isoflurane. Although the therapeutic advantage achieved with our hypothetical example *may* be clinically significant, this degree of stereoselectivity is unlikely to be attained with relatively simple halogenated ethers (see above). Although the stereoisomers of inhalation agents provide valuable tools to examine the molecular basis of anesthesia, obtaining a significantly safer agent through stereochemistry does not appear to be a practical approach for the immediate future.

REFERENCES

1. Calderone, S.N., Rothman, R.B., Porreca, F., Flippen-Anderson, J.L., McNutt, R.W., Xu, H., Smith, L. E., Bilsky, E.J., Davis, P., and Rice, K.C., Probes for narcotic receptor mediated phenomena. 19. Synthesis of (+)-4[αR)-α-((2S,5R)-4-allyl-2,5-dimethyl-1-piperazinyl)-3-methoxybenzyl]-N,N-diethylbenzamide (SNC 80): A high selective, nonpeptide delta opioid receptor agonist. *J. Med. Chem.* 37, 2125, 1994.
2. Dickinson, R., Franks, N.P., and Lieb, W.R., Can the stereoselective effects of the anesthetic isoflurane be accounted for by lipid solubility? *Biophys. J.* 66, 2019, 1994.
3. Drenger, B., Quigg, M., and Blanck, T.J.J., Volatile anesthetics depress calcium channel blocker binding to bovine cardiac sarcolemna. *Anesthesiology* 74, 155, 1991.
4. Eckenhoff, R.G., Amino acid resolution of halothane binding sites in serum albumin. *J. Biol. Chem.* 271, 15521, 1996.
5. Eckenhoff, R.G., and Shuman, H., Halothane binding to soluble proteins determined by photoaffinity labeling. *Anesthesiology* 79, 96, 1993.
6. Eger, E.I., II, Saidman, L.J., and Brandstater, B., Minimum alveolar anesthetic concentration: A standard of anesthetic potency. *Anesthesiology* 26, 756, 1965.
7. El-Maghrabi, E.A., and Eckenhoff, R.G., Inhibition of dopamine transport in rat brain synaptosomes by volatile anesthetics. *Anesthesiology* 78, 750, 1994.
8. Firestone, S., Ferguson, C., and Firestone, L.L., Isoflurane's optical isomers are equipotent in *Rana pipiens* tadpoles. *Anesthesiology* 77, A758, 1992.
9. Franks, N.P., and Lieb, W.R., What is the molecular nature of general anesthetic target sites? *Trends Pharmacol. Sci.* 8, 169, 1987.
10. Franks, N.P., and Lieb, W.R., Selective actions of volatile general anaesthetics at molecular and cellular levels. *Anesthesiology* 71, 65, 1993.
11. Franks, N.P., and Lieb, W.R., Molecular and cellular mechanisms of general anesthesia. *Nature* 367, 607, 1994.
12. Fritschy, J.M., and Mohler, H., $GABA_A$-receptor heterogeneity in the adult rat brain: Differential regional and cellular distribution of seven major subunits. *J. Comp. Neurol.* 359, 154, 1995.
13. Graf, B.M., Boban, M., Stowe, D.F., Kampine, J.P., and Bosnajak, Z.J., Lack of stereoselective effects of isoflurane and desflurane isomers in isolated guinea pig hearts. *Anesthesiology* 81, 129, 1994.
14. Graham, S.H., Shimizu, H., Newman, A., Weinstein, P., and Faden, A.I., Opioid receptor antagonist nalmefene stereospecifically inhibits glutamate release during global cerebral ischemia. *Brain Res.* 632, 346, 1993.
15. Hall, A.C., Lieb, W.R., and Franks, N.P., Stereoselective and non-stereoselective actions of isoflurane on the $GABA_A$ receptor. *Br. J. Pharmacol.* 112, 906, 1994.

16. Harris, B.D., Moody, E.J., Basile, A.S., and Skolnick, P., Volatile anesthetics bidirectionally and stereospecifically modulate ligand binding to GABA receptors. *Eur. J. Pharmacol.: Mol. Pharmacol. Section* 267, 269, 1994.
17. Harris, B., Moody, E.J., and Skolnick, P., Isoflurane anesthesia is stereoselective. *Eur. J. Pharmacol.* 217, 215, 1992.
18. Harris, B.D., Wong, G., Moody, E. J., and Skolnick, P., Different subunit requirements for volatile and nonvolatile anesthetics at gamma-aminobutyric acid type A receptors. *Mol. Pharmacol.* 47, 363, 1995.
19. Hoehner, P.J., Quigg, M.C., and Blanck, T.J., Halothane depresses D600 binding to bovine heart sarcolemna. *Anesthesiology* 75, 1019, 1991.
20. Huang, C.G., Rozov, L.A., Halpern, D.F., and Vernice, C.G., Preparation of the isoflurane enantiomers. *J. Biol. Chem.* 58, 7382, 1993.
21. Huidobro-Toro, J.P., Bleck, V., Allan, M., and Harris, R.A., Neurochemical actions of anesthetic drugs on the aminobutyric acid receptor-chloride channel complex. *J. Pharmacol. Exp. Ther.* 242, 963, 1987.
22. Iijima, I., Minamilkawa, J.I., Jacobson, A.E., Brossi, A., Rice, K.C., and Klee, W.A., Structure in the (+)-morphanan series. V. Synthesis and biological properties of (+)-naloxone. *J. Med. Chem.* 21, 298, 1978.
23. Jones, M.V., and Harrison, N.L., Effects of volatile anesthetics on kinetics of inhibitory postsynaptic currents in cultured rat hippocampal neurons. *J. Neurophys.* 70, 1340, 1993.
24. Kendig, J.J., Trudell, J.R., and Cohen, E.N., Halothane stereoisomer: Lack of stereospecificity in two model systems. *Anesthesiology* 39, 518, 1973.
25. Lin, L.H., Chen, L.L., and Harris, R.A., Enflurane inhibitis NMDA, AMPA, and kainate-induced currents in *Xenopus* oocytes expressing mouse and human brain mRNA. *FASEB J.* 7, 479, 1993.
26. Lysko, G.S., Robinson, J.L., Casto, R., and Ferrone, R.A., The stereospecific effects of isoflurane isomer *in vivo*. *Eur. J. Pharmacol.* 263, 25, 1994.
27. Meinwald, J., Thompson, W.R., Pearson, F.L., Konig, W.A., Runge, T., and Franke, W., Inhalational anesthetics stereochemistry: Optical resolution of halothane, enflurane, and isoflurane. *Science* 251, 560, 1991.
28. Mohler, H., and Okada, T., Benzodiazepine receptor: Demonstration in the central nervous system. *Science* 198, 849, 1977.
29. Moody, E.J., Harris, B.D., Hoehner, P., and Skolnick, P., Isoflurane stereoisomers inhibit [^{3}H]-isradipine binding to brain and heart: Lack of stereoselectivity. *Anesthesiology* 81, 124, 1994.
30. Moody, E.J., Harris, B., and Skolnick, P., Stereospecific actions of the inhalation anesthetic isoflurane at the $GABA_A$ receptor complex. *Brain Res.* 615, 101, 1993.
31. Moody, E.J., Mattson, M., Newman, A.H., Rice, K.C., and Skolnick, P., Stereospecific reversal of nitrous oxide analgesia by naloxone. *Life Sci.* 44, 703, 1989.
32. Moody, E.J., Suzdak, P.D., Paul, S.M., and Skolnick, P., Modulation of the benzodiazepine/γ-aminobutyric acid receptor complex by inhalational anesthetics. *J. Neurochem.* 51, 1386, 1988.
33. Moody, E.J., Yeh, H.J.C., and Skolnick, P., The $GABA_A$ receptor complex: Is it a locus of action for inhalational anesthetics. In: *Neuropharmacology of Ethanol,* R.E. Miller, Koob, G.F., Lewis, M.J., and Paul, S.M., Eds., Boston: Birkhauser, 1991, p. 77–92.
34. Quinlan, J.J., Firestone, S., and Firestone, L., Isoflurane's enhancement of chloride flux through rat brain g-aminobutyric acid type A receptors is stereoselective. *Anesthesiology* 83, 611, 1995.

35. Rauh, J.J., Lummis, S.C.R., and Sattelle, D.B., Pharmacological and biochemical properties of insect GABA receptors. *Trends Pharmacol. Sci.* 11, 325, 1990.
36. Sedensky, M.M., Cascrobi, H.F., Meinwald, J., Radford, P., and Morgan, P.G., Genetic differences affecting the potency of stereoisomers of halothane. *Proc. Natl. Acad. Sci. USA* 91, 10054, 1994.
37. Segal, D.S., Behavioral characterization of d- and l-amphetamine: Neurochemical implications. *Science* 190, 75, 1975.
38. Skolnick, P., and Paul, S., The benzodiazepine/GABA receptor chloride channel complex. *ISI Atlas Pharmacol.* 2, 19, 1988.
39. Wieland, H., and Lüddens, H., Four amino acid exchanges convert a diazepam-insensitive, inverse agonist-preferring $GABA_A$ receptor into a diazepam-preferring $GABA_A$ receptor. *J. Med. Chem.* 37, 4576, 1994.
40. Wieland, H.A., Lüddens, H., and Seeburg, P.H., A single histidine in $GABA_A$ receptors is essential for benzodiazepine agonist binding. *J. Biol. Chem.* 267, 1426, 1992.
41. Williamson, M., Paul, S.M., and Skolnick, P., Labelling of benzodiazepine receptors *in vivo*. *Nature* 275, 551, 1978.
42. Wolfson, B., Hetrick, W.D., Lake, C.L., and Siker, E.S., Anesthetic indices — further data. *Anesthesiology* 48, 187, 1978.
43. Wong, G., Sei, Y., and Skolnick, P., Stable expression of type I γ-aminobutyric $acid_A$/benzodiazepine receptors in a transfected cell line. *Mol. Pharmacol.* 42, 996, 1992.

13 Effects of Volatile Anesthetics at Nicotinic Acetylcholine Receptors

Pamela Flood

CONTENTS

Despite more than 150 years of increasing clinical expertise with general anesthetics, their mechanism of action remains enigmatic. It is clear, however, that general anesthetics inhibit synaptic transmission in sensory, motor, and limbic areas of the central nervous system (CNS).[13,14,3] As is frequently the case, the devil is in the details. Evidence has converged on the idea that general anesthetics act potently and specifically at several receptors of the ligand-gated ion channel family in the CNS, including the inhibitory $GABA_A$ and glycine receptors (see Ref. 12 and Chapters 10 and 11 of this book). Recent work has demonstrated that volatile anesthetics have potent activity at several neuronal-type nicotinic acetylcholine receptors (neuronal nicotinic acetylcholine receptors, n-nAChRs) as well.[9,31] The n-nAChRs are from an important family of excitatory ion channel receptors in the CNS and autonomic nervous system. Thus, general anesthetics cause synaptic inhibition by both augmenting inhibitory and inhibiting excitatory input.

The recent work on n-nAChRs was preceded by many studies on general anesthetic activity at the muscle-type AChRs (m-nAChRs) and those from the electric organ of Torpedo, which are easily isolated and purified in large quantity.[7,8] Although the muscle and invertebrate nAChRs share sequence homology with their neuronal cousins, they are distinct in terms of subunit composition, pharmacology, and physiology (Figure 13.1). In fact, some subunits that form neuronal nicotinic receptors are evolutionarily closer to subunits that form $5HT_3$ and $GABA_A$ receptors than those that form muscle nicotinic receptors[24] (see Figure 13.2).

0-8493-8555-5/01/$0.00+$.50

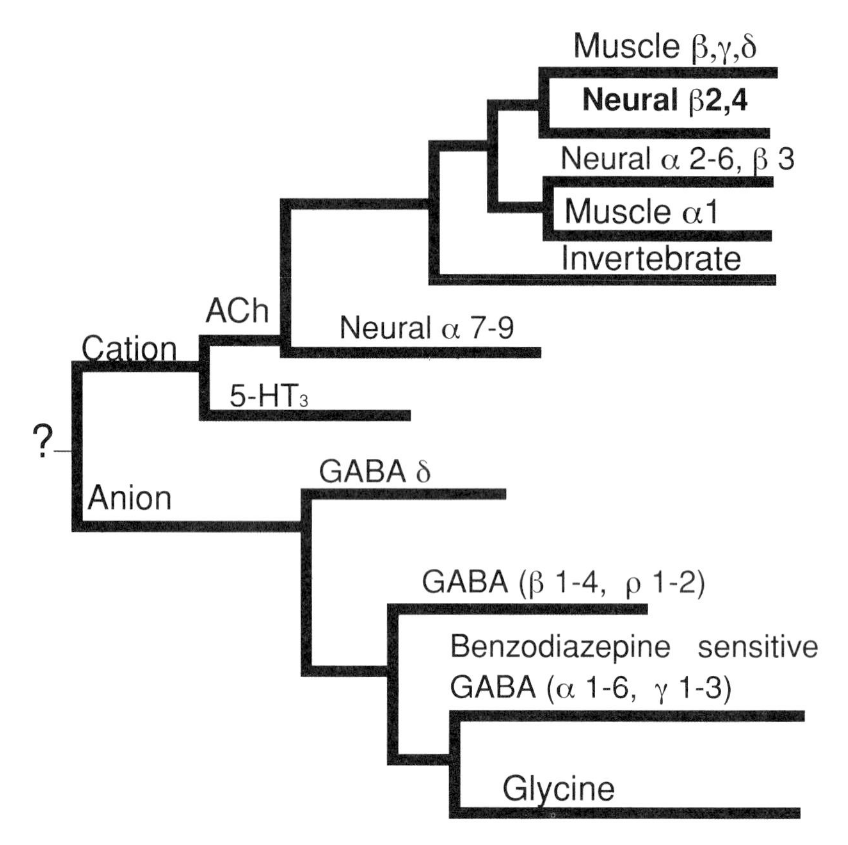

FIGURE 13.1 The evolutionary relationship between ligand-gated ion channels with four transmembrane-spanning segments.

Candidates for receptors involved in the transaction of general anesthetic action need to fulfill certain criteria: (1) they must be potently and specifically effected by clinically relevant concentrations of general anesthetics; and (2) they must be reasonably anatomically located to have a role in a specific behavioral effect. The nAChRs fill these criteria. Volatile anesthetic inhibition of excitatory input in the central and autonomic nervous systems fulfill these criteria because inhibitory anesthetic concentrations are lower than the clinical concentrations in both recombinant and native systems.[9,21,31] Because n-nAChRs are expressed throughout the CNS, where they are thought to act presynaptically to augment release of glutamate, GABA, serotonin, dopamine, and ACh itself, it is easy to imagine that reduction of nAChR activity could effect the inhibition in synaptic transmission caused by general anesthetics.

The recent cloning of the genes encoding the nAChR subunits found in neurons and the explosion of studies on the pharmacology, physiology, and behavioral consequences of the n-nAChRs have made possible the study of volatile anesthetic activity on this family of excitatory ion channel/receptors. The remainder of this chapter will present a brief overview of what is currently known about the molecular biology, physiology, and pharmacology of these receptors and the experimental evidence that demonstrates that volatile anesthetics do indeed have potent activity at specific members of the n-nAChR family.

13.1 NICOTINIC ACETYLCHOLINE RECEPTORS AND THEIR HOMOLOGUES

The ligand-gated ion channels are a family of pentameric proteins that form an ion channel when inserted into a membrane. Each of the five subunits that make up these receptors traverses the membrane four times with the N terminal and C terminal portions of the protein in the extracellular space. The N terminal, extracellular portions of these proteins are highly conserved and contain two cysteine residues separated by 13 amino acids, which are disulfide linked in the ACh receptor.[12] The extracellular domain of a subset of these subunits contains the agonist binding site. The second and part of the first two transmembrane domains (M2 and M1) form the ion conducting pore.[1,2] The third, fourth, and part of the first transmembrane domain make contact with the membrane lipid, forming a hydrophobic "barrel" around the ion conducting pore.[3]

Neurotransmitter-gated ion channels can conduct anions, as is the case for the inhibitory $GABA_A$ and glycine receptors, or cations as neuronal and muscle-type nAChRs and the 5-HT_3 receptor.[15] Figure 13.1 demonstrates the phylogenetic relationship between ligand-gated ion channels containing four transmembrane domains.[24] It is of interest that volatile anesthetics have activity at each member of this family that is much more potent than other types of ion channels. This clustering of affected channels raises the likelihood that the portion perturbed by the volatile anesthetic is conserved among these receptors. Recent studies using both chimeric receptor subunits and site-directed mutagenesis have identified specific amino acid residues in the extracellular portion of the M2 and M3 domains of the α subunit of $GABA_A$ and glycine receptors involved in the augmentation of the agonist response by volatile anesthetics.[22] Homologous amino acids in the β subunit of the $GABA_A$ receptor have been identified as being responsible for augmentation by etomidate.[20,23] Anesthetic inhibition of m-nAChR has been shown to occur via two modalities: direct channel block and favoring of desensitized states.[11,27] Specific residues within the M2 domain of the α subunit of m-nAChR have been demonstrated to influence channel block of the muscle receptor by isoflurane.[11] It is unclear if these residues also effect desensitization, or if another site is involved. The Hill number derived from these experiments is 2-3, indicating a potential for multiple sites of anesthetic interaction. Work is underway to determine the site of inhibition of n-nAChRs by general anesthetics.

13.2 SUBUNIT IDENTITY

The muscle-type n-AChRs are composed of α, β, γ, and δ subunits in the embryo and α, β, ε, and δ subunits in the adult. The α subunit contains the ligand binding site, although the γ and δ subunits in the embryo and the ε and δ subunits in the adult also contribute to ligand binding.[5]

In contrast, n-nAChRs are composed of only α and β subunits. The simplicity ends here: eight neuronal α-type genes have been cloned and named α_{2-9} by analogy to the muscle α subunit, now called α_1. α subunits contain the vicinal cysteines, which mark the agonist binding site. In addition, three non-α (lacking the cysteines) or β subunits have been cloned and are named β_{2-4} by analogy to the muscle β subunit, now called β_1. The neuronal subunits do not share great homology with the muscle β subunit, but are called β on the basis of lacking the vicinal cysteines of the α subunits. Unlike the muscle receptors, which presumably form a homogeneous population of receptors containing two α subunits, one β subunit, one δ subunit, or one ε subunit in adult life, the n-nAChR family is similar in variety and complexity to the homologous $GABA_A$ receptor family. Many α and β subunits can form heteromeric channels in which one, two (or perhaps three) α subunits coassemble with two β subunits to form functional receptors.[17] The nAChR subunits α_{7-9}, which are thought to have split off very early in evolution from the other nAChR subunits, can form homomeric receptors that preferentially conduct calcium with a P_{ca}/P_{na} more than twice that of the NMDA receptor.[29] The calcium current activated by ACh may be responsible for some of the activities for which calcium is thought to be a second messenger. Figure 13.2 demonstrates some of the n-nAChRs that may be formed *in vivo*.

13.3 NEURONAL nAChR EXPRESSION PATTERNS

Although cholinergic neurons represent a minute fraction of the neurons that make up the CNS, they regulate essentially every muscle, organ, gland, and neural region.[32] The cholinergic systems in the CNS are diverse, diffuse, and thus certainly appealing as putative targets of volatile anesthetic action. The cholinergic basal forebrain contains cells that are thought to be involved with memory and arousal.[32] Pontomesencephalic cholinergic neurons appear to be involved with mechanisms of sleep, memory, and locomotor activity. n-nAChRs can serve either postsynaptic or presynaptic functions.[32]

The classical postsynaptic role of n-nAChRs is in the autonomic ganglia, where they are responsible for direct synaptic transmission. Cholinergic somatic and autonomic nerves control smooth and cardiac muscle as well as glandular secretion. Thus, inhibition of n-nAChRs may result in alterations of heart rate and blood pressure, common side effects of general anesthetics.

The n-nAChRs are thought to have an important presynaptic role in the CNS. There are many cells throughout the CNS that express n-nAChRs both on the soma and on their axonal terminals. Stimulation of these nAChRs results in the presynaptic augmentation of release of glutamate, GABA, dopamine, serotonin, norepinephrine, and ACh itself.[28] This increase in release has been shown to augment synaptic transmission in many areas of the CNS, including medial habenula, diagonal band, laterodorsal tegmental nucleus, prefrontal cortex, primary visual cortex, and hippocampus.[32] The behavioral effects of general anesthetics are varied and include hypnosis, amnesia, analgesia, immobility, as well as hemodynamic, gastrointestinal, and thermoregulatory side effects. It is likely that multiple neurophysiologic mechanisms

Muscle-Nicotinic Acetylcholine Receptor

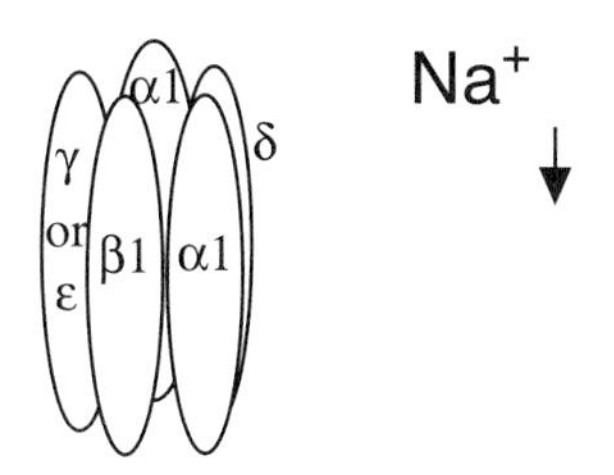

Neuronal Nicotinic Acetylcholine Receptors

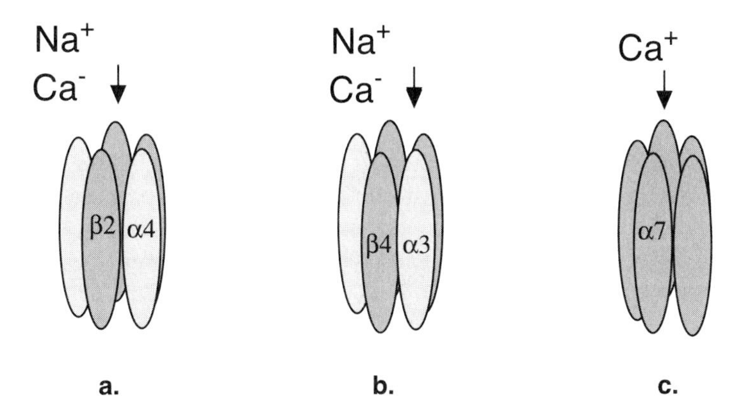

FIGURE 13.2 Three types of neuronal nicotinic acetylcholine receptors that have been expressed in heterologous systems, thought to typify native receptors. (A) $\alpha_4\beta_2$, most common subunits found in the central nervous system. Conducts sodium and calcium to a lesser degree. (B) $\alpha_3\beta_4$, most prevalent subunits in postsynaptic sympathetic ganglia neurons. Mediates synaptic transmission in the autonomic nervous system. Conducts sodium and calcium to a lesser degree. (C) α_7, found in many presynaptic terminals in central and autonomic nervous systems. Conducts significant amount of calcium.

underlie these diverse behaviors. Potential responses in which n-nAChR inhibition may be involved include any of the aformentioned behaviors because the nAChRs are presynaptic in all of the neuronal areas thought to be involved in the mediation of these behaviors. Supporting this possibility, the nAChRs have been shown to be presynaptic in hippocampus, amygdala, hypothalamus, and nucleus solitarius.[28] For example, nicotinic inhibition may be involved in hypnosis. The classical nicotinic agonist, nicotine, has been shown to improve short-term memory.[18] Thus, blockade of the central receptor for nicotine may decrease acquisition of short-term memory under anesthesia.

Volatile anesthetics in particular are well known to inhibit sympathetic tone. Recent evidence indicates that isoflurane inhibits postsynaptic n-nAChRs in lumbar sympathetic ganglia neurons, but is relatively ineffective at presynaptic inhibition at up to five times MAC (1.6 m*M*).[10]

13.4 VOLATILE ANESTHETIC ACTIVITY AT NEURONAL nAChRs

In contrast to the voluminous information about volatile anesthetic actions at the m-nAChR, little is known about their effects on n-nAChRs. Although the first n-nAChR subunit (α_3) was cloned in 1986,[4] there have been few studies on their modulation by volatile anesthetics. This may be due, in part, to the complexity of the native receptor composition. In addition, after many years of searching for examples of direct synaptic transmission in the CNS, it has only recently become clear that presynaptic activity may predominate.[18] In the autonomic nervous system, neuronal nicotinic receptors both mediate synaptic transmission and play a presynaptic modulatory role.[16]

There is strong evidence for direct inhibition of n-nAChRs in recombinant systems. Because the n-nAChR family is large and diverse and it is thought that native neurons express multiple receptor types, most of this work has been conducted in the *Xenopus laevis* oocyte. The α_4 and β_2 subunits are highly expressed throughout the brain. In *X. laevis* oocytes, the $\alpha_4\beta_2$ nAChR is inhibited by isoflurane and halothane with IC_{50} values of 85 and 27 μM, respectively (Figure 13.3). These concentrations are far below the half-maximal concentrations for surgical anesthesia with these drugs. This inhibition is concentration dependent and in some studies incomplete. The inability to completely inhibit the response of the receptor to ACh may be explained by the fact that isoflurane alone is able to gate the $\alpha_4\beta_2$ nAChR. Because these studies were conducted in *Xenopus* oocytes, using slow perfusion techniques and different protocols, it is unclear whether the inhibition is competitive or noncompetitive.[9,31] The mechanism of inhibition will be clarified by further studies with better kinetic resolution. Receptors that are highly expressed in the sympathetic nervous system, composed of α_3 and β_4 receptors, are inhibited in a dose-dependent manner by isoflurane and halothane.[31] Inhibition of nAChRs by volatile anesthetics is by no means global. Inhibition of n-nAChRs expressed in the autonomic nervous system is recapitulated by lumbar sympathetic and adrenal suppression as discussed below.[26] The α_7 homomeric receptors are not affected by up to two times MAC isoflurane. Neither do all general anesthetics inhibit nAChRs in a clinically relevant range. Propofol inhibits the α_7 nAChR at only 1000 times its clinical EC_{50}.

There have been few studies on the effects of general anesthetics on n-nAChRs in vertebrate neurons. However, in the invertebrate snail, *Lymnea stagnalis*, a neuronal nicotinic receptor has been shown to be inhibited by isoflurane.[21] In bovine chromaffin cells, the release of norepinephrine has been shown to be inhibited by halothane via nicotinic inhibition.[26] In addition, the ACh response of lumbar sympathetic ganglia neurons is inhibited by isoflurane in a concentration-dependent manner, with an IC_{50} near 0.5 MAC (P. Flood, unpublished observations). This inhibition may contribute to the sympathectomy seen clinically with these drugs.

There have been several studies on the effect on volatile anesthetics on synaptic transmission and transmitter release that have implicated a role in presynaptic inhibition. This presynaptic inhibition may have several etiologies, including the inhibition of an excitatory nicotinic response or the augmentation of an inhibitory GABA or glycinergic input. Schlame and Hemmings have demonstrated that volatile

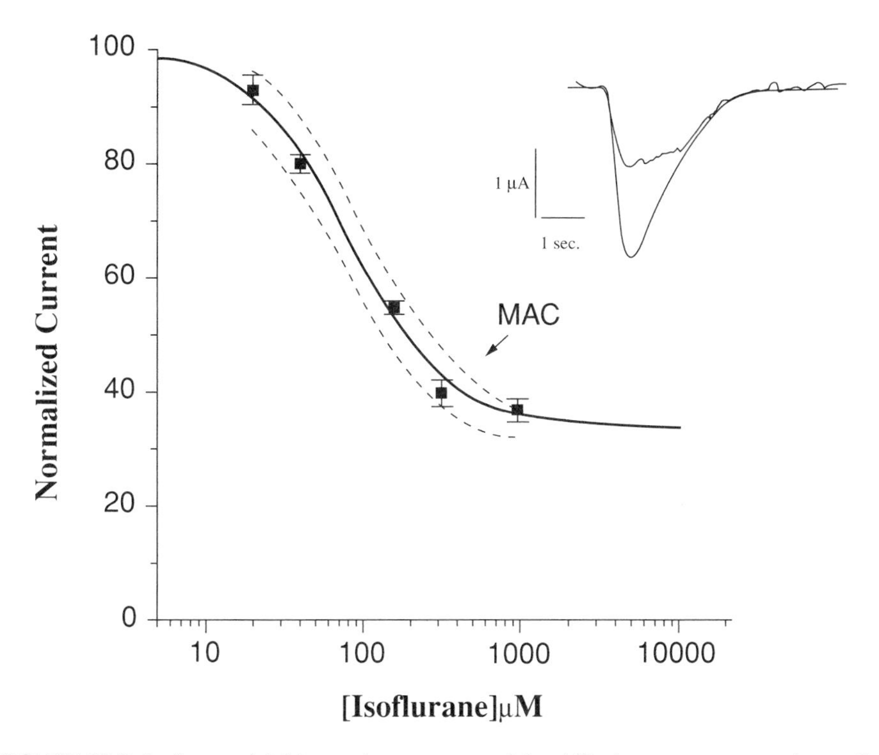

FIGURE 13.3 Isoflurane inhibits peak current gated by ACh in oocytes expressing $\alpha_4\beta_2$ nAChRs. Concentration–response curve — increasing concentrations of isoflurane progressively inhibit peak ACh-gated current. MAC refers to minimum alveolar concentration necessary to immobilize 50% of subjects. Inset — raw current where larger current is control response of $\alpha_4\beta_2$ nAChR to acetylcholine at EC_{50}, and smaller current is in the presence of 640 μM isoflurane (two times MAC).

anesthetics inhibit glutamate release in cortical synaptosomes. Although they did not identify the molecular basis of their findings, anesthetic inhibition of n-nAChRs that are present on cortical synaptosomes might produce similar results. In the hippocampal slice, volatile anesthetics have been shown to inhibit synaptic transmission by a presynaptic mechanism insensitive to bicuculline (a $GABA_A$ receptor inhibitor).[25] Inhibition of excitatory nicotinic input could be responsible for this inhibition. In work with striatal synaptosomes, it has been shown that the release of glutamate, GABA, and dopamine is inhibited and that this inhibition is at least partially due to presynaptic nicotinic inhibition. Because the striatum is involved in motor control and planning, this inhibition may result in the absence of movement to surgical stimulus caused by volatile anesthetics.

13.5 MUSCLE-TYPE nAChR

Although earlier work on the muscle-type (and torpedo) nAChRs was originally conceived as work on an ion channel model system, the muscle relaxant sparing effect of larger doses of general anesthetics may well be secondary to direct inhibition at this receptor. The volatile anesthetic isoflurane inhibits the m-nAChRs at about two to three times MAC.[6] Thus, one would expect partial blockade of the muscle receptor at clinical concentrations, which would be additive with the competitive inhibition caused by nondepolarizing muscle relaxants.

Although the concentrations of volatile anesthetics that cause inhibition of m-nAChR are higher than clinical concentrations, more information is available about the mechanism of inhibition of the m-nAChR by volatile anesthetics than is available for n-nAChRs. From measurements of the effects of isoflurane on the kinetics of single-muscle AChRs in the presence of isoflurane it appears that this inhibition is, at least in part, secondary to blockade ion conduction through the channel pore.[6] Experiments utilizing site-directed mutagenesis have suggested that the portion of the receptor that participates in channel block is located in the pore of the channel.[11] Propofol may also inhibit m-nAChRs by channel block, although this effect is not seen until concentrations of greater than 100 times the clinical EC_{50}.[6] There is evidence from experiments using spin labeling or agonists that rates of desensitization of the m-nAChR are effected by isoflurane as well.[27]

In summary, the n-nAChRs are a family of important excitatory receptor/ion channels in both the central and autonomic nervous systems. The neuronal nAChRs have recently joined the $GABA_A$ receptor family as exquisitely sensitive targets of volatile anesthetics. Some neuronal forms of nAChRs are inhibited by volatile anesthetics in a clinically relevant range. The physiologic and clinical consequences of this inhibition to cerebral function and clinical symptoms are only now being explored. Future research will reveal the mechanism and sites of anesthetic inhibition of the n-nAChRs. A further challenge will be to tease out which anesthetic behaviors result from an anesthetic effect on a particular ion channel target. Only with this information will the mechanisms of general anesthetic action be elucidated.

REFERENCES

1. Akabas, M. H. and A. Karlin. (1995). "Identification of acetylcholine receptor channel-lining residues in the M1 segment of the alpha-subunit." *Biochemistry* 34(39): 12496–500.
2. Akabas, M. H., C. Kaufmann, et al. (1994). "Identification of acetylcholine receptor channel-lining residues in the entire M2 segment of the alpha subunit." *Neuron* 13(4): 919–27.
3. Blanton, M. P. and J. B. Cohen. (1992). "Mapping the lipid-exposed regions in the Torpedo californica nicotinic acetylcholine receptor" [published erratum appears in *Biochemistry* 1992; 31(25):5951]. *Biochemistry* 31(15): 3738–50.
4. Boulter, J., K. Evans, et al. (1986). "Isolation of a cDNA clone coding for a possible neural nicotinic acetylcholine receptor alpha-subunit." *Nature* 319(6052): 368–74.
5. Czajkowski, C. and A. Karlin. (1995). "Structure of the nicotinic receptor acetylcholine-binding site. Identification of acidic residues in the delta subunit within 0.9 nm of the 5 alpha subunit-binding." *J. Biol. Chem.* 270(7): 3160–4.
6. Dilger, J., A. Vidal, et al. (1994). "Evidence for direct actions of general anesthetics on an ion channel protein." *Anesthesiology* 81: 431.

7. Dilger, J. P., R. S. Brett, et al. (1993). "The effects of isoflurane on acetylcholine receptor channels. 2. Currents elicited by rapid perfusion of acetylcholine." *Mol. Pharmacol.* 44(5): 1056–63.
8. Firestone, L. L., J. F. Sauter, et al. (1986). "Actions of general anesthetics on acetylcholine receptor-rich membranes from Torpedo californica." *Anesthesiology* 64(6): 694–702.
9. Flood, P., J. Ramirez-Latorre, et al. (1997). "a4b2 neuronal nicotinic acetylcholine receptors in the central nervous system are inhibited by isoflurane and propofol, but a7-type nicotinic acetylcholine receptors are unaffected." *Anesthesiology* 86(4): 859–865.
10. Flood, P. and L. Role. (1997). "Effects of general anesthetics on ACh evoked currents in autonomic neurons in vitro." *Neurosci. Abstr.* 23: 915.
11. Forman, S. A., K. W. Miller, et al. (1995). "A discrete site for general anesthetics on a postsynaptic receptor." *Mol. Pharmacol.* 48(4): 574–81.
12. Franks, N. P. and Lieb, W. R. (1994). "Molecular and Cellular Mechanisms of General Anesthesia." *Nature* 367: 607–614.
13. Karlin, A., M. H. Akabas, et al. (1994). "Structures involved in binding, gating, and conduction in nicotinic acetylcholine receptors." *Ren. Physiol. Biochem.* 17(3–4): 184–6.
14. Larsen, M., E. Hegstad, et al. (1997). "Isoflurane increases the uptake of glutamate in synaptosomes from rat cerebral cortex." *Br. J. Anaesth.* 78(1): 55–9.
15. Maclver, M. B., A. A. Mikulec, et al. (1996). "Volatile anesthetics depress glutamate transmission via presynaptic actions." *Anesthesiology* 85(4): 823–34.
16. Maricq, A. V., A. S. Peterson, et al. (1991). "Primary structure and functional expression of the 5HT3 receptor, a serotonin-gated ion channel." *Science* 254(5030): 432–7.
17. McGehee, D. S., M. J. Heath, et al. (1995). "Nicotine enhancement of fast excitatory synaptic transmission in CNS by presynaptic receptors [see comments]." *Science* 269(5231): 1692–6.
18. McGehee, D. S. and L. W. Role. (1995). "Physiological diversity of nicotinic acetylcholine receptors expressed by vertebrate neurons." *Annu. Rev. Physiol.* 57: 521–46.
19. McGehee, D. S. and L. W. Role. (1996). "Neurobiology: Memories of nicotine [news; comment]." *Nature* 383(6602): 670–1.
20. McGehee, D. S. and L. W. Role. (1996). "Presynaptic ionotropic receptors." *Curr. Opin. Neurobiol.* 6(3): 342–9.
21. McGurk, K., M. Pistis, et al. (1998). "The effect of a transmembrane amino acid on etomidate sensetivity of a invertebrate GABA receptor." *Br. J. Pharmacol.* 123: 1–8.
22. McKenzie, D., N. Franks, et al. (1995). "Actions of general anaesthetics on a neuronal nicotinic acetylcholine receptor in isolated identified neurones of Lymnea stagnalis." *Br. J. Pharmacol.* 115: 275–282.
23. Mihic, S. J., Q. Ye, et al. (1997). "Sites of alcohol and volatile anaesthetic action on $GABA_A$ and glycine receptors [see comments]." *Nature* 389(6649): 385–9.
24. Moody, E. J., C. Knauer, et al. (1997). "Distinct loci mediate the direct and indirect actions of the anesthetic etomidate at $GABA_A$ receptors." *J. Neurochem.* 69(3): 1310–3.
25. Ortells, M. O. and G. G. Lunt. (1995). "Evolutionary history of the ligand-gated ion-channel superfamily of receptors [see comments]." *Trends Neurosci.* 18(3): 121–7.
26. Perouansky, M., D. Baranov, et al. (1995). "Effects of halothane on glutamate receptor-mediated excitatory postsynaptic currents." *Anesthesiology* 83: 109–119.
27. Pocock, G. and C. D. Richards. (1988). "The action of volatile anaesthetics on stimulus-secretion coupling in bovine adrenal chromaffin cells." *Br. J. Pharmacol.* 95(1): 209–17.

28. Raines, D. E., S. E. Rankin, et al. (1995). "General anesthetics modify the kinetics of nicotinic acetylcholine receptor desensitization at clinically relevant concentrations." *Anesthesiology* 82(1): 276–87; discussion 31A–32A.
29. Role, L. W. and D. K. Berg. (1996). "Nicotinic receptors in the development and modulation of CNS synapses." *Neuron* 16(6): 1077–85.
30. Schlame, M. and Hemmings, H. C. (1995). "Inhibition by Volative Anesthetics of Endogenous Glutamate Release from Synaptosomes by a Presynaptic Mechanism." *Anesthesiology* 82: 1406–1416.
31. Seguela, P., J. Wadiche, et al. (1993). "Molecular cloning, functional properties, and distribution of rat brain alpha 7: A nicotinic cation channel highly permeable to calcium." *J. Neurosci.* 13(2): 596–604.
32. Spencer, G. E., N. I. Syed, et al. (1995). "Halothane-induced synaptic depression at both in vivo and in vitro reconstructed synapses between identified Lymnaea neurons." *J. Neurophysiol.* 74(6): 2604–13.
33. Violet, J. M., D. L. Downie, et al. (1997). "Differential sensitivities of mammalian neuronal and muscle nicotinic acetylcholine receptors to general anesthetics [see comments]." *Anesthesiology* 86(4): 866–74.
34. Woolf, N. (1991). "Cholinergic systems in mammalian brain and spinal cord." *Prog. Neurobiol.* 37: 475–524.

Index

A

B

C

D

E

F

G

H

I

K

L

M

N

O

P

Q

R

S

T

U

V

W

X

Y

Z